Engineering Analysis with ANSYS Software

Engineering Analysis with ANSYS Software

Second Edition

Tadeusz Stolarski

Professor in the Department of Mechanical
and Aerospace Engineering, Brunel University
London, UK and also at Gdansk University
of Technology, Poland

Yuji Nakasone

Professor in the Department of Mechanical
Engineering, Tokyo University of Science, Tokyo,
Japan

Shigeka Yoshimoto

Professor in the Department of Mechanical
Engineering, Tokyo University of Science, Tokyo,
Japan

Butterworth-Heinemann
An imprint of Elsevier

Butterworth-Heinemann is an imprint of Elsevier
The Boulevard, Langford Lane, Kidlington, Oxford OX5 1GB, United Kingdom
50 Hampshire Street, 5th Floor, Cambridge, MA 02139, United States

Notices
Knowledge and best practice in this field are constantly changing. As new research and
experience broaden our understanding, changes in research methods, professional practices,
or medical treatment may become necessary.

Practitioners and researchers must always rely on their own experience and knowledge in
evaluating and using any information, methods, compounds, or experiments described herein.
In using such information or methods they should be mindful of their own safety and the safety
of others, including parties for whom they have a professional responsibility.

To the fullest extent of the law, neither the Publisher nor the authors, contributors, or editors,
assume any liability for any injury and/or damage to persons or property as a matter of products
liability, negligence or otherwise, or from any use or operation of any methods, products,
instructions, or ideas contained in the material herein.

Library of Congress Cataloging-in-Publication Data
A catalog record for this book is available from the Library of Congress

British Library Cataloguing-in-Publication Data
A catalogue record for this book is available from the British Library

ISBN: 978-0-08-102164-4

For information on all Butterworth-Heinemann publications
visit our website at https://www.elsevier.com/books-and-journals

Working together
to grow libraries in
developing countries

www.elsevier.com • www.bookaid.org

Publisher: Mathew Deans
Acquisition Editor: Brian Guerin
Editorial Project Manager: Leticia Lima
Production Project Manager: Surya Narayanan Jayachandran
Cover Designer: Mark Rogers

Typeset by SPi Global, India

Contents

Preface

The aims and scope of this text are described later. The book is very much the result of collaboration between the three co-authors: Professors Nakasone and Yoshimoto of Tokyo University of Science, Japan and Professor Stolarski of Brunel University, United Kingdom. This collaboration started some time ago and initially covered only research topics of interest to the authors. Exchange of academic staff and research students has taken place and archive papers have been published. However, being an academic does not mean research only. The other important activity of any academic is to teach students on degree courses. Since the authors are involved in teaching students various aspects of engineering analyses using ANSYS, it is only natural that the need for a textbook about aiding students to solve problems with ANSYS has been identified.

The ethos of the book, worked out during a number of discussion meetings, aims at assisting in learning the use of ANSYS through examples taken from engineering practice. It is hoped that the book will meet its primary aim and provide practical help to those who embark on the road leading to effective use of ANSYS in solving engineering problems.

This second edition of the book has been thoroughly revised in the light of the latest version of ANSYS. New solved examples have been added and the use of the Workbench module is explained through illustrative engineering problems.

J. Nakasone
T.A. Stolarski
S. Yoshimoto

Preface

The aims and scope of the book

It is true to say that in many instances, the best way to learn complex behaviour is by means of imitation. For instance, most of us learned to walk, to talk, and to throw a ball solely by imitating the actions and speech of those around us. To a considerable extent, the same approach can be adopted to learn to use the ANSYS programme by imitation, using the examples provided in this book. This is the essence of the philosophy and innovative approach used in the book. The authors have attempted to provide readers with a comprehensive cross-section of analysis types in a variety of engineering areas, in order to provide a broad choice of examples to be imitated in one's own work. In developing these examples, the authors' intent has been to demonstrate many programme features and refinements. By displaying these features in an assortment of disciplines and analysis types, the authors hope to give readers the confidence to employ these programme enhancements in their own ANSYS applications.

The primary aim of the book is to assist in learning the use of ANSYS through examples taken from various areas of engineering. The content and treatment of the subject matter are considered to be most appropriate for university students studying engineering and for practising engineers who wish to learn the use of ANSYS. The book is exclusively structured around ANSYS, and no other finite element (FE) software currently available is considered. In this second edition of the book, revised and updated, ANSYS version 17 is used throughout. Although the fundamentals underlying ANSYS have not changed, nevertheless there are some subtle differences between the latest version and the version used in the first edition of the book, and these differences are incorporated in the second edition of the book. Also, instead of FLOTRAN used in the first edition, FLUENT—a new module of the programme—is utilised to solve illustrative examples. Moreover, some new solved examples have been introduced in order to bring the contents of the book closer to practical engineering applications.

The book is organised into seven chapters. Chapter 1 introduces a reader to fundamental concepts of the FE method. In Chapter 2 an overview of ANSYS is presented. Chapter 3 deals with example problems concerning stress analysis. Chapter 4 contains solved problems pertinent to dynamics of machinery. Chapter 5 is devoted to fluid dynamics problems, while Chapter 6 shows how to use ANSYS to solve problems typical in the field of thermos-mechanics. Finally, Chapter 7 outlines the approach, through solved examples, to problems related to contact and surface mechanics of machine elements.

The authors are of the opinion that the book is very timely as there is a considerable demand, primarily from university engineering courses, for a book that could be used to teach, in a practical way, ANSYS—a premiere finite element analysis computer programme. In addition, practising engineers increasingly use ANSYS for computer-based analyses of various systems, hence there is a need for a book that they could use in a self-learning mode.

The strategy used in this book, i.e. to enable readers to learn ANSYS by means of imitation, is quite unique and thus differs significantly from that in other books where ANSYS is also involved.

Fundamentals of the finite element method

1

1.1 Method of weighted residuals

Differential equations are generally formulated so as to be satisfied at any points that belong to regions of interest. The method of weighted residuals determines the approximate solution $\bar{u}$ to a differential equation such that the integral of the weighted error of the differential equation of the approximate function $\bar{u}$ over the region of interest be zero. In other words, this method determines the approximate solution that satisfies the differential equation of interest on average over the region of interest.

$$\begin{cases} L[u(x)] = f(x) \quad (a \le x \le b) \\ \text{B.C.(Boundary conditions)}: \ u(a) = u_a, \ u(b) = u_b \end{cases} \tag{1.1}$$

Now, let us suppose an approximate solution to the function u to be

$$\bar{u}(x) = \phi_0(x) + \sum_{i=1}^{n} a_i \phi_i(x). \tag{1.2}$$

where ϕ_i are called trial functions $(i = 1, 2, \ldots, n)$ which are chosen arbitrarily as any function $\phi_0(x)$ and a_i some parameters which are computed so as to obtain a good 'fit'.

The substitution of $\bar{u}$ into Eq. (1.1) makes the right-hand side nonzero, but some error R:

$$L[\bar{u}(x)] - f(x) = R. \tag{1.3}$$

The method of weighted residuals determines $\bar{u}$ such that the integral of the error R over the region of interest weighted by arbitrary functions w_i $(i = 1, 2, \ldots, n)$ be zero, i.e. the coefficients a_i in Eq. (1.2) are determined so as to satisfy the following equation:

$$\int_D w_i R \, dv = 0 \tag{1.4}$$

where D is the region considered.

Engineering Analysis with ANSYS Software. https://doi.org/10.1016/B978-0-08-102164-4.00001-7

1.1.1 Subdomain method (finite volume method)

The choice of the following weighting function brings about the subdomain method or finite volume method.

$$w_i(x) = \begin{cases} 1 & (\text{for } x \in D) \\ 0 & (\text{for } x \notin D) \end{cases} \tag{1.5}$$

Example 1.1. Consider a boundary-value problem described by the following one-dimensional differential equation:

$$\begin{cases} \dfrac{d^2u}{dx^2} - u = 0 & (0 \le x \le 1) \\ \text{B.C.}: u(0) = 0, \ u(1) = 1 \end{cases} \tag{1.6}$$

where L is a linear differential operator, $f(x)$ a function of x, and u_a and u_b the values of a function $u(x)$ of interest at the endpoints, or the one-dimensional boundaries of the region D. The linear operator is defined as follows:

$$L[\cdot] \equiv \frac{d^2(\cdot)}{dx^2}, \quad f(x) \equiv u(x). \tag{1.7}$$

For simplicity, let us choose the power series of x as the trial functions ϕ_i, i.e.

$$\overline{u}(x) = \sum_{i=0}^{n+1} c_i x^i. \tag{1.8}$$

For satisfying the required boundary conditions,

$$c_0 = 0, \quad \sum_{i=1}^{n+1} c_i = 1 \tag{1.9}$$

so that

$$\overline{u}(x) = x + \sum_{i=1}^{n} A_i (x^{i+1} - x). \tag{1.10}$$

If only the second term of the right-hand side of Eq. (1.10) is chosen as a first-order approximate solution,

$$\overline{u}_1(x) = x + A_1 (x^2 - x), \tag{1.11}$$

the error, or residual is obtained as

$$R = \frac{d^2\bar{u}}{dx^2} - \bar{u} = -A_1 x^2 + (A_1 - 1)x + 2A_1 \neq 0. \tag{1.12}$$

$$\int_0^1 w_i R dx = \int_0^1 1 \cdot \left[-A_1 x^2 + (A_1 - 1)x + 2A_1 \right] dx = \frac{13}{6}A_1 - \frac{1}{2} = 0 \tag{1.13}$$

Consequently, the first-order approximate solution is obtained as

$$\bar{u}_1(x) = x + \frac{3}{13}x(x-1) \tag{1.14}$$

which agrees well with the exact solution

$$u(x) = \frac{e^x - e^{-x}}{e - e^{-1}} \tag{1.15}$$

as shown by the solid line in Fig. 1.1.

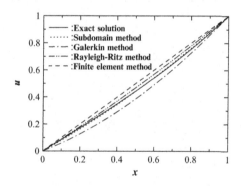

Fig. 1.1 Comparison of the results obtained by various kinds of discrete analyses.

1.1.2 Galerkin method

When the weighting function w_i in Eq. (1.4) is equal to the trial function ϕ_i, this is called the Galerkin method, i.e.,

$$w_i(x) = \phi_i(x) \quad (i = 1, 2, ..., n) \tag{1.16}$$

and thus Eq. (1.4) is changed to

$$\int_D \phi_i R dv = 0. \tag{1.17}$$

This method determines the coefficients a_i by directly using Eq. (1.17) or by integrating it by parts.

Example 1.2. Let us solve the same boundary-value problem described by Eq. (1.6) as in Section 1.1.1 by the Galerkin method.

The trial function ϕ_i is chosen as the weighting function w_i in order to find the first-order approximate solution.

$$w_1(x) = \phi_1(x) = x(x-1) \tag{1.18}$$

Integrating Eq. (1.4) by parts,

$$\int_0^1 w_i R dx = \int_0^1 w_i \left(\frac{d^2 \bar{u}}{dx^2} - \bar{u} \right) dx = \left[w_i \frac{d\bar{u}}{dx} \right]_0^1 - \int_0^1 \frac{dw_i}{dx} \frac{d\bar{u}}{dx} dx - \int_0^1 w_i \bar{u} dx = 0 \tag{1.19}$$

is obtained. Choosing $\bar{u}_1$ in Eq. (1.11) as the approximate solution $\bar{u}$, the substitution of Eq. (1.18) into Eq. (1.19) gives

$$\begin{aligned}
\int_0^1 \phi_1 R dx &= -\int_0^1 \frac{d\phi_1}{dx} \frac{d\bar{u}}{dx} dx - \int_0^1 \phi_1 \bar{u} dx \\
&= -\int_0^1 (2x-1)[1 + A_1(2x-1)] dx \\
&\quad -\int_0^1 (x^2 - x)\left[1 + A_1(x^2 - x)\right] dx \\
&= -\frac{A_1}{3} + \frac{1}{12} - \frac{A_1}{30} = 0. \tag{1.20}
\end{aligned}$$

Thus, the following approximate solution is obtained

$$\bar{u}_1(x) = x + \frac{5}{22}x(x-1). \tag{1.21}$$

Fig. 1.1 shows that the approximate solution obtained by the Galerkin method also agrees well with the exact solution throughout the region of interest.

1.2 Rayleigh–Ritz method

When there exists the functional which is equivalent to a given differential equation, the Rayleigh–Ritz method can be used.

Let us consider an example of functional illustrated in Fig. 1.2, where a particle having a mass of M slide from a point P_0 to a lower point P_1 along a curve in a vertical plane under the force of gravity. The time t that the particle needs for sliding from the points P_0 to P_1 varies with the shape of a curve denoted by $y(x)$ which connects the two points. Namely, the time t can be considered as a kind of function $t = F[y]$, which is determined by a function $y(x)$ of an independent variable x. The function of a function $F[y]$ is called a functional. The method for determining the maximum or the minimum of a given functional is called the variational method. In the case of Fig. 1.2, the method determines the shape of the curve $y(x)$ which gives the possible minimum time t_{min} in which the particle slides from P_0 to P_1.

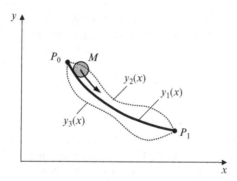

Fig. 1.2 Particle M sliding from point P_0 to lower point P_1 under gravitational force.

The principle of the virtual work or the minimum potential energy in the field of the solid mechanics is one of the variational principles that guarantee the existence of the function which makes the functional minimum or maximum. For unsteady thermal conductivity problems and for viscous flow problems, no variational principle can be established; in such a case, the method of weighting residuals can be applied instead.

Now, let $\Pi[u]$ be the functional which is equivalent to the differential equation in Eq. (1.1). The Rayleigh–Ritz method assumes that an approximate solution $\bar{u}(x)$ of $u(x)$ is a linear combination of trial functions ϕ_i, as shown in the following equation:

$$\bar{u}(x) = \sum_{i=1}^{n} a_i \phi_i(x) \tag{1.22}$$

where a_i $(i = 1, 2, ..., n)$ are arbitrary constants, ϕ_i are C^0-class functions which have continuous first-order derivatives for $a \leq x \leq b$ and are chosen such that the following boundary conditions are satisfied:

$$\sum_{i=1}^{n} a_i \phi_i(a) = u_a, \quad \sum_{i=1}^{n} a_i \phi_i(b) = u_b. \tag{1.23}$$

The approximate solution $\bar{u}(x)$ in Eq. (1.22) is the function which makes the functional $\Pi[u]$ take stationary value, and is called the admissible function.

Next, integrating the functional Π after substituting Eq. (1.22) into the functional, the constants a_i are determined by the stationary conditions:

$$\frac{\partial \Pi}{\partial a_i} = 0 \quad (i = 1, 2, ..., n). \tag{1.24}$$

The Rayleigh–Ritz method determines the approximate solution $\bar{u}(x)$ by substituting the constants a_i into Eq. (1.22). The method is generally understood to be a method which determines the coefficients a_i so as to make the distance between the approximate solution $\bar{u}(x)$ and the exact one $u(x)$ minimum.

Example 1.3. Let us solve again the boundary-value problem described by Eq. (1.6) by the Rayleigh–Ritz method. The functional equivalent to the first equation of Eq. (1.6) is written as

$$\Pi[u] = \int_0^1 \left[\frac{1}{2} \left(\frac{du}{dx} \right)^2 + \frac{1}{2} u^2 \right] dx. \tag{1.25}$$

Eq. (1.25) is obtained by intuition, but Eq. (1.25) is shown to really give the functional of the first equation of Eq. (1.6) as follows: first, let us take the first variation of Eq. (1.25) in order to obtain the stationary value of the equation:

$$\delta \Pi = \int_0^1 \left[\frac{du}{dx} \delta \left(\frac{du}{dx} \right) + u \delta u \right] dx. \tag{1.26}$$

Then, integrating the above equation by parts, we have

$$\delta \Pi = \int_0^1 \left(\frac{du}{dx} \frac{d\delta u}{dx} + u \delta u \right) dx = \left[\frac{du}{dx} \delta u \right] - \int_0^1 \left[\frac{d}{dx} \left(\frac{du}{dx} \right) \delta u - u \delta u \right] dx$$

$$= -\int_0^1 \left(\frac{d^2 u}{dx^2} - u \right) \delta u \, dx. \tag{1.27}$$

For satisfying the stationary condition that $\delta \Pi = 0$, the rightmost-hand side of Eq. (1.27) should be identically zero over the interval considered ($a \le x \le b$), so that

$$\frac{d^2 u}{dx^2} - u = 0. \tag{1.28}$$

This is exactly the same as the first equation of Eq. (1.6).

Now, let us consider the following first-order approximate solution $\bar{u}_1$ which satisfies the following boundary conditions:

$$\bar{u}_1(x) = x + a_1 x(x - 1). \tag{1.29}$$

Substitution of Eq. (1.29) into Eq. (1.25) and integration of Eq. (1.25) lead to

$$\Pi[\bar{u}_1] = \int_0^1 \left\{ \frac{1}{2}[1 + a_1(2x-1)]^2 + \frac{1}{2}[x + a_1(x^2-x)]^2 \right\} dx$$
$$= \frac{2}{3} - \frac{1}{12}a_1 + \frac{1}{3}a_1^2.$$

(1.30)

Since the stationary condition for Eq. (1.30) is written by

$$\frac{\partial \Pi}{\partial a_1} = -\frac{1}{12} + \frac{2}{3}a_1 = 0,$$

(1.31)

the first-order approximate solution can be obtained as follows:

$$\bar{u}_1(x) = x + \frac{1}{8}x(x-1).$$

(1.32)

Fig. 1.1 shows that the approximate solution obtained by the Rayleigh–Ritz method agrees well with the exact solution throughout the region considered.

1.3 Finite element method

There are two ways of the formulation of the finite element method; one is based on the direct variational method such as the Rayleigh–Ritz method and the other on the method of weighted residuals such as the Galerkin method. In the formulation based on the variational method, the fundamental equations are derived from the stationary conditions of the functional for the boundary-value problems. This formulation has an advantage that the process of deriving functionals is not necessary, so it is easy to formulate the finite element method based on the method of the weighted residuals. In the formulation based on the variational method, however, it is generally difficult to derive the functional except for the case where the variational principles are already established, as in the case of the principle of the virtual work or the principle of the minimum potential energy in the field of the solid mechanics.

This section will explain how to derive the fundamental equations for the finite element method based on the Galerkin method.

Let us consider again the boundary-value problem stated by Eq. (1.1).

$$\begin{cases} L[u(x)] = f(x) \quad (a \le x \le b) \\ \text{B.C.(Boundary conditions)}: \ u(a) = u_a, \ u(b) = u_b \end{cases}$$

(1.33)

First, divide the region of interest $(a \le x \le b)$ into n subregions, as illustrated in Fig. 1.3. These subregions are called 'elements' in the finite element method.

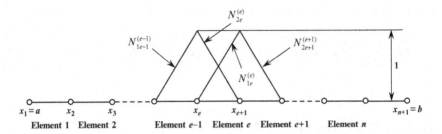

Fig. 1.3 Discretisation of the domain to analyse by finite elements and their interpolation functions.

Now, let us assume that an approximate solution $\bar{u}$ of u can be expressed by piecewise linear functions which form a straight line in each subregion, i.e.,

$$\bar{u}(x) = \sum_{i=1}^{n+1} u_i N_i(x) \tag{1.34}$$

where u_i represents the value of u in element 'e' at a boundary point, or a nodal point 'i' between two one-dimensional elements. Functions $N_i(x)$ are the following piecewise linear functions and are called interpolation or shape functions of the nodal point 'i':

$$\begin{cases} N_{1e}^{(e)} = \dfrac{x_{e+1} - x}{x_{e+1} - x_e} = \dfrac{x_{2e} - x}{x_{2e} - x_{1e}} = \dfrac{h^{(e)} - \xi}{h^{(e)}} \\[3mm] N_{2e}^{(e)} = \dfrac{x - x_e}{x_{e+1} - x_e} = \dfrac{x - x_{1e}}{x_{2e} - x_{1e}} = \dfrac{\xi}{h^{(e)}} \end{cases} \tag{1.35}$$

where e $(e = 1, 2, ..., n)$ denotes the element number, x_i the global coordinate of the nodal point i $(i = 1, ..., e - 1, e, ...n, n + 1)$, $N_{ie}^{(e)}$ is the value of the interpolation function at the nodal point i_e $(i_e = 1_e, 2_e)$ which belongs to the eth element, and 1_e and 2_e are the numbers of two nodal points of the eth element. The symbol ξ is the local coordinate of an arbitrary point in the eth element, $\xi = x - x_e = x - x_{1e}$ $(0 \leq \xi \leq h^{(e)})$, $h^{(e)}$ is the length of the eth element, and $h^{(e)}$ is expressed as $h^{(e)} = x - x_e = x - x_{1e}$.

As the interpolation function, the piecewise linear or quadric function is often used. Generally speaking, the quadric interpolation function gives better solutions than the linear one.

The Galerkin-method-based finite element method adopts the weighting functions $w_i(x)$ equal to the interpolation functions $N_i(x)$, i.e.,

$$w_i(x) = N_i(x) \quad (i = 1, 2, ..., n + 1). \tag{1.36}$$

Thus, Eq. (1.4) becomes

$$\int_D N_i R dv = 0.$$

(1.37)

In the finite element method, a set of simultaneous algebraic equations for unknown variables of $u(x)$ at the ith nodal point u_i and those of its derivatives du/dx, $(du/dx)_i$ are derived by integrating Eq. (1.37) by parts and then by taking boundary conditions into consideration. The simultaneous equations can be easily solved by digital computers to determine the unknown variables u_i and $(du/dx)_i$ at all the nodal points in the region considered.

Example 1.4. Let us solve the boundary-value problem stated in Eq. (1.6) by the finite element method.

First, the integration of Eq. (1.37) by parts gives

$$\int_0^1 w_i R dx = \int_0^1 w_i \left(\frac{d^2\bar{u}}{dx^2} - \bar{u}\right) dx = \left[w_i \frac{d\bar{u}}{dx}\right]_0^1 - \int_0^1 \left(\frac{dw_i}{dx}\frac{d\bar{u}}{dx} + w_i\bar{u}\right) dx$$
$$= 0 \quad (i = 1, 2, ..., n+1).$$

(1.38)

Then, the substitution of Eqs (1.34), (1.36) into Eq. (1.38) gives

$$\sum_{j=1}^{n+1} \int_0^1 \left(\frac{dN_i}{dx}\frac{dN_j}{dx} + N_iN_j\right) u_j dx - \left[N_i\frac{d\bar{u}}{dx}\right]_0^1 = 0 \quad (i = 1, 2, ..., n+1).$$

(1.39)

Eq. (1.39) is a set of simultaneous linear algebraic equations composed of $(n + 1)$ nodal values u_i of the solution u and also $(n + 1)$ nodal values $(du/dx)_i$ of its derivative du/dx. The matrix notation of the simultaneous equations above is written in a simpler form as follows:

$$[K_{ij}]\{u_j\} = \{f_i\}$$

(1.40)

where $[K_{ij}]$ is a square matrix of $(n + 1)$ by $(n + 1)$, $\{f_i\}$ is a column vector of $(n + 1)$ by 1 and each component of the matrix K_{ij} and the vector f_i are expressed as

$$\begin{cases} K_{ij} \equiv \int_0^1 \left(\frac{dN_i}{dx}\frac{dN_j}{dx} + N_iN_j\right) dx & (1 \leq i, j \leq n+1), \\ f_i \equiv \left[N_i\frac{d\bar{u}}{dx}\right]_0^1 & (1 \leq i \leq n+1). \end{cases}$$

(1.41)

1.3.1 One-element case

As a first example, let us compute Eq. (1.37) by regarding the whole region as one finite element as shown in Fig. 1.4. From Eqs (1.34), (1.35), since $x_1 = 0$ and $x_2 = 1$, the approximate solution $\bar{u}$ and the interpolation functions N_i ($i = 1, 2$) become

$$\bar{u}(x) = u_1 N_1 + u_2 N_2 \tag{1.42}$$

and

$$\begin{cases} N_1 = N_{11}^{(1)} = \dfrac{x_2 - x}{x_2 - x_1} = 1 - x \\[2mm] N_2 = N_{21}^{(1)} = \dfrac{x - x_1}{x_2 - x_1} = x \end{cases} \tag{1.43}$$

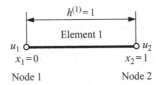

Fig. 1.4 One-element model of one-dimensional finite element method.

Thus, from Eq. (1.41),

$$K_{ij} \equiv K_{ij}^{(1)} \equiv \int_0^1 \left(\frac{dN_i}{dx}\frac{dN_j}{dx} + N_i N_j \right) dx = \begin{cases} 4/3 & (i = j) \\ -5/6 & (i \neq j) \end{cases},$$

$$f_i \equiv \left[N_i \frac{d\bar{u}}{dx} \right]_0^1 = \begin{cases} -\dfrac{d\bar{u}}{dx}\Big|_{x=0} \\[3mm] \dfrac{d\bar{u}}{dx}\Big|_{x=1} \end{cases}. \tag{1.44}$$

The global simultaneous equations are obtained as

$$\begin{bmatrix} \dfrac{4}{3} & -\dfrac{5}{6} \\[3mm] -\dfrac{5}{6} & \dfrac{4}{3} \end{bmatrix} \begin{Bmatrix} u_1 \\ u_2 \end{Bmatrix} = \begin{Bmatrix} -\dfrac{d\bar{u}}{dx}\Big|_{x=0} \\[3mm] \dfrac{d\bar{u}}{dx}\Big|_{x=1} \end{Bmatrix}. \tag{1.45}$$

According to the boundary conditions, $u_1 = 0$ and $u_2 = 1$ in the left-hand side of the above equations are known variables, whereas $(du/dx)_{x=0}$ and $(du/dx)_{x=1}$ in the left-hand side are unknown variables. The substitution of the boundary conditions into Eq. (1.45) directly gives the nodal values of the approximate solution, i.e.,

$$\begin{cases} \dfrac{d\bar{u}}{dx}\Big|_{x=0} = 0.8333 \\[3mm] \dfrac{d\bar{u}}{dx}\Big|_{x=1} = 1.3333 \end{cases} \tag{1.46}$$

which agree well with those of the exact solution:

$$
\begin{cases}
\dfrac{du}{dx}\Big|_{x=0} = \dfrac{2}{e-e^{-1}} = 0.8509 \\[3mm]
\dfrac{du}{dx}\Big|_{x=1} = \dfrac{e+e^{-1}}{e-e^{-1}} = 1.3130
\end{cases}
\tag{1.47}
$$

The approximate solution in this example is determined as

$$
\bar{u}(x) = x \tag{1.48}
$$

and agrees well with the exact solution throughout the whole region of interest as depicted in Fig. 1.1.

1.3.2 Three-element case

In this subsection, let us compute the approximate solution $\bar{u}$ by dividing the whole region considered into three subregions having the same length as shown in Fig. 1.5. From Eqs (1.34), (1.35), the approximate solution $\bar{u}$ and the interpolation functions N_i ($i = 1, 2$) are written as

$$
\bar{u}(x) = \sum_{i=1}^{4} u_i N_i \tag{1.49}
$$

and

$$
\begin{cases}
N_{1e} = \dfrac{x_{2e} - x}{x_{2e} - x_{1e}} = \dfrac{h^{(e)} - \xi}{h^{(e)}} \\[3mm]
N_{2e} = \dfrac{x - x_{1e}}{x_{2e} - x_{1e}} = \dfrac{\xi}{h^{(e)}}
\end{cases}
\tag{1.50}
$$

where $h^{(e)} = 1/3$ and $0 \le \xi \le 1/3 (e = 1, 2, 3)$.

Calculating all the components of the K-matrix in Eq. (1.41), the following equation is obtained:

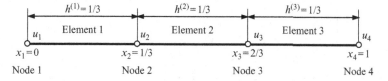

Fig. 1.5 Three-element model of one-dimensional finite element method.

$$K_{ij}^{(e)} \equiv \int_0^1 \left(\frac{dN_i^{(e)}}{dx} \frac{dN_j^{(e)}}{dx} + N_i^{(e)} N_j^{(e)} \right) dx$$

$$= \begin{cases} \dfrac{1}{h^{(e)}} + \dfrac{h^{(e)}}{3} = \dfrac{28}{9} & (i = mj \text{ and } i,j = 1e, 2e) \\[3mm] -\dfrac{1}{h^{(e)}} + \dfrac{h^{(e)}}{6} = -\dfrac{53}{18} & (i \neq j \text{ and } i,j = 1e, 2e) \\[3mm] 0 & (i, j \neq 1e, 2e) \end{cases} \qquad (1.51\text{a})$$

The components relating to the first derivative of the function u in Eq. (1.41) are calculated as follows:

$$f_i \equiv \left[N_i \frac{d\bar{u}}{dx} \right]_0^1 = \begin{cases} -\dfrac{d\bar{u}}{dx}\bigg|_{x=0} & (i = 1) \\[3mm] 0 & (i = 2, 3) \\[3mm] \dfrac{d\bar{u}}{dx}\bigg|_{x=1} & (i = 4) \end{cases} \qquad (1.51\text{b})$$

The coefficient matrix in Eq. (1.51a) calculated for each element is called the 'element matrix', and the components of the matrix are obtained as follows:

$$\left[K_{ij}^{(1)} \right] = \begin{bmatrix} \dfrac{1}{h^{(1)}} + \dfrac{h^{(1)}}{3} & -\dfrac{1}{h^{(1)}} + \dfrac{h^{(1)}}{6} & 0 & 0 \\[3mm] -\dfrac{1}{h^{(1)}} + \dfrac{h^{(1)}}{6} & \dfrac{1}{h^{(1)}} + \dfrac{h^{(1)}}{3} & 0 & 0 \\[3mm] 0 & 0 & 0 & 0 \\[3mm] 0 & 0 & 0 & 0 \end{bmatrix} \qquad (1.52\text{a})$$

$$\left[K_{ij}^{(2)} \right] = \begin{bmatrix} 0 & 0 & 0 & 0 \\[3mm] 0 & \dfrac{1}{h^{(2)}} + \dfrac{h^{(2)}}{3} & -\dfrac{1}{h^{(2)}} + \dfrac{h^{(2)}}{6} & 0 \\[3mm] 0 & -\dfrac{1}{h^{(2)}} + \dfrac{h^{(2)}}{6} & \dfrac{1}{h^{(2)}} + \dfrac{h^{(2)}}{3} & 0 \\[3mm] 0 & 0 & 0 & 0 \end{bmatrix} \qquad (1.52\text{b})$$

$$\left[K_{ij}^{(3)} \right] = \begin{bmatrix} 0 & 0 & 0 & 0 \\[3mm] 0 & 0 & 0 & 0 \\[3mm] 0 & 0 & \dfrac{1}{h^{(3)}} + \dfrac{h^{(3)}}{3} & -\dfrac{1}{h^{(3)}} + \dfrac{h^{(3)}}{6} \\[3mm] 0 & 0 & -\dfrac{1}{h^{(3)}} + \dfrac{h^{(3)}}{6} & \dfrac{1}{h^{(3)}} + \dfrac{h^{(3)}}{3} \end{bmatrix} \qquad (1.52\text{c})$$

From Eqs (1.52a)–(1.52c), it is concluded that only the components of the element matrix relating to the nodal points which belong to the corresponding element is non-zero and that the other components are zero. Namely, for example, element 2 is composed of nodal points 2 and 3 and among the components of the element matrix only $K_{22}^{(2)}$, $K_{23}^{(2)}$, $K_{32}^{(2)}$, and $K_{33}^{(2)}$ are nonzero and the other components are zero. The superscript (2) of the element matrix components above indicates that the components are computed in element 2, and the subscripts indicate that the components are calculated for nodal points 2 and 3 in element 2.

A matrix that relates all the known and unknown variables for the problem concerned is called the global matrix. The global matrix can be obtained only by summing up Eqs (1.52a)–(1.52c) as follows:

$$[K_{ij}] = \left[\sum_{e=1}^{3} K_{ij}^{(e)}\right]$$

$$= \begin{bmatrix} \frac{1}{h^{(1)}}+\frac{h^{(1)}}{3} & -\frac{1}{h^{(1)}}+\frac{h^{(1)}}{6} & 0 & 0 \\ -\frac{1}{h^{(1)}}+\frac{h^{(1)}}{6} & \sum_{e=1}^{2}\left(\frac{1}{h^{(e)}}+\frac{h^{(e)}}{3}\right) & -\frac{1}{h^{(2)}}+\frac{h^{(2)}}{6} & 0 \\ 0 & -\frac{1}{h^{(2)}}+\frac{h^{(2)}}{6} & \sum_{e=2}^{3}\left(\frac{1}{h^{(e)}}+\frac{h^{(e)}}{3}\right) & -\frac{1}{h^{(3)}}+\frac{h^{(3)}}{6} \\ 0 & 0 & -\frac{1}{h^{(3)}}+\frac{h^{(3)}}{6} & \frac{1}{h^{(3)}}+\frac{h^{(3)}}{3} \end{bmatrix} \quad (1.53)$$

Consequently, the global simultaneous equation becomes

$$\begin{bmatrix} \frac{1}{h^{(1)}}+\frac{h^{(1)}}{3} & -\frac{1}{h^{(1)}}+\frac{h^{(1)}}{6} & 0 & 0 \\ -\frac{1}{h^{(1)}}+\frac{h^{(1)}}{6} & \sum_{e=1}^{2}\left(\frac{1}{h^{(e)}}+\frac{h^{(e)}}{3}\right) & -\frac{1}{h^{(2)}}+\frac{h^{(2)}}{6} & 0 \\ 0 & -\frac{1}{h^{(2)}}+\frac{h^{(2)}}{6} & \sum_{e=2}^{3}\left(\frac{1}{h^{(e)}}+\frac{h^{(e)}}{3}\right) & -\frac{1}{h^{(3)}}+\frac{h^{(3)}}{6} \\ 0 & 0 & -\frac{1}{h^{(3)}}+\frac{h^{(3)}}{6} & \frac{1}{h^{(3)}}+\frac{h^{(3)}}{3} \end{bmatrix} \begin{Bmatrix} u_1 \\ u_2 \\ u_3 \\ u_4 \end{Bmatrix}$$

$$= \begin{Bmatrix} -\frac{d\bar{u}}{dx}\Big|_{x=0} \\ 0 \\ 0 \\ \frac{d\bar{u}}{dx}\Big|_{x=1} \end{Bmatrix} \quad (1.54)$$

Note that the coefficient matrix $[K_{ij}]$ in the left-hand side of Eq. (1.54) is symmetric with respect to the nondiagonal components $(i \neq j)$, i.e., $K_{ij} = K_{ji}$. Only the components in the band region around the diagonal of the matrix are nonzero, and the others are zero. Due to this nature, the coefficient matrix is called the sparse or band matrix.

From the boundary conditions, the values of u_1 and u_4 in the left-hand side of Eq. (1.54) are known, i.e. $u_1 = 0$ and $u_4 = 1$ and, from Eq. (1.51), the values of f_2 and f_3 in the right-hand side are also known, i.e. $f_2 = 0$ and $f_3 = 0$. On the other hand, u_2 and u_3 in the left-hand side, and $\left.\dfrac{d\bar{u}}{dx}\right|_{x=0}$ and $\left.\dfrac{d\bar{u}}{dx}\right|_{x=1}$ in the right-hand side are unknown variables. By changing unknown variables $\left.\dfrac{d\bar{u}}{dx}\right|_{x=0}$ and $\left.\dfrac{d\bar{u}}{dx}\right|_{x=1}$ with the first and the fourth components of the vector in the left-hand side of Eq. (1.54), and by substituting $h^{(1)} = h^{(2)} = h^{(3)} = 1/3$ into Eq. (1.54), after rearrangement of the equation, the global simultaneous equation is rewritten as follows:

$$
\begin{bmatrix}
-1 & -\dfrac{53}{18} & 0 & 0 \\
0 & \dfrac{56}{9} & -\dfrac{53}{18} & 0 \\
0 & -\dfrac{53}{18} & \dfrac{56}{9} & 0 \\
0 & 0 & -\dfrac{53}{18} & -1
\end{bmatrix}
\left\{
\begin{array}{c}
\left.\dfrac{d\bar{u}}{dx}\right|_{x=0} \\
u_2 \\
u_3 \\
\left.\dfrac{d\bar{u}}{dx}\right|_{x=1}
\end{array}
\right\}
=
\left\{
\begin{array}{c}
0 \\
0 \\
\dfrac{53}{18} \\
-\dfrac{28}{9}
\end{array}
\right\}
\tag{1.55}
$$

where the new vector in the left-hand side of the equation is an unknown vector and the one in the right-hand side is a known vector.

After solving Eq. (1.55), it is found that $u_2 = 0.2885$, $u_3 = 0.6098$, $\left.\dfrac{d\bar{u}}{dx}\right|_{x=0} = 0.8496$, and $\left.\dfrac{d\bar{u}}{dx}\right|_{x=1} = 1.3157$. The exact solutions for u_2 and u_3 can be calculated as $u_2 = 0.2889$ and $u_3 = 0.6102$ from Eq. (1.15). The relative errors for u_2 and u_3 are as small as 0.1% and 0.06%, respectively. The calculated values of the derivative, $\left.\dfrac{d\bar{u}}{dx}\right|_{x=0}$ and $\left.\dfrac{d\bar{u}}{dx}\right|_{x=1}$ are improved when compared to those by the one-element finite element method.

In this subsection, only the one-dimensional finite element method was described. The finite element method can be applied to two- and three-dimensional continuum problems of various kinds, which are described in terms of ordinary and partial differential equations. There is no essential difference between the formulation for one-dimensional problems and the formulations for higher dimensions, except for the intricacy of formulation.

1.4   FEM in two-dimensional elastostatic problems

Generally speaking, elasticity problems are reduced to solving the partial differential equations known as the equilibrium equations together with the stress–strain relations or the constitutive equations, the strain–displacement relations, and the compatibility equation under given boundary conditions. The exact solutions can be obtained in quite limited cases only, and exact solutions in general cannot be solved in closed forms. In order to conquer these difficulties, the finite element method (FEM) has been developed as a powerful numerical method to obtain approximate solutions for various kinds of elasticity problems. The finite element method assumes an object of analysis as an aggregate of elements having arbitrary shapes and finite sizes (called finite elements), approximates partial differential equations by simultaneous algebraic equations, and numerically solves various elasticity problems. Finite elements take the form of a line segment in one-dimensional problems as shown in the preceding section, a triangle or rectangle in two-dimensional problems and tetrahedron, and a cuboid or prism in three-dimensional problems. Since the procedure of the finite element method is mathematically based on the variational method, it can be applied not only to elasticity problems of structures, but also to various problems relating to thermodynamics, fluid dynamics, and vibrations, which are described by partial differential equations.

1.4.1 Elements of finite-element procedures in the analysis of plane elastostatic problems

Limited to static (without time variation) elasticity problems, the procedure described in the preceding section is essentially the same as that of the stress analyses by the finite element method. The procedure is summarised as follows:

Procedure 1: *Discretisation:* Divide an object of analysis into a finite number of finite elements.

Procedure 2: *Selection of the interpolation function*: Select the element type or the interpolation function which approximates displacements and strains in each finite element.

Procedure 3: *Derivation of element stiffness matrices*: Determine the element stiffness matrix which relates forces and displacements in each element.

Procedure 4: *Assembly of stiffness matrices into the global stiffness matrix*: Assemble the element stiffness matrices into the global stiffness matrix which relates forces and displacements in the whole elastic body to analyse.

Procedure 5: *Rearrangement of the global stiffness matrix*: Substitute prescribed applied forces (mechanical boundary conditions) and displacements (geometrical boundary conditions) into the global stiffness matrix, and rearrange the matrix by collecting unknown variables for forces and displacements, say, in the left-hand side and known values of the forces and displacements in the right-hand side in order to set up the simultaneous equations.

Procedure 6: *Derivation of unknown forces and displacements*: Solve the simultaneous equations set up in Procedure 5 above to solve the unknown variables for forces and displacements. The solutions for unknown forces are reaction forces and those for unknown displacements are deformations of the elastic body of interest for given geometrical and mechanical boundary conditions, respectively.

Procedure 7: *Computation of strains and stresses*: Compute the strains and stresses from the displacements obtained in Procedure 6 by using the strain–displacement relations and the stress–strain relations explained later.

1.4.2   Fundamental formulae in plane elastostatic problems

1.4.2.1   Equations of equilibrium

Consider the static equilibrium state of an infinitesimally small rectangle with its sides parallel to the coordinate axes in a two-dimensional elastic body as shown in Fig. 1.6. If the body forces F_x and F_y act in the directions of the x- and y-axes, respectively, the equations of equilibrium in the elastic body can be derived as follows:

$$\begin{cases} \dfrac{\partial \sigma_x}{\partial x} + \dfrac{\partial \tau_{xy}}{\partial y} + F_x = 0 \\ \dfrac{\partial \tau_{yx}}{\partial x} + \dfrac{\partial \sigma_y}{\partial y} + F_y = 0 \end{cases} \tag{1.56}$$

where σ_x and σ_y are normal stresses in the x- and y-axes, respectively, and τ_{xy} and τ_{yx} shear stresses acting in the x–y plane. The shear stresses τ_{xy} and τ_{yx} are generally equal to each other due to the rotational equilibrium of the two-dimensional elastic body around its centre of gravity.

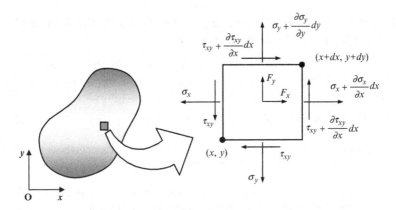

Fig. 1.6 Stress states in an infinitesimal element of two-dimensional elastic body.

1.4.2.2   Strain–displacement relations

If the deformation of a two-dimensional elastic body is infinitesimally small under the applied load, the normal strains ε_x and ε_y in the directions of the x- and the y-axes and the engineering shearing strain γ_{xy} in the x–y plane are expressed by the following equations:

$$\begin{cases} \varepsilon_x = \dfrac{\partial u}{\partial x} \\ \varepsilon_y = \dfrac{\partial v}{\partial y} \\ \gamma_{xy} = \dfrac{\partial v}{\partial x} + \dfrac{\partial u}{\partial y} \end{cases} \tag{1.57}$$

where u and v are infinitesimal displacements in the directions of the x- and the y-axes, respectively.

1.4.2.3 Stress–strain relations (constitutive equations)

The stress–strain relations describe states of deformation, or strains induced by the internal forces, or stresses resisting against applied loads. Unlike the other fundamental equations shown in Eqs (1.56), (1.57), which can be determined mechanistically or geometrically, these relations depend on the properties of the material and they are determined experimentally and often called as constitutive relations or constitutive equations. One of the most popular relations is the generalised Hooke's law, which relates three components of the two-dimensional stress tensor with those of strain tensor through the following simple linear expressions:

$$
\begin{cases}
\sigma_x = \dfrac{\nu E}{(1+\nu)(1-2\nu)}e_v + 2G\varepsilon_x \\[2mm]
\sigma_y = \dfrac{\nu E}{(1+\nu)(1-2\nu)}e_v + 2G\varepsilon_y \\[2mm]
\sigma_y = \dfrac{\nu E}{(1+\nu)(1-2\nu)}e_v + 2G\varepsilon_z \\[2mm]
\tau_{xy} = G\gamma_{xy} = \dfrac{E}{2(1+\nu)}\gamma_{xy} \\[2mm]
\tau_{yz} = G\gamma_{yz} = \dfrac{E}{2(1+\nu)}\gamma_{yz} \\[2mm]
\tau_{zx} = G\gamma_{zx} = \dfrac{E}{2(1+\nu)}\gamma_{zx}
\end{cases}
\tag{1.58a}
$$

or inversely

$$
\begin{cases}
\varepsilon_x = \dfrac{1}{E}\left[\sigma_x - \nu(\sigma_y + \sigma_z)\right] \\[2mm]
\varepsilon_y = \dfrac{1}{E}\left[\sigma_y - \nu(\sigma_z + \sigma_x)\right] \\[2mm]
\varepsilon_z = \dfrac{1}{E}\left[\sigma_z - \nu(\sigma_x + \sigma_y)\right] \\[2mm]
\gamma_{xy} = \dfrac{\tau_{xy}}{G} \\[2mm]
\gamma_{yz} = \dfrac{\tau_{yz}}{G} \\[2mm]
\gamma_{zx} = \dfrac{\tau_{zx}}{G}
\end{cases}
\tag{1.58b}
$$

where E is Young's modulus, ν is Poisson's ratio, G is the shear modulus, and e_v is the volumetric strain expressed by the sum of the three normal components of strain, i.e., $e_V = \varepsilon_x + \varepsilon_y + \varepsilon_z$. In other words, the volumetric strain e_V can be written as $e_v = \Delta V/V$ where V is the initial volume of the elastic body of interest in an undeformed state and ΔV the change of the volume after deformation.

In the two-dimensional elasticity theory, the three-dimensional Hooke's law is converted into two-dimensional form by using the two types of approximations, i.e. plane stress and plane strain approximations.

1. Plane stress approximation: For thin plates, for example, one can assume the plane stress approximation that all the stress components in the direction perpendicular to the plate surface vanish, i.e. $\sigma_z = \tau_{zx} = \tau_{yz} = 0$. The stress–strain relations in this approximation are written by the following two-dimensional Hooke's law:

$$\begin{cases} \sigma_x = \dfrac{E}{1-\nu^2}\left(\varepsilon_x + \nu\varepsilon_y\right) \\[2mm] \sigma_y = \dfrac{E}{1-\nu^2}\left(\varepsilon_y + \nu\varepsilon_x\right) \\[2mm] \tau_{xy} = G\gamma_{xy} = \dfrac{E}{2(1+\nu)}\gamma_{xy} \end{cases} \qquad (1.59a)$$

or

$$\begin{cases} \varepsilon_x = \dfrac{1}{E}\left(\sigma_x - \nu\sigma_y\right) \\[2mm] \varepsilon_y = \dfrac{1}{E}\left(\sigma_y - \nu\sigma_x\right) \\[2mm] \gamma_{xy} = \dfrac{\tau_{xy}}{G} = \dfrac{2(1+\nu)}{E}\tau_{xy} \end{cases} \qquad (1.59b)$$

However, the normal strain component ε_z in the thickness direction is not zero, but $\varepsilon_z = -\nu(\sigma_x + \sigma_y)/E$.

The plane stress approximation satisfies the equations of equilibrium (1.56), nevertheless the normal strain in the direction of the z-axis, ε_z must take a special form, i.e., ε_z must be a linear function of coordinate variables x and y in order to satisfy the compatibility condition which ensures the single-valuedness and continuity conditions of strains. Since this approximation impose a special requirement for the form of the strain ε_z and thus the forms of the normal stresses σ_x and σ_y, this approximation cannot be considered as a general rule. Strictly speaking, the plane stress state does not exist in reality.

2. Plane strain approximation: In the case where plate thickness (in the direction of the z-axis) is so large that the displacement is subjected to large constraints in the direction of the z-axis such that $\varepsilon_z = \gamma_{zx} = \gamma_{yz} = 0$. This case is called the plane stress approximation. The generalised Hooke's law can be written as follows:

$$\begin{cases} \sigma_x = \dfrac{E}{(1+\nu)(1-2\nu)}\left[(1-\nu)\varepsilon_x + \nu\varepsilon_y\right] \\[2mm] \sigma_y = \dfrac{E}{(1+\nu)(1-2\nu)}\left[\nu\varepsilon_x + (1-\nu)\varepsilon_y\right] \\[2mm] \tau_{xy} = G\gamma_{xy} = \dfrac{E}{2(1+\nu)}\gamma_{xy} \end{cases} \qquad (1.60a)$$

or

$$\begin{cases} \varepsilon_x = \dfrac{1+\nu}{E}\left[(1-\nu)\sigma_x - \nu\sigma_y\right] \\ \varepsilon_y = \dfrac{1+\nu}{E}\left[-\nu\sigma_x + (1-\nu)\sigma_y\right] \\ \gamma_{xy} = \dfrac{\tau_{xy}}{G} = \dfrac{2(1+\nu)}{E}\tau_{xy} \end{cases} \tag{1.60b}$$

The normal stress component σ_z in the thickness direction is not zero, but $\sigma_z = \nu E(\sigma_x + \sigma_y)/[(1+\nu)(1-2\nu)]$. Since the plane strain state satisfies the equations of equilibrium (1.56) and the compatibility condition, this state can exist in reality.

If we redefine Young's modulus and Poisson's ratio by the following formulae:

$$E' = \begin{cases} E & \text{(plane stress)} \\ \dfrac{E}{1-\nu^2} & \text{(plane strain)} \end{cases} \tag{1.61a}$$

$$\nu' = \begin{cases} \nu & \text{(plane stress)} \\ \dfrac{\nu}{1-\nu} & \text{(plane strain)} \end{cases} \tag{1.61b}$$

the two-dimensional Hooke's law can be expressed in a unified form:

$$\begin{cases} \sigma_x = \dfrac{E'}{1-\nu'^2}\left(\varepsilon_x + \nu'\varepsilon_y\right) \\ \sigma_y = \dfrac{E'}{1-\nu'^2}\left(\varepsilon_y + \nu'\varepsilon_x\right) \\ \tau_{xy} = G\gamma_{xy} = \dfrac{E'}{2(1+\nu')}\gamma_{xy} \end{cases} \tag{1.62a}$$

or

$$\begin{cases} \varepsilon_x = \dfrac{1}{E'}\left(\sigma_x - \nu'\sigma_y\right) \\ \varepsilon_y = \dfrac{1}{E'}\left(\sigma_y - \nu'\sigma_x\right) \\ \gamma_{xy} = \dfrac{\tau_{xy}}{G} = \dfrac{2(1+\nu')}{E'}\tau_{xy} \end{cases} \tag{1.62b}$$

The shear modulus G is invariant under this transformations as shown in Eqs (1.61a), (1.61b), i.e.,

$$G = \dfrac{E}{2(1+\nu)} = \dfrac{E'}{2(1+\nu')} = G'$$

1.4.2.4 Boundary conditions

When solving the partial differential equations (1.56), there remains indefiniteness in the form of integral constants. In order to eliminate this indefiniteness, prescribed conditions on stress and/or displacements must be imposed on the bounding surface of the elastic body. These conditions are called boundary conditions. There are two types of boundary conditions: mechanical boundary conditions prescribing stresses or surface tractions, and geometrical boundary conditions prescribing displacements.

Let us denote a portion of the surface of the elastic body where stresses are prescribed by S_σ and the remaining surface where displacements are prescribed by S_u. The whole surface of the elastic body is denoted by $S = S_\sigma + S_u$. Note that it is not possible to prescribe both stresses and displacements on a portion of the surface of the elastic body.

The mechanical boundary conditions on S_σ are given by the following equations:

$$\begin{cases} t_x^* = \bar{t}_x^* \\ t_y^* = \bar{t}_y^* \end{cases} \tag{1.63}$$

where t_x^* and t_y^* are the x- and y-components of the traction force $\mathbf{t}^*$, respectively, and the bar over t_x^* and t_y^* indicates that those quantities are prescribed on that portion of the surface. Taking $\mathbf{n} = [\cos\alpha, \sin\alpha]$ as the outward unit normal vector at a point of a small element of the surface portion S_σ, the Cauchy relations which represent the equilibrium conditions for surface traction forces and internal stresses:

$$\begin{cases} \bar{t}_x^* = \sigma_x \cos\alpha + \tau_{xy} \sin\alpha \\ \bar{t}_y^* = \tau_{xy} \cos\alpha + \sigma_y \sin\alpha \end{cases} \tag{1.64}$$

where α is the angle between the normal vector $\mathbf{n}$ and the x-axis. For free surfaces where no forces are applied, $t_x^* = 0$ and $t_y^* = 0$.

The geometrical boundary conditions on S_u are given by the following equations:

$$\begin{cases} u = \bar{u} \\ v = \bar{v} \end{cases} \tag{1.65}$$

where $\bar{u}$ and $\bar{v}$ are the x- and y-components of prescribed displacements $\mathbf{u}$ on S_u. One of the most popular geometrical boundary conditions, i.e., clamped end condition, is denoted by $u = 0$ and/or $v = 0$ as shown in Fig. 1.7.

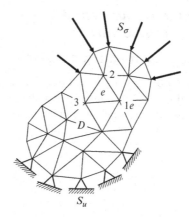

Fig. 1.7 Finite-element discretisation of two-dimensional elastic body by triangular elements.

1.4.3 Variational formulae in elastostatic problems: The principle of virtual work

The variational principle used in two-dimensional elasticity problems is the principle of virtual work, which is expressed by the following integral equation:

$$\iint_D \left(\sigma_x \delta\varepsilon_x + \sigma_y \delta\varepsilon_y + \tau_{xy}\delta\gamma_{xy}\right) t dx dy - \iint_D \left(F_x \delta u + F_y \delta v\right) t dx dy$$

$$- \int_{S_\sigma} \left(\bar{t}_x^* \delta u + \bar{t}_y^* \delta v\right) t ds = 0 \tag{1.66}$$

where D denotes the whole region of a two-dimensional elastic body of interest, S the whole portion of the surface of the elastic body $S(=S_\sigma \cup S_u)$ where the boundary conditions are prescribed, and t the thickness.

The first term in the left-hand side of Eq. (1.66) represents the increment of the strain energy of the elastic body, the second term the increment of the work done by the body forces, and the third term the increment of the work done by the surface traction forces. Therefore, Eq. (1.66) claims that the increment of the strain energy of the elastic body is equal to the work done by the forces applied.

The fact that the integrand in each integral in the left-hand side of Eq. (1.66) is identically equal to zero brings about the equations of equilibrium (1.56) and the boundary conditions (1.63) and/or (1.65). Therefore, instead of solving the partial differential equations (1.56) under the boundary conditions of Eqs (1.63) and/or (1.65), two-dimensional elasticity problems can be solved by using the integral equation (1.66).

1.4.4 Formulation of the fundamental finite-element equations in plane elastostatic problems

1.4.4.1 Strain–displacement matrix or [B] matrix

Let us use the constant-strain triangular element (see Fig. 1.8A) to derive the fundamental finite element equations in plane elastostatic problems. The constant-strain triangular element assumes the displacements within the element to be expressed by the following linear functions of the coordinate variables (x, y):

$$\begin{cases} u = \alpha_0 + \alpha_1 x + \alpha_2 y \\ v = \beta_0 + \beta_1 x + \beta_2 y \end{cases} \tag{1.67}$$

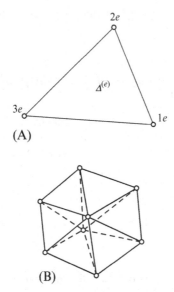

Fig. 1.8 (A) Triangular constant strain element and (B) the continuity of displacements.

The above interpolation functions for displacements convert straight lines joining arbitrary two points in the element into straight lines after deformation. Since the boundaries between neighbouring elements are straight lines joining the apices or nodal points of triangular elements, any incompatibility does not occur along the boundaries between adjacent elements, and displacements are continuous everywhere in the domain to analyse as shown in Fig. 1.8B. For the eth triangular element consisting of three apices or nodal points $(1_e, 2_e, 3_e)$ having the coordinates (x_{1e}, y_{1e}), (x_{2e}, y_{2e}), and (x_{3e}, y_{3e}) and the nodal displacements (u_{1e}, v_{1e}), (u_{2e}, v_{2e}), and (u_{3e}, v_{3e}), the coefficients $\alpha_0, \alpha_1, \alpha_2, \beta_0, \beta_1,$ and β_2 in Eq. (1.67) are obtained by the following equations:

$$\left\{ \begin{array}{l} \left\{ \begin{array}{c} \alpha_0 \\ \alpha_1 \\ \alpha_2 \end{array} \right\} = \left(\begin{array}{ccc} a_{1e} & a_{2e} & a_{3e} \\ b_{1e} & b_{2e} & b_{3e} \\ c_{1e} & c_{2e} & c_{3e} \end{array} \right) \left\{ \begin{array}{c} u_{1e} \\ u_{2e} \\ u_{3e} \end{array} \right\} \\ \left\{ \begin{array}{c} \beta_0 \\ \beta_1 \\ \beta_2 \end{array} \right\} = \left(\begin{array}{ccc} a_{1e} & a_{2e} & a_{3e} \\ b_{1e} & b_{2e} & b_{3e} \\ c_{1e} & c_{2e} & c_{3e} \end{array} \right) \left\{ \begin{array}{c} v_{1e} \\ v_{2e} \\ v_{3e} \end{array} \right\} \end{array} \right. \tag{1.68}$$

where

$$\left\{ \begin{array}{l} a_{1e} = \dfrac{1}{2\Delta^{(e)}} (x_{2e}y_{3e} - x_{3e}y_{2e}) \\ b_{1e} = \dfrac{1}{2\Delta^{(e)}} (y_{2e} - y_{3e}) \\ c_{1e} = \dfrac{1}{2\Delta^{(e)}} (x_{3e} - x_{2e}) \end{array} \right. \tag{1.69a}$$

$$\left\{ \begin{array}{l} a_{2e} = \dfrac{1}{2\Delta^{(e)}} (x_{3e}y_{1e} - x_{1e}y_{3e}) \\ b_{2e} = \dfrac{1}{2\Delta^{(e)}} (y_{3e} - y_{1e}) \\ c_{2e} = \dfrac{1}{2\Delta^{(e)}} (x_{1e} - x_{3e}) \end{array} \right. \tag{1.69b}$$

$$\left\{ \begin{array}{l} a_{3e} = \dfrac{1}{2\Delta^{(e)}} (x_{1e}y_{2e} - x_{2e}y_{1e}) \\ b_{3e} = \dfrac{1}{2\Delta^{(e)}} (y_{1e} - y_{2e}) \\ c_{3e} = \dfrac{1}{2\Delta^{(e)}} (x_{2e} - x_{1e}) \end{array} \right. \tag{1.69c}$$

The numbers with subscript 'e'—$1_e, 2_e,$ and 3_e—in the above equations are called element nodal numbers and denote the numbers of three nodal points of the eth element. Nodal points should be numbered counter-clockwise. These three numbers are used only in the eth element. Nodal numbers of the other type called global nodal numbers are also assigned to the three nodal points of the eth element numbered throughout the whole model of the elastic body. The symbol $\Delta^{(e)}$ represent the area of the eth element and can be expressed only by the coordinates of the nodal points of the element, i.e.,

$$\Delta^{(e)} = \frac{1}{2}[(x_{1e} - x_{3e})(y_{2e} - y_{3e}) - (y_{3e} - y_{1e})(x_{3e} - x_{2e})] = \frac{1}{2} \begin{vmatrix} 1 & x_{1e} & y_{1e} \\ 1 & x_{2e} & y_{2e} \\ 1 & x_{3e} & y_{3e} \end{vmatrix} \tag{1.69d}$$

Consequently, the components of the displacement vector $[u, v]$ can be expressed by the components of the nodal displacement vectors $[u_{1e}, v_{1e}]$, $[u_{2e}, v_{2e}]$, and $[u_{3e}, v_{3e}]$ as follows:

$$\begin{cases} u = (a_{1e}+b_{1e}x+c_{1e}y)u_{1e} + (a_{2e}+b_{2e}x+c_{2e}y)u_{2e} + (a_{3e}+b_{3e}x+c_{3e}y)u_{3e} \\ v = (a_{1e}+b_{1e}x+c_{1e}y)v_{1e} + (a_{2e}+b_{2e}x+c_{2e}y)v_{2e} + (a_{3e}+b_{3e}x+c_{3e}y)v_{3e} \end{cases} \quad (1.70)$$

The matrix notation of Eq. (1.70) is

$$\begin{Bmatrix} u \\ v \end{Bmatrix} = \begin{bmatrix} N_{1e}^{(e)} & 0 & N_{2e}^{(e)} & 0 & N_{3e}^{(e)} & 0 \\ 0 & N_{1e}^{(e)} & 0 & N_{2e}^{(e)} & 0 & N_{3e}^{(e)} \end{bmatrix} \begin{Bmatrix} u_{1e} \\ v_{1e} \\ u_{2e} \\ v_{2e} \\ u_{3e} \\ v_{3e} \end{Bmatrix} = [\mathbf{N}]\{\boldsymbol{\delta}\}^{(e)} \quad (1.71)$$

where

$$\begin{cases} N_{1e}^{(e)} = a_{1e}+b_{1e}x+c_{1e}y \\ N_{2e}^{(e)} = a_{2e}+b_{2e}x+c_{2e}y \\ N_{3e}^{(e)} = a_{3e}+b_{3e}x+c_{3e}y \end{cases} \quad (1.72)$$

and the superscript of $\{\boldsymbol{\delta}\}^{(e)}$, (e), indicates that $\{\boldsymbol{\delta}\}^{(e)}$ is the displacement vector determined by the three displacement vectors at the three nodal points of the eth triangular element. Eq. (1.72) formulates the definitions of the interpolation functions or shape functions $N_{ie}^{(e)}(i = 1, 2, 3)$ for the triangular constant-strain element.

Now, let us consider strains derived from the displacements given by Eq. (1.71). Substitution of Eq. (1.71) into Eq. (1.57) gives.

$$\{\boldsymbol{\varepsilon}\} = \begin{Bmatrix} \varepsilon_x \\ \varepsilon_y \\ \gamma_{xy} \end{Bmatrix} = \begin{Bmatrix} \dfrac{\partial u}{\partial x} \\ \dfrac{\partial v}{\partial y} \\ \dfrac{\partial v}{\partial x}+\dfrac{\partial u}{\partial y} \end{Bmatrix} = \begin{bmatrix} \dfrac{\partial N_{1e}^{(e)}}{\partial x} & 0 & \dfrac{\partial N_{2e}^{(e)}}{\partial x} & 0 & \dfrac{\partial N_{3e}^{(e)}}{\partial x} & 0 \\ 0 & \dfrac{\partial N_{1e}^{(e)}}{\partial y} & 0 & \dfrac{\partial N_{2e}^{(e)}}{\partial y} & 0 & \dfrac{\partial N_{3e}^{(e)}}{\partial y} \\ \dfrac{\partial N_{1e}^{(e)}}{\partial x} & \dfrac{\partial N_{1e}^{(e)}}{\partial y} & \dfrac{\partial N_{2e}^{(e)}}{\partial x} & \dfrac{\partial N_{2e}^{(e)}}{\partial y} & \dfrac{\partial N_{3e}^{(e)}}{\partial x} & \dfrac{\partial N_{3e}^{(e)}}{\partial y} \end{bmatrix} \begin{Bmatrix} u_{1e} \\ v_{1e} \\ u_{2e} \\ v_{2e} \\ u_{3e} \\ v_{3e} \end{Bmatrix}$$

$$= \begin{bmatrix} b_{1e} & 0 & b_{2e} & 0 & b_{3e} & 0 \\ 0 & c_{1e} & 0 & c_{2e} & 0 & c_{3e} \\ c_{1e} & b_{1e} & c_{2e} & b_{2e} & c_{3e} & b_{3e} \end{bmatrix} \begin{Bmatrix} u_{1e} \\ v_{1e} \\ u_{2e} \\ v_{2e} \\ u_{3e} \\ v_{3e} \end{Bmatrix} = [\mathbf{B}]\{\boldsymbol{\delta}\}^{(e)}$$

$$(1.73)$$

where [**B**] establishes the relationship between the nodal displacement vector $\{\boldsymbol{\delta}\}^{(e)}$ and the element strain vector $\{\boldsymbol{\varepsilon}\}$, and is called the strain–displacement matrix or [**B**] matrix. All the components of the [**B**] matrix are expressed only by the coordinate values of the three nodal points consisting of the element.

From the discussion above, it can be concluded that strains are constant throughout a three-node triangular element since its interpolation functions are linear functions of the coordinate variables within the element. For this reason, a triangular element with three nodal points is called a 'constant strain' element. Hence, three-node triangular elements cannot satisfy the compatibility condition in the strict sense, since strains are discontinuous among elements. It is demonstrated, however, that the results obtained by elements of this type converge to exact solutions as the size of the elements becomes smaller.

It is known that elements must fulfil the following three criteria for the finite element solutions to converge to the exact solutions as the subdivision into even smaller elements is attempted. Namely, the elements must:

1. represent rigid body displacements;
2. represent constant strains; and
3. ensure the continuity of displacements among elements.

1.4.4.2 Stress–strain matrix or [D] matrix

Substitution of Eq. (1.73) into Eq. (1.62a) gives

$$\{\boldsymbol{\sigma}\} = \left\{ \begin{array}{c} \sigma_x \\ \sigma_y \\ \tau_{xy} \end{array} \right\} = \frac{E'}{1-v'^2} \begin{bmatrix} 1 & v' & 0 \\ v' & 1 & 0 \\ 0 & 0 & \dfrac{1-v'}{2} \end{bmatrix} \left\{ \begin{array}{c} \varepsilon_x \\ \varepsilon_y \\ \gamma_{xy} \end{array} \right\} = [\mathbf{D}^e]\{\boldsymbol{\varepsilon}\}$$

$$= [\mathbf{D}^e][\mathbf{B}]\{\boldsymbol{\delta}\}^{(e)}$$

(1.74)

where [**D**^{e}] establishes the relationship between stresses and strains, or the constitutive relations. The matrix [**D**^{e}] is for elastic bodies and thus is called the elastic stress–strain matrix or just [**D**] matrix. In the case where initial strains $\{\boldsymbol{\varepsilon}_0\}$ such as plastic strains, thermal strains, and residual strains exist, $\{\boldsymbol{\varepsilon}\} - \{\boldsymbol{\varepsilon}_0\}$ is used instead of $\{\boldsymbol{\varepsilon}\}$.

1.4.4.3 Element stiffness equations

First, let $\{\mathbf{P}\}^{(e)}$ define the equivalent nodal forces which are statically equivalent to the traction forces $\mathbf{t}^* = [t_x^*, t_y^*]$ on the element boundaries and the body forces $\{\mathbf{F}\}^{(e)}$ in the element:

$$\{\mathbf{F}\}^{(e)T} = \begin{bmatrix} F_x, & F_y \end{bmatrix}$$

(1.75)

$$\{\mathbf{P}\}^{(e)T} = \begin{bmatrix} X_{1e}, & Y_{1e}, & X_{2e}, & Y_{2e}, & X_{3e}, & Y_{3e} \end{bmatrix}$$

(1.76)

In the above equations, $\{\cdot\}$ represents a column vector, $[\cdot]$ a row vector, and the superscript T transpose of a vector or a matrix.

In order to make differentiations shown in Eq. (1.57), displacements assumed by Eq. (1.71) must be continuous everywhere in an elastic body of interest. The remaining conditions to satisfy are the equations of equilibrium (1.56) and the mechanical boundary conditions (1.63); nevertheless, these equations generally cannot be satisfied in the strict sense. Hence, defining the equivalent nodal forces, for instance, (X_{1e}, Y_{1e}), (X_{2e}, Y_{2e}), and (X_{3e}, Y_{3e}) on the three nodal points of the eth element and determining these forces by the principle of the virtual work in order to satisfy the equilibrium and boundary conditions element by element. Namely, the principle of the virtual work to be satisfied for arbitrary virtual displacements $\{\delta^*\}^{(e)}$ of the eth element is derived from Eq. (1.66) as

$$\{\delta^*\}^{(e)T}\{\mathbf{P}\}^{(e)} = \iint\limits_{D} \left(\{\varepsilon^*\}^T\{\delta\} - \{\mathbf{f}^*\}^T\{\mathbf{F}\}^{(e)} \right) t\,dx\,dy \qquad (1.77)$$

where

$$\{\varepsilon^*\} = [\mathbf{B}]\{\delta^*\}^{(e)} \qquad (1.78)$$

$$\{\mathbf{f}^*\} = [\mathbf{N}]\{\delta^*\}^{(e)} \qquad (1.79)$$

Substitution of Eqs (1.78), (1.79) into Eq. (1.77) gives.

$$\{\delta^*\}^{(e)T}\{\mathbf{P}\}^{(e)} = \{\delta^*\}^{(e)T} \left(\iint\limits_{D} [\mathbf{B}]^T\{\sigma\}t\,dx\,dy - \iint\limits_{D} [\mathbf{N}]^T\{\mathbf{F}\}^{(e)}t\,dx\,dy \right) \qquad (1.80)$$

Since Eq. (1.80) holds true for any virtual displacements $\{\delta^*\}^{(e)}$, the equivalent nodal forces can be obtained by the following equation:

$$\{\mathbf{P}\}^{(e)} = \iint\limits_{D} [\mathbf{B}]^T\{\sigma\}t\,dx\,dy - \iint\limits_{D} [\mathbf{N}]^T\{\mathbf{F}\}^{(e)}t\,dx\,dy \qquad (1.81)$$

From Eqs (1.73), (1.74),

$$\{\sigma\} = [\mathbf{D}^e](\{\varepsilon\} - \{\varepsilon_0\}) = [\mathbf{D}^e][\mathbf{B}]\{\delta\}^{(e)} - [\mathbf{D}^e]\{\varepsilon_0\} \qquad (1.82)$$

Substitution of Eq. (1.82) into Eq. (1.81) gives

$$\{\mathbf{P}\}^{(e)} = \left(\iint\limits_{D} [\mathbf{B}]^T[\mathbf{D}^e][\mathbf{B}]t\,dx\,dy \right)\{\delta\}^{(e)} - \iint\limits_{D} [\mathbf{B}]^T[\mathbf{D}^e]\{\varepsilon_0\}t\,dx\,dy$$

$$- \iint\limits_{D} [\mathbf{N}]^T\{\mathbf{F}\}^{(e)}t\,dx\,dy \qquad (1.83)$$

Eq. (1.83) is rewritten in the form

$$\{P\}^{(e)} = \left[k^{(e)}\right]\{\delta\}^{(e)} + \{F_{\varepsilon_0}\}^{(e)} + \{F_F\}^{(e)} \tag{1.84}$$

where

$$\left[k^{(e)}\right] \equiv \iint_D [B]^T [D^e][B] t \, dx dy = \Delta^{(e)} [B]^T [D^e][B] t \tag{1.85}$$

$$\{F_{\varepsilon_0}\}^{(e)} \equiv - \iint_D [B]^T [D^e]\{\varepsilon_0\} t \, dx dy \tag{1.86}$$

$$\{F_F\}^{(e)} \equiv - \iint_D [N]^T \{F\}^{(e)} t \, dx dy \tag{1.87}$$

Eq. (1.84) is called the element stiffness equation for the eth triangular finite element and $[k^{(e)}]$ defined by Eq. (1.85) the element stiffness matrix. Since the matrices $[B]$ and $[D^e]$ are constant throughout the element, they can be taken out of the integral and the integral is simply equal to the area of the element $\Delta^{(e)}$ so that the rightmost side of Eq. (1.85) is obtained. The forces $\{F_{\varepsilon_0}\}^{(e)}$ and $\{F_F\}^{(e)}$ are the equivalent nodal forces due to initial strains and to body forces, respectively. Since the integrand in Eq. (1.85) is generally a function of the coordinate variables x and y except for the case of three-node triangular elements, the integrals appearing in Eq. (1.85) are often evaluated by numerical integration scheme such as the Gaussian quadrature.

The element stiffness matrix $[k^{(e)}]$ in Eq. (1.85) is a 6×6 square matrix which can be decomposed into nine 2×2 submatrices, as shown in the following equation:

$$\left[k^{(e)}_{ieje}\right] = \begin{bmatrix} k^{(e)}_{1e1e} & k^{(e)}_{1e2e} & k^{(e)}_{1e3e} \\ k^{(e)}_{2e1e} & k^{(e)}_{2e2e} & k^{(e)}_{2e3e} \\ k^{(e)}_{3e1e} & k^{(e)}_{3e2e} & k^{(e)}_{3e3e} \end{bmatrix} \tag{1.88}$$

$$k^{(e)}_{ieje} = k^{(e)T}_{jeie} \tag{1.89}$$

$$k^{(e)}_{ieje} = \int_{\Delta^{(e)}} [B_{ie}]^T [D^e][B_{je}] t \, dx dy \begin{cases} (2 \times 2) \text{ asymmetric matrix } (i_e \neq j_e) \\ (2 \times 2) \text{ symmetric matrix } (i_e = j_e) \end{cases} \tag{1.90}$$

where

$$[B_{ie}] \equiv \frac{1}{2\Delta^{(e)}} \begin{bmatrix} b_{ie} & 0 \\ 0 & c_{ie} \\ c_{ie} & b_{ie} \end{bmatrix} \quad (i_e = 1, 2, 3) \tag{1.91}$$

and the subscripts i_e and j_e of $\mathbf{k}_{i_e j_e}^{(e)}$ refers to element nodal numbers, and

$$\mathbf{k}_{i_e j_e}^{(e)} = [\mathbf{B}_{ie}]^T [\mathbf{D}^e] [\mathbf{B}_{je}] t\Delta^{(e)} \tag{1.92}$$

In the above discussion, the formulae have been obtained for just one triangular element, but are available for any types of elements, if necessary, with some modifications.

1.4.4.4 Global stiffness equations

Element stiffness equations as shown in Eq. (1.84) are determined element by element, and then they are assembled into the global stiffness equations for the whole elastic body of interest. Since nodal points which belong to different elements but have the same coordinates are the same points, the following points during the assembly procedure of the global stiffness equations should be noted:

1. The displacement components u and v of the same nodal points which belong to different elements are the same; i.e. there exist no incompatibilities such as cracks between elements.
2. For nodal points on the bounding surfaces and for those in the interior of the elastic body to which no forces are applied, the sums of the nodal forces are to be zero.
3. Similarly, for nodal points to which forces are applied, the sums of the nodal forces are equal to the sums of the forces applied to those nodal points.

The same global nodal numbers are to be assigned to the nodal points which have the same coordinates. Taking the points described above into consideration, let us rewrite the element stiffness matrix [$\mathbf{k}^{(e)}$] in Eq. (1.88) by using the global nodal numbers I, J and K ($I, J, K = 1, 2, ..., 2n$) instead of the element nodal numbers i_e, j_e and k_e ($i_e, j_e, k_e = 1, 2, 3$); i.e.,

$$\left[\mathbf{k}^{(e)}\right] = \begin{bmatrix} \mathbf{k}_{II}^{(e)} & \mathbf{k}_{IJ}^{(e)} & \mathbf{k}_{IK}^{(e)} \\ \mathbf{k}_{JI}^{(e)} & \mathbf{k}_{JJ}^{(e)} & \mathbf{k}_{JK}^{(e)} \\ \mathbf{k}_{KI}^{(e)} & \mathbf{k}_{KJ}^{(e)} & \mathbf{k}_{KK}^{(e)} \end{bmatrix} \tag{1.93}$$

Then, let us embed the element stiffness matrix in a square matrix having the same size as the global stiffness matrix of $2n \times 2n$, as shown in Eq. (1.94):

$$(1.94)$$

where n denotes the number of degrees of freedom. This procedure is called the method of extended matrix. The number of degrees of freedom here means the number of unknown variables. In two-dimensional elasticity problems, since two displacements and forces in the x- and y-directions are unknown variables for one nodal point, every nodal point has two degrees of freedom. Hence, the number of degrees of freedom for a finite element model consisting of n nodal points is $2n$.

By summing up the element stiffness matrices for all the n_e elements in the finite element model, the global stiffness matrix $[\mathbf{K}]$ is obtained as shown in the following equation:

$$[\mathbf{K}] \equiv [K_{ij}] = \sum_{e=1}^{ne} [\mathbf{K}^{(e)}] \quad (i, j = 1, 2, ..., 2n \text{ and } e = 1, 2, ..., n_e) \tag{1.95}$$

Since the components of the global nodal displacement vector $\{\boldsymbol{\delta}\}$ are common for all the elements, they remain unchanged during the assembly of the global stiffness equations. By rewriting the components of $\{\boldsymbol{\delta}\}$, $u_1, u_2, ..., u_n$ as $u_1, u_3, ..., u_{2i-1}, ..., u_{2n-1}$ and $v_1, v_2, ..., v_n$ as $u_2, u_4, ..., u_{2i}, ..., u_{2n}$, the following expression for the global nodal displacement vector $\{\boldsymbol{\delta}\}$ is obtained:

$$\{\boldsymbol{\delta}\} = [u_1, u_2, \cdots, u_{2I-1}, u_{2I}, \cdots, u_{2J-1}, u_{2J}, \cdots, u_{2K-1}, u_{2K}, \cdots, u_{2n-1}, u_{2n}]^T \tag{1.96}$$

The global nodal force of a node is the sum of the nodal forces for all the elements to which the node belongs. Hence, the global nodal force vector $\{\mathbf{P}\}$ can be written as

$$\{\mathbf{P}\} = [X_1, Y_1, \cdots, X_I, Y_I, \cdots, X_J, Y_J, \cdots, X_K, Y_K, \cdots, X_n, Y_n]^T \tag{1.97}$$

where

$$X_I = \sum X_I^{(e)}, \quad Y_I = \sum Y_I^{(e)} \quad (I = 1, 2, ..., n) \tag{1.98}$$

By rewriting the global nodal force vector $\{\mathbf{P}\}$ in a similar way to $\{\boldsymbol{\delta}\}$ in Eq. (1.96) as

$$\{\mathbf{P}\} = [X_1, X_2, \cdots, X_{2I-1}, X_{2I}, \cdots, X_{2J-1}, X_{2J}, \cdots, X_{2K-1}, X_{2K}, \cdots, X_{2n-1}, X_{2n}]^T \tag{1.99}$$

where

$$X_I = \sum X_I^{(e)} \quad (I = 1, 2, ..., 2n) \tag{1.100}$$

The symbol Σ in Eqs (1.98), (1.100) indicate that the summation is taken over all the elements that possess the node in common. The values of X_I in Eq. (1.100), however, are zero for the nodes inside of the elastic body and for those on the bounding surfaces which are subjected to no applied loads.

Consequently, the following formula is obtained as the governing global stiffness equation:

$$[\mathbf{K}]\{\boldsymbol{\delta}\} = \{\mathbf{P}\} \tag{1.101}$$

which is the $2n$th degree simultaneous linear equations for $2n$ unknown variables of nodal displacements and/or forces.

1.4.4.5 Example: Finite-element calculations for a square plate subjected to uniaxial uniform tension

Procedures 5–7 described in Section 1.4.1 will be explained by taking an example of the finite element calculations for a square plate subjected to uniaxial uniform tension, as illustrated in Fig. 1.9. The square plate model has a side of unit length 1 and a thickness of unit length 1, and consists of two constant-strain triangular elements, i.e., the model has four nodes and thus eight degrees of freedom.

Let us determine the element stiffness matrix for Element 1. From Eqs (1.73), (1.69a)–(1.69c), the $[\mathbf{B}]$ matrix of Element 1 is calculated as

$$
\begin{aligned}
[\mathbf{B}] &= \begin{bmatrix} b_{1_1} & 0 & b_{2_1} & 0 & b_{3_1} & 0 \\ 0 & c_{1_1} & 0 & c_{2_1} & 0 & c_{3_1} \\ c_{1_1} & b_{1_1} & c_{2_1} & b_{2_1} & c_{3_1} & b_{3_1} \end{bmatrix} \\[2mm]
&= \frac{1}{2\Delta^{(1)}} \begin{bmatrix} y_{2_1} - y_{3_1} & 0 & y_{3_1} - y_{1_1} & 0 & y_{1_1} - y_{2_1} & 0 \\ 0 & x_{3_1} - x_{2_1} & 0 & x_{1_1} - x_{3_1} & 0 & x_{2_1} - x_{1_1} \\ x_{3_1} - x_{2_1} & y_{2_1} - y_{3_1} & x_{1_1} - x_{3_1} & y_{3_1} - y_{1_1} & x_{2_1} - x_{1_1} & y_{1_1} - y_{2_1} \end{bmatrix}
\end{aligned} \tag{1.102}
$$

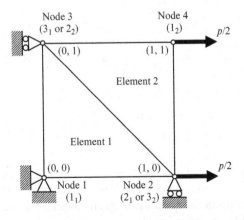

Fig. 1.9 Finite element model of a square plate subjected to uniaxial uniform tension.

Since the area of Element 1 $\Delta^{(e)}$ in the above equation can be easily obtained as 1/2 without using Eq. (1.69d),

$$[\mathbf{B}] = \begin{bmatrix} -1 & 0 & 1 & 0 & 0 & 0 \\ 0 & -1 & 0 & 0 & 0 & 1 \\ -1 & -1 & 0 & 1 & 1 & 0 \end{bmatrix} \tag{1.103}$$

Hence, from Eq. (1.85), the element stiffness matrix for Element 1 $[\mathbf{k}^{(e)}]$ is calculated as

$$\left[\mathbf{k}^{(1)}\right] = \frac{t}{2} \cdot \frac{E'}{1-\nu'^2} \begin{bmatrix} -1 & 0 & -1 \\ 0 & -1 & -1 \\ 1 & 0 & 0 \\ 0 & 0 & 1 \\ 0 & 0 & 1 \\ 0 & 1 & 0 \end{bmatrix} \begin{bmatrix} 1 & \nu' & 0 \\ \nu' & 1 & 0 \\ 0 & 0 & \alpha \end{bmatrix} \begin{bmatrix} -1 & 0 & 1 & 0 & 0 & 0 \\ 0 & -1 & 0 & 0 & 0 & 1 \\ -1 & -1 & 0 & 1 & 1 & 0 \end{bmatrix} \tag{1.104}$$

where $\alpha = (1 - \nu')/2$. After multiplications of the matrices in Eq. (1.104), the element stiffness equation is obtained from Eq. (1.84) as

$$\begin{Bmatrix} X_{1_1}^{(1)} \\ Y_{1_1}^{(1)} \\ X_{2_1}^{(1)} \\ Y_{2_1}^{(1)} \\ X_{3_1}^{(1)} \\ Y_{3_1}^{(1)} \end{Bmatrix} = \frac{t}{2} \cdot \frac{E'}{1-\nu'^2} \begin{bmatrix} 1+\alpha & \nu'+\alpha & -1 & -\alpha & -\alpha & -\nu' \\ \nu'+\alpha & 1+\alpha & -\nu' & -\alpha & -\alpha & -1 \\ -1 & -\nu' & 1 & 0 & 0 & \nu' \\ -\alpha & -\alpha & 0 & \alpha & \alpha & 0 \\ -\alpha & -\alpha & 0 & \alpha & \alpha & 0 \\ -\nu' & -1 & \nu' & 0 & 0 & 1 \end{bmatrix} \begin{Bmatrix} u_{1_1} \\ v_{1_1} \\ u_{2_1} \\ v_{2_1} \\ u_{3_1} \\ v_{3_1} \end{Bmatrix} \tag{1.105}$$

since the equivalent nodal forces due to initial strains ε_0 and body forces F_x and F_y, $\{\mathbf{F}_{e0}\}^{(1)}$ and $\{\mathbf{F}_F\}^{(1)}$ are zero. The components of the nodal displacement and force vectors are written by the element nodal numbers. By rewriting these components by the global nodal numbers as shown in Fig. 1.9, Eq. (1.105) is rewritten as

$$\begin{Bmatrix} X_1 \\ Y_1 \\ X_2 \\ Y_2 \\ X_3 \\ Y_3 \end{Bmatrix} = \frac{t}{2} \cdot \frac{E'}{1-\nu'^2} \begin{bmatrix} 1+\alpha & \nu'+\alpha & -1 & -\alpha & -\alpha & -\nu' \\ \nu'+\alpha & 1+\alpha & -\nu' & -\alpha & -\alpha & -1 \\ -1 & -\nu' & 1 & 0 & 0 & \nu' \\ -\alpha & -\alpha. & 0 & \alpha & \alpha & 0 \\ -\alpha & -\alpha & 0 & \alpha & \alpha & 0 \\ -\nu' & -1 & \nu' & 0 & 0 & 1 \end{bmatrix} \begin{Bmatrix} u_1 \\ v_1 \\ u_2 \\ v_2 \\ u_3 \\ v_3 \end{Bmatrix} \tag{1.106}$$

In a similar way, the element stiffness equation for Element 2 is obtained as

$$\begin{Bmatrix} X_{1_2}^{(2)} \\ Y_{1_2}^{(2)} \\ X_{2_2}^{(2)} \\ Y_{2_2}^{(2)} \\ X_{3_2}^{(2)} \\ Y_{3_2}^{(2)} \end{Bmatrix} = \frac{t}{2} \cdot \frac{E'}{1-\nu'^2} \begin{bmatrix} 1+\alpha & \nu'+\alpha & -1 & -\alpha & -\alpha & -\nu' \\ \nu'+\alpha & 1+\alpha & -\nu' & -\alpha & -\alpha & -1 \\ -1 & -\nu' & 1 & 0 & 0 & \nu' \\ -\alpha & -\alpha & 0 & \alpha & \alpha & 0 \\ -\alpha & -\alpha & 0 & \alpha & \alpha & 0 \\ -\nu' & -1 & \nu' & 0 & 0 & 1 \end{bmatrix} \begin{Bmatrix} u_{1_2} \\ v_{1_2} \\ u_{2_2} \\ v_{2_2} \\ u_{3_2} \\ v_{3_2} \end{Bmatrix} \tag{1.107}$$

and is rewritten by using the global nodal numbers as

$$\begin{Bmatrix} X_4 \\ Y_4 \\ X_3 \\ Y_3 \\ X_2 \\ Y_2 \end{Bmatrix} = \frac{t}{2} \cdot \frac{E'}{1-\nu'^2} \begin{bmatrix} 1+\alpha & \nu'+\alpha & -1 & -\alpha & -\alpha & -\nu' \\ \nu'+\alpha & 1+\alpha & -\nu' & -\alpha & -\alpha & -1 \\ -1 & -\nu' & 1 & 0 & 0 & \nu' \\ -\alpha & -\alpha & 0 & \alpha & \alpha & 0 \\ -\alpha & -\alpha & 0 & \alpha & \alpha & 0 \\ -\nu' & -1 & \nu' & 0 & 0 & 1 \end{bmatrix} \begin{Bmatrix} u_4 \\ v_4 \\ u_3 \\ v_3 \\ u_2 \\ v_2 \end{Bmatrix} \tag{1.108}$$

By assembling the two element stiffness matrices, the following global stiffness equation for the square plate subjected to uniform tension is obtained (Procedure 4):

$$\frac{t}{2} \cdot \frac{E'}{1-\nu'^2} \begin{bmatrix} 1+\alpha & \nu'+\alpha & -1 & -\alpha & -\alpha & -\nu' & 0 & 0 \\ \nu'+\alpha & 1+\alpha & -\nu' & -\alpha & -\alpha & -1 & 0 & 0 \\ -1 & -\nu' & 1+\alpha & 0 & 0 & \nu'+\alpha & -\alpha & -\alpha \\ -\alpha & -\alpha & 0 & 1+\alpha & \nu'+\alpha & 0 & -\nu' & -1 \\ -\alpha & -\alpha & 0 & \nu'+\alpha & 1+\alpha & 0 & -1 & -\nu' \\ -\nu' & -1 & \nu'+\alpha & 0 & 0 & 1+\alpha & -\alpha & -\alpha \\ 0 & 0 & -\alpha & -\nu' & -1 & -\alpha & 1+\alpha & \nu'+\alpha \\ 0 & 0 & -\alpha & -1 & -\nu' & -\alpha & \nu'+\alpha & 1+\alpha \end{bmatrix} \begin{Bmatrix} u_1 \\ v_1 \\ u_2 \\ v_2 \\ u_3 \\ v_3 \\ u_4 \\ v_4 \end{Bmatrix} = \begin{Bmatrix} X_1 \\ Y_1 \\ X_2 \\ Y_2 \\ X_3 \\ Y_3 \\ X_4 \\ Y_4 \end{Bmatrix} \tag{1.109}$$

where the left-hand and right-hand sides of the equation are replaced with each other.

Let us now impose boundary conditions on the nodes. Namely, node 1 is clamped in both the x- and y-directions, node 2 is clamped only in the x-direction, and nodes 2 and 4 are subjected to equal nodal forces $X_2 = X_4 = (p \times 1 \times 1)/2 = p/2$, respectively, in the x-direction. A pair of the equal nodal forces $p/2$ applied to nodes 2 and 4 in the x-direction is the finite element model of a uniformly distributed tension force p per unit area exerted on the side $\overline{24}$ in the x-direction, as illustrated in Fig. 1.9. The geometrical and mechanical boundary conditions for the present case are

$$u_1 = v_1 = v_2 = u_3 = 0 \tag{1.110}$$

and

$$X_2 = X_4 = (pt)/2 = p/2, \quad Y_3 = Y_4 = 0 \tag{1.111}$$

respectively. Substitution of Eqs (1.110), (1.111) into Eq. (1.109) gives the global stiffness equation, i.e.,

$$\frac{t}{2}\cdot\frac{E'}{1-\nu'^2}
\begin{bmatrix}
1+\alpha & \nu'+\alpha & -1 & -\alpha & -\alpha & -\nu' & 0 & 0 \\
\nu'+\alpha & 1+\alpha & -\nu' & -\alpha & -\alpha & -1 & 0 & 0 \\
-1 & -\nu' & 1+\alpha & 0 & 0 & \nu'+\alpha & -\alpha & -\alpha \\
-\alpha & -\alpha & 0 & 1+\alpha & \nu'+\alpha & 0 & -\nu' & -1 \\
-\alpha & -\alpha & 0 & \nu'+\alpha & 1+\alpha & 0 & -1 & -\nu' \\
-\nu' & -1 & \nu'+\alpha & 0 & 0 & 1+\alpha & -\alpha & -\alpha \\
0 & 0 & -\alpha & -\nu' & -1 & -\alpha & 1+\alpha & \nu'+\alpha \\
0 & 0 & -\alpha & -1 & -\nu' & -\alpha & \nu'+\alpha & 1+\alpha
\end{bmatrix}
\begin{Bmatrix}
0 \\ 0 \\ u_2 \\ 0 \\ 0 \\ v_3 \\ u_4 \\ v_4
\end{Bmatrix}
=
\begin{Bmatrix}
X_1 \\ Y_1 \\ p/2 \\ Y_2 \\ X_3 \\ 0 \\ p/2 \\ 0
\end{Bmatrix} \tag{1.112}$$

Rearrangement of Eq. (1.112) by collecting unknown variables for forces and displacements in the left-hand side and known values of the forces and displacements in the right-hand side brings about the following simultaneous equations (Procedure 5):

$$\frac{t}{2}\cdot\frac{E'}{1-\nu'^2}
\begin{bmatrix}
-\dfrac{2}{t}\cdot\dfrac{E'}{1-\nu'^2} & 0 & -1 & 0 & 0 & -\nu' & 0 & 0 \\[2mm]
0 & -\dfrac{2}{t}\cdot\dfrac{E'}{1-\nu'^2} & -\nu' & 0 & 0 & -1 & 0 & 0 \\[2mm]
0 & 0 & 1+\alpha & 0 & 0 & \nu'+\alpha & -\alpha & -\alpha \\[2mm]
0 & 0 & 0 & -\dfrac{2}{t}\cdot\dfrac{E'}{1-\nu'^2} & 0 & 0 & -\nu' & -1 \\[2mm]
0 & 0 & 0 & 0 & -\dfrac{2}{t}\cdot\dfrac{E'}{1-\nu'^2} & 0 & -1 & -\nu' \\[2mm]
0 & 0 & \nu'+\alpha & 0 & 0 & 1+\alpha & -\alpha & -\alpha \\[2mm]
0 & 0 & -\alpha & 0 & 0 & -\alpha & 1+\alpha & \nu'+\alpha \\[2mm]
0 & 0 & -\alpha & 0 & 0 & -\alpha & \nu'+\alpha & 1+\alpha
\end{bmatrix}$$

$$\begin{Bmatrix}
X_1 \\ Y_1 \\ u_2 \\ Y_2 \\ X_3 \\ v_3 \\ u_4 \\ v_4
\end{Bmatrix}
=
\begin{Bmatrix}
0 \\ 0 \\ p/2 \\ 0 \\ 0 \\ 0 \\ p/2 \\ 0
\end{Bmatrix} \tag{1.113}$$

Eq. (1.113) can be solved numerically by, for instance, the Gauss elimination procedure. The solutions for Eq. (1.113) are

$$u_2 = u_4 = p/E', \quad v_3 = v_4 = -v'p/E', \quad X_1 = X_3 = -p/2, \quad Y_1 = Y_2 = 0. \qquad (1.114)$$

(Procedure 6).

The strains and stresses in the square palate can be calculated by substituting the solutions (1.114) into Eqs (1.73), (1.74), respectively (Procedure 7). The resultant strains and stresses are given by the following equations:

$$
\begin{Bmatrix} \varepsilon_x \\ \varepsilon_y \\ \gamma_{xy} \end{Bmatrix} = [\mathbf{B}] \begin{Bmatrix} u_4 \\ v_4 \\ u_3 \\ v_3 \\ u_2 \\ v_2 \end{Bmatrix} = \frac{p}{E'} \begin{bmatrix} 1 & 0 & -1 & 0 & 0 & 0 \\ 0 & 1 & 0 & 0 & 0 & -1 \\ 1 & 1 & 0 & 1 & -1 & 0 \end{bmatrix} \begin{Bmatrix} 1 \\ -v' \\ 0 \\ v' \\ 1 \\ 0 \end{Bmatrix}
$$

$$
= \frac{p}{E'} \begin{Bmatrix} 1 \\ -v' \\ 1 - v' + v' - 1 \end{Bmatrix} = \frac{p}{E'} \begin{Bmatrix} 1 \\ -v' \\ 0 \end{Bmatrix} \qquad (1.115)
$$

and

$$
\begin{Bmatrix} \sigma_x \\ \sigma_y \\ \tau_{xy} \end{Bmatrix} = \frac{E'}{1 - v'^2} \begin{bmatrix} 1 & v' & 0 \\ v' & 1 & 0 \\ 0 & 0 & \frac{1-v'}{2} \end{bmatrix} \begin{Bmatrix} \varepsilon_x \\ \varepsilon_y \\ \gamma_{xy} \end{Bmatrix}
$$

$$
= \frac{p}{1 - v'^2} \begin{bmatrix} 1 & v' & 0 \\ v' & 1 & 0 \\ 0 & 0 & \frac{1-v'}{2} \end{bmatrix} \begin{Bmatrix} 1 \\ -v' \\ 0 \end{Bmatrix} = \frac{p}{1 - v'^2} \begin{Bmatrix} 1 - v'^2 \\ 0 \\ 0 \end{Bmatrix} \qquad (1.116)
$$

$$
= p \begin{Bmatrix} 1 \\ 0 \\ 0 \end{Bmatrix}
$$

The results obtained by the present finite element calculations imply that a square plate subjected to uniaxial uniform tension in the x-direction is elongated by a uniform strain of p/E' in the loading direction, whereas it is contacted by a uniform strain of $-v'p/E'$ in the direction perpendicular to the loading direction and that only a uniform normal stress of $\sigma_x = p$ is induced in the plate. The result that the nodal reaction forces at nodes 1 and 3 are equal to $-p/2$, i.e. $X_1 = X_3 = -p/2$ implies that a uniform reaction force of $-p$ is produced along the side $\overline{13}$. It is concluded that the above results obtained by the finite element method agree well with the physical interpretations of the present problem.

References

[1] O.C. Zienkiewicz, K. Morgan, Finite Elements and Approximation, John Wiley & Sons, New York, 1983.

[2] O. C. Zienkiewicz and R. Taylor, The Finite Element Method (fourth ed.), vol. 1, 1989, McGraw Hill Book Co., London.

[3] G.W. Rowe, et al., Finite-Element Plasticity and Metalforming Analysis, Cambridge University Press, Cambridge, 1991.

[4] C.L. Dym, I.H. Shames, Solid Mechanics: A Variational Approach, McGraw-Hill, New York, 1973.

[5] K. Washizu, Variational Methods in Elasticity and Plasticity, second ed., Pergamon, New York, 1975.

[6] K.-J. Bathe, Finite Element Procedures, Prentice-Hall International, New Jersey, 1975.

[7] B.A. Finlayson, The Method of Weighted Residuals and Variational Principles, Academic Press, New York, 1972.

[8] M. Gotoh, Engineering Finite Element Method—For Analysis of Large Elastic-Plastic Deformation, Corona Publishing Co., Ltd., Tokyo, 1995 (in Japanese).

[9] H. Togawa, Introduction to the Finite Element Method, Baifukan Co., Ltd., Tokyo, 1984 (in Japanese).

[10] G. Yagawa, et al., Computational Mechanics, Iwanami Shoten Publishers, Tokyo, 2000 (in Japanese).

[11] K. Washizu, et al., Handbook of the Finite Element Method (Basics), Baifukan Co., Ltd., Tokyo, 1994 (in Japanese).

Overview of ANSYS structure and its graphic capabilities

2

2.1 Introduction

ANSYS is a general-purpose finite element modelling package for numerically solving a wide variety of mechanical problems. These problems include: static/dynamic, structural analysis (both linear and nonlinear), heat transfer and fluid problems, as well as acoustic and electro-magnetic problems.

In general, a finite element solution may be broken into the following three stages.

1. **Preprocessing—defining the problem:** The major steps in preprocessing are: (i) define keypoints/lines/areas/volumes; (ii) define element type and material/geometric properties; and (iii) mesh lines/areas/volumes as required. The amount of detail required will depend on the dimensionality of the analysis—that is, 1D, 2D, axi-symmetric, 3D.
2. **Solution—assigning loads, constraints, and solving:** Here, it is necessary to specify the loads (point or pressure) and constraints (translational and rotational), and solve the resulting set of equations.
3. **Postprocessing—further processing and viewing of the results:** In this stage, one may wish to see: (i) lists of nodal displacements; (ii) element forces and moments; (iii) deflection plots; and (iv) stress contour diagrams or temperature maps.

2.2 Starting the programme

2.2.1 Preliminaries

There are two methods to use ANSYS. The first is by means of the graphical user interface or GUI. This method follows the conventions of popular Windows- and X-Windows-based programmes. The GUI method is exclusively used in this book.

The second is by means of command files. The command file approach has a steeper learning curve for many, but it has the advantage that the entire analysis can be described in a small text file, typically in fewer than 50 lines of commands. This approach enables easy model modifications and has minimal file space requirements.

The ANSYS environment contains two windows: the Main Window and an Output Window. Within the Main Window, there are five divisions (see Fig. 2.1).

1. The **Utility Menu** [A] contains functions that are available throughout the ANSYS session, such as file controls, selections, graphic controls, and parameters.
2. The **Input Line** [B] shows programme prompt messages and allows the user to type in commands directly.
3. The **Toolbar** [C] contains push buttons that execute commonly used ANSYS commands. More push buttons can be made available if desired.

Engineering Analysis with ANSYS Software. https://doi.org/10.1016/B978-0-08-102164-4.00002-9

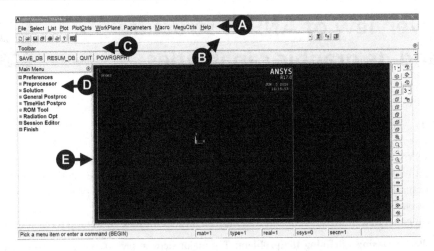

Fig. 2.1 Main window of ANSYS.

4. The **Main Menu** [D] contains the primary ANSYS functions, organised by preprocessor, solution, general postprocessor, and design optimiser. It is from this menu that the vast majority of modelling commands are issued.
5. The **Graphics Window** [E] is where graphics are shown and graphical picking can be made. It is here where the model in its various stages of construction and the ensuing results from the analysis can be viewed.

The Output Window, shown in Fig. 2.2, displays text output from the programme, such as listing of data. It is usually positioned behind the Graphics Window and can be brought to the front if necessary.

2.2.2 Saving and restoring jobs

It is good practice to save the model at various stages during its creation. Very often, the stage in the modelling is reached where things have gone well, and the model ought to be saved at this point. In that way, if mistakes are made later on, it will be possible to return to this point.

To save the model, from the ANSYS Utility Menu, select **File → Save as Jobname.db**.

The model will be saved in a file called Jobname.db, where Jobname is the name that was specified in the Launcher when ANSYS was first started. It is a good idea to save the job at different times throughout the building and analysis of the model, to back up the work in case of a system crash or other unforeseen problems.

Alternatively, select **File → Save as**. In response to the second option, the frame shown in Fig. 2.3 is produced.

Select the [A] appropriate drive and [B] give the file a name. Clicking the [C] **OK** button saves the model as a database with the name given.

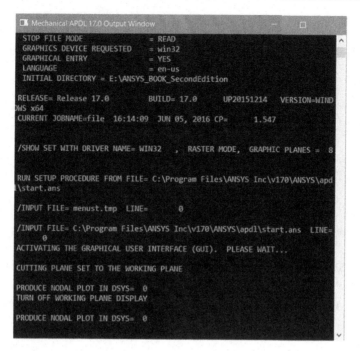

Fig. 2.2 Output window of ANSYS.

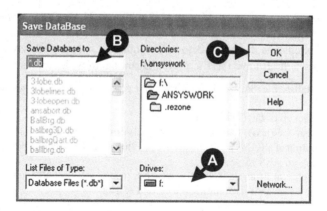

Fig. 2.3 Save database of the problem.

Frequently there is a need to start up ANSYS and recall and continue a previous job. There are two methods to do this:

1. Using the Launcher: (i) In the ANSYS Launcher, select Interactive and specify the previously defined jobname; (ii) When ANSYS is running, select **Utility Menu: File: Resume Jobname.db**. This will restore as much of the database (geometry, loads, solution, etc.) as was previously saved.

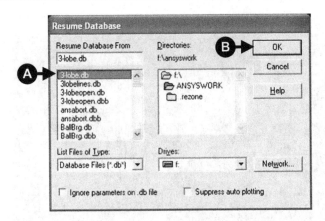

Fig. 2.4 Resume problem from saved database.

2. Start ANSYS and select from the Utility Menu **File → Resume from**, then click on the job from the list that appears. Fig. 2.4 shows the resulting frame.

Select the [A] appropriate file from the list and click the [B] **OK** button to resume the analysis.

2.2.3 Organisation of files

A large number of files are created when ANSYS is run. If ANSYS is started without specifying a jobname, the name of all files created will be File.*, where the * represents various extensions described below. If a jobname is specified, say Beam, then the created files will all have the file prefix, Beam again with various extensions:

beam.db—database file (binary). This file stores the geometry, boundary conditions, and any solutions.

beam.dbb—backup of the database file (binary).

beam.err—error file (text). Listing of all error and warning messages.

beam.out—output of all ANSYS operations (text). This is what normally scrolls in the output window during an ANSYS session.

beam.log—logfile or listing of ANSYS commands (text). Listing of all equivalent ANSYS command line commands used during the current session.

Depending on the operations carried out, other files may have been written. These files may contain results, for example. It is important to know what to save—for instance, when there is a need to clean up a directory or to move things from the /scratch directory. If the GUI is always used, then only the .db file is required. This file stores the geometry, boundary conditions, and any solutions. Once the ANSYS programme has started, and the jobname has been specified, only the resume command has to be activated to proceed from where the model was last left off. If, however, ANSYS command files are intended to be used, then only the command file and/or the log file have to be stored. The log file contains a complete listing of the ANSYS commands used to get the model to its current stage. That file may be run as is, or edited and rerun as desired.

2.2.4 Printing and plotting

ANSYS produces lists and tables of many types of results that are normally displayed on the screen. However, it is often desirable to save the results to a file to be analysed later or included in a report.

In case of stresses, instead of using **Plot Results** to plot the stresses, choose **List Results**. Select **Elem Table Data**, and choose the appropriate item(s) from the menu. Multiple items can be picked. When the list appears on the screen in its own window, select **File: Save As** and give a file name to store the results. Any other solution can be done in the same way. For example, select **Nodal Solution** from the **List Results** menu to get displacements. Preprocessing and Solution data can be listed and saved from the **List** menu in the ANSYS Utility Menu. Save the resulting list in the same way described above.

When there is a need to quickly save an image of the entire screen or the current Graphics Window, from the ANSYS Utility Menu select **Plot Ctrls → Hard Copy**. There are two options: **To Printer** (see Fig. 2.5) and **To File** (see Fig. 2.6).

In the frame shown in Fig. 2.5, select the [A] printer from the list of printers and press the [B] **Print** button.

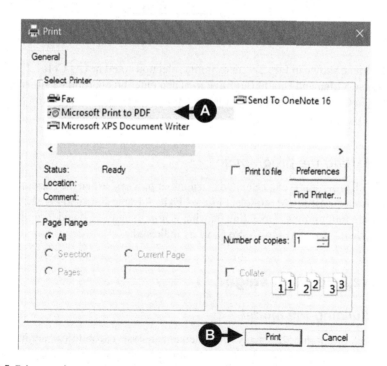

Fig. 2.5 Print to printer.

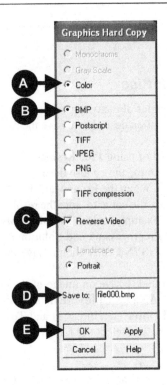

Fig. 2.6 Print to file.

In the frame shown in Fig. 2.6, an example selection could be [A] **Color**, [B] **BMP**, [C] **Reverse Video** and finally [D] **Save to**. Then enter an appropriate file name and click the [E] **OK** button. This image file may now be printed on a PostScript printer or included in a document.

2.2.5 Exiting the programme

The ANSYS programme can be exited in a number of ways. When the current analysis is saved, from the Utility Menu select **File → Exit**. A frame shown in Fig. 2.7 appears.

There are four options, depending on what is important to be saved. If nothing is to be saved, then select [A] **Quit - No Save** as indicated.

2.3 Preprocessing stage

2.3.1 Building the model

The ANSYS programme has many finite element analysis capabilities, ranging from a simple linear static analysis to a complex nonlinear transient dynamic analysis. Building a finite element model requires more time than any other part of the analysis. First a jobname and analysis title have to be specified. Next, the PREP7 preprocessor is

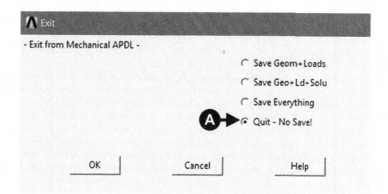

Fig. 2.7 Exit from ANSYS.

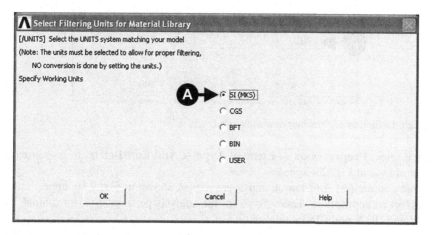

Fig. 2.8 Selection of units for the problem.

used to define the element types, element real constants, material properties, and model geometry. It is important to remember that ANSYS does not assume a system of units for intended analysis. Except in magnetic field analyses, any system of units can be used, as long as it is ensured that units are consistent for all input data. Units cannot be set directly from the GUI. In order to set units as the international system of units (SI), from the ANSYS Main Menu select **Preprocessor → Material Props → Material Library → Select Units**. Fig. 2.8 shows the resulting frame.

Activate the [A] **SI (MKS)** button to inform the ANSYS programme that this system of units is going to be used in the analysis.

2.3.1.1 Defining element types and real constants

The ANSYS element library contains more than 100 different element types. Each element type has a unique number and a prefix that identifies the element category. In order to define element types, the user must be in PREP7. From the ANSYS Main

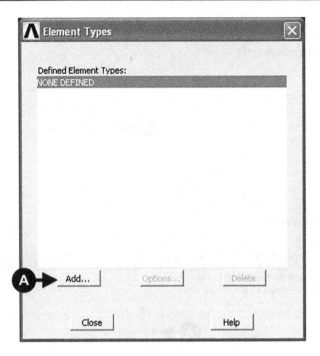

Fig. 2.9 Definition of element types to be used.

Menu, select **Preprocessor → Element Type → Add/Edit/Delete**. In response, the frame shown in Fig. 2.9 appears.

Click on the [A] **Add** button and a new frame, shown in Fig. 2.10, appears.

Select an appropriate element type for the analysis performed—for example, [A] **Solid** and [B] **8 node 183** as shown in Fig. 2.10.

Element real constants are properties that depend on the element type, such as cross-sectional properties of a beam element. As with element types, each set of real constants has a reference number, and the table of reference number versus the real

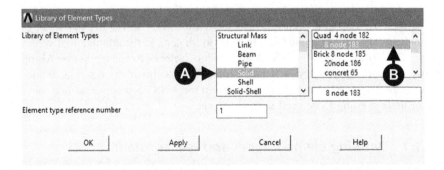

Fig. 2.10 Selection of element types from library.

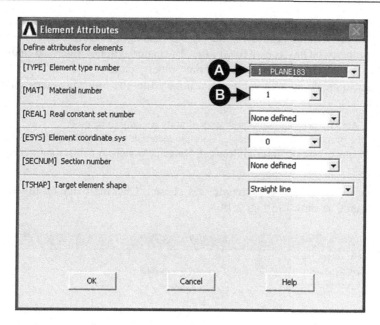

Fig. 2.11 Element attributes.

constants set is called the real constant table. Not all element types require real constants, and different elements of the same type may have different real constant values. The ANSYS Main Menu command **Preprocessor → Modelling → Create → Elements → Elem Attributes** can be used to define an element real constant. Fig. 2.11 shows a frame in which one can select the element type. According to Fig. 2.11, the element type already selected is [A] **Plane 183**, for which the real constant is being defined. A corresponding [B] **Material number**, allocated by ANSYS when material properties are defined (see section below), is also shown in the frame.

Other element attributes can be defined as required by the type of analysis performed. Chapter 7 contains sample problems where elements attributes are defined in accordance with the requirements of the problem.

2.3.1.2 Defining material properties

Material properties are required for most element types. Depending on the application, material properties may be linear or nonlinear, isotropic, orthotropic or anisotropic, constant temperature or temperature-dependent. As with element types and real constants, each set of material properties has a material reference number. The table of material reference numbers versus material property sets is called the material table. In one analysis there may be multiple material property sets corresponding with multiple materials used in the model. Each set is identified using a unique reference number. Although material properties can be defined separately for each finite element analysis, the ANSYS programme enables storing a material property set in an archival material library file, then retrieving the set and reusing it in multiple analyses. Each material property set has its own library file. The material library files also enable several users to share commonly used material property data.

In order to create an archival material library file, the following steps should be followed:

1. Tell the ANSYS programme what system of units is going to be used.
2. Define the properties of, for example, isotropic material; use the ANSYS Main Menu and select **Preprocessor → Material Props → Material Models**. A frame, shown in Fig. 2.12, appears.

As shown in Fig. 2.12, [A] **Isotropic** was chosen. Clicking twice on this calls up another frame, as shown in Fig. 2.13.

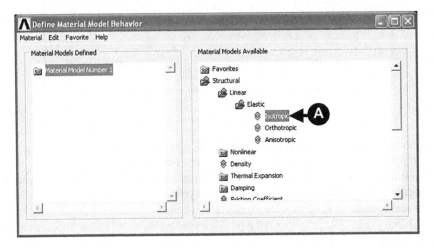

Fig. 2.12 Define material model behaviour.

Fig. 2.13 Linear isotropic material properties.

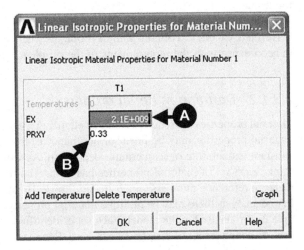

Enter data characterising the material to be used in the analysis into the appropriate field: for example, [A] **EX** = 2.1e9 and [B] **PRXY** = 0.33, as shown in Fig. 2.13. If the problem requires a number of different materials to be used, then the above procedure should be repeated and another material model is created with appropriate material number allocated by the programme.

2.3.2 Construction of the model

2.3.2.1 Creating the model geometry

Once material properties are defined, the next step in an analysis is generating a finite element model-nodes and element adequately describing the model geometry. There are two methods to create the finite element model: solid modelling and direct generation. With solid modelling, the geometry of shape of the model is described and then the ANSYS programme automatically meshes the geometry with nodes and elements. The size and shape of the elements that the programme creates can be controlled. With direct generation, the location of each node and the connectivity of each element are manually defined. Several convenience operations, such as copying patterns of existing nodes and elements, symmetry reflection, etc., are available.

Solved example problems in this book amply illustrate, in a step-by-step manner, how to create the model geometry.

2.3.2.2 Applying loads

Loads can be applied using either the **PREP7** preprocessor or the **SOLUTION** processor. Regardless of the strategy chosen, it is necessary to define the analysis type and analysis options, apply loads, specify load step options, and initiate the finite element solution.

The analysis type to be used is based on the loading conditions and the response which is wished to be calculated. For example, if natural frequencies and mode shapes are to be calculated, then a modal analysis should be chosen. The ANSYS programme offers the following analysis types: static (or steady-state), transient, harmonic, modal, spectrum, buckling, and substructuring. Not all analysis types are valid for all disciplines. Modal analysis, for instance, is not valid for a thermal model. Analysis options allow for customisation of analysis type. Typical analysis options are the method of solution, stress stiffening on or off, and Newton-Raphson options. In order to define the analysis type and analysis options, use the ANSYS Main Menu and select **Main Menu: Preprocessor → Loads → Analysis Type → New Analysis**. In response to the selection, the frame shown in Fig. 2.14 appears.

Select the type of analysis appropriate for the problem at hand by activating the [A] **Static** button, for example.

The word 'loads' used here includes boundary conditions: constraints, supports, or boundary field specifications. It also includes other externally and internally applied loads. Loads in the ANSYS programme are divided into six categories: DOF constraints, forces, surface loads, body loads, inertia loads, and coupled-field loads.

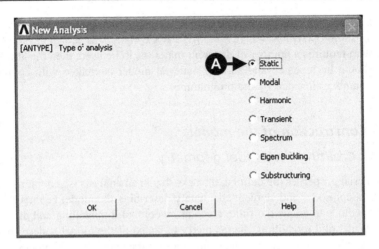

Fig. 2.14 Type of analysis definition.

Most of these loads can be applied either on the solid model (keypoints, lines, and areas) or the finite element model (nodes and elements).

There are two important load-related terms. A load step is simply a configuration of loads for which the solution is obtained. In a structural analysis, for instance, wind loads may be applied in one load step and gravity in a second load step. Load steps are also useful in dividing a transient load history curve into several segments.

Substeps are incremental steps taken within a load step. They are mainly used for accuracy and convergence purposes in transient and nonlinear analyses. Substeps are also known as time steps which are taken over a period of time.

Load step options are alternatives that can be changed from load step to load step, such as number of substeps, time at the end of a load step, and output controls. Depending on the type of analysis performed, load step options may or may not be required. Sample problems solved here provide a practical guide to appropriate load step options as necessary.

2.4 Solution stage

To initiate solution calculations, use the ANSYS Main Menu, selecting **Solution →** **Solve → Current LS**. Fig. 2.15 shows the resulting frame.

After reviewing summary information about the model, click the [A] **OK** button to start the solution. When this command is issued, the ANSYS programme takes the model and loading information from the database, and calculates the results. Results are written to the results file and also to the database. The only difference is that just one set of results can reside in the database at one time, while a number of result sets can be written to the results file.

Once the solution has been calculated, the ANSYS postprocessors can be used to review the results.

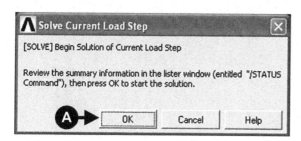

Fig. 2.15 Start solution of current problem.

2.5 Postprocessing stage

Two postprocessors are available: POST1 and POST26. POST1 (the general post-processor) is used to review results at one substep (time step) over the entire model or selected portion of the model. The command to enter POST1 requires selecting from the ANSYS Main Menu **General Postproc**. Using this postprocessor, contour displays, deformed shapes, and tabular listings to review and interpret the results of the analysis can be obtained. POST1 offers many other capabilities, including error estimation, load case combinations, calculations among results data, and path operations.

POST26 (the time history postprocessor) is used to review results at specific points in the model over all time steps. The command to enter POST26 is as follows: from the ANSYS Main Menu select **TimeHist Postpro**. Graph plots of results data versus time (or frequency) and tabular listings can be obtained. Other POST26 capabilities include arithmetic calculations and complex algebra.

Application of ANSYS to stress analysis

3

3.1 Cantilever beam

Beams are important fundamental structural and/or machine elements; they are found in buildings and in bridges. Beams are also used as shafts in cars and trains (see Fig. 3.1), as wings in aircrafts, and as book shelves in bookstores. Arms and femurs of human beings and branches of trees are good examples of portions of living creatures that support their bodies like cantilever beams as illustrated in Fig. 3.2. Beams play important roles not only in inorganic but in organic structures.

The mechanics of beams is one of the most important subjects in engineering.

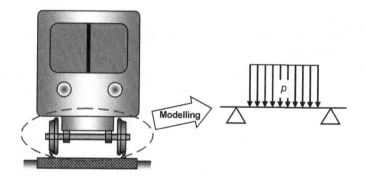

Fig. 3.1 Modelling of an axle shaft by a simply supported beam.

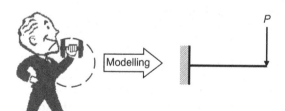

Fig. 3.2 Modelling of an arm of a human being by a cantilever beam.

3.1.1 Example problem: A cantilever beam

Perform an FEM analysis of a 2D cantilever beam shown in Fig. 3.3 and calculate the deflection of the beam at the loading point and the longitudinal stress distribution in the beam.

Engineering Analysis with ANSYS Software. https://doi.org/10.1016/B978-0-08-102164-4.00003-0

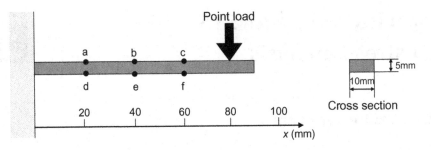

Fig. 3.3 Bending of a cantilever beam to solve.

3.1.2 Problem description

Geometry: length $l = 90$ mm, height $h = 5$ mm, thickness $b = 10$ mm.
Material: mild steel having Young's modulus $E = 210$ GPa and Poisson's ratio $\nu = 0.3$.
Boundary conditions: The beam is clamped to a rigid wall at the left end and loaded at $x = 80$ mm by a point load of $P = 100$ N.

3.1.3 Review of the solutions obtained by the elementary beam theory

Before proceeding to the FEM analysis of the beam, let us review the solutions to the example problem obtained by the elementary beam theory. The maximum deflection of the beam δ_{max} can be calculated by the following equation:

$$\delta_{max} = \frac{Pl_1^3}{3EI}\left(1 + \frac{3l_1}{l_2}\right) \tag{3.1}$$

where l_1 (=80 mm) is the distance of the application point of the load from the rigid wall and $l_2 = l - l_1$.

The maximum tensile stress $\sigma_{max}(x)$ at x in the longitudinal direction appears at the upper surface of the beam in a cross section at x from the wall.

$$\sigma_{max}(x) = \begin{cases} \dfrac{P(l_1 - x)h}{I}\dfrac{h}{2} & (0 \leq x \leq l_1) \\ 0 & (0 \leq x) \end{cases} \tag{3.2}$$

where l (=90 mm) is the length, h (=5 mm) is the height, b (=10 mm) is the thickness, E (=210 GPa) is Young's modulus, and I is the area moment of inertia of the cross section of the beam. For a beam having a rectangular cross section of a height h by a thickness b, the value of I can be calculated by the following equation:

$$I = \frac{bh^3}{12} \tag{3.3}$$

3.1.4 Analytical procedures

Fig. 3.4 shows how to make structural analyses by using FEM. In this chapter, the analytical procedures will be explained following the flowchart illustrated in Fig. 3.4.

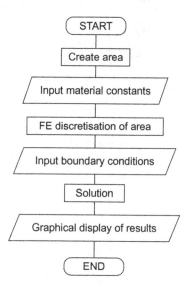

Fig. 3.4 Flowchart of the structural analyses by ANSYS.

3.1.4.1 Creation of an analytical model

(1) Creation of a beam shape to analyse

Here we shall analyse a rectangular slender beam of 5 mm (0.005 m) in height by 90 mm (0.09 m) in length by 10 mm (0.01 m) in width, as illustrated in Fig. 3.3. Fig. 3.5 shows the ANSYS Main Menu window where we can find layered command options imitating folders and files in the Microsoft Explorer folder window.

In order to prepare for creating the beam, the following operations should be made:

1. Click [A] **Preprocessor** to open its submenus in **ANSYS Main Menu** window.
2. Click [B] **Modeling** to open its submenus and select **Create** menu.
3. Click [C] **Areas** to open its submenus and select **Rectangle** menu.
4. Click to select the [D] **By 2 corners** menu.

After carrying out the operations above, a window called **Rectangle by 2 corners** appears, as shown in Fig. 3.6, for the input of the geometry of a 2D rectangular beam.

(2) Input of the beam geometry to analyse

The **Rectangle by 2 corners** window has four boxes for inputting the coordinates of the lower left corner point of the rectangular beam and the width and height of the beam to create. The following operations complete the creation procedure of the beam.

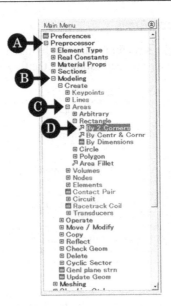

Fig. 3.5 'ANSYS Main Menu' window.

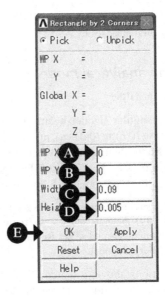

Fig. 3.6 'By 2 Corners' window.

1. Input two **0**s into [A] **WP X** and [B] **WP Y** to determine the lower left corner point of the beam on the Cartesian coordinates of the working plane.
2. Input **0.09** and **0.005** (m) into [C] **Width** and [D] **Height**, respectively, to determine the shape of the beam model.

3. Click the [E] **OK** button to create the rectangular area, or beam on the **ANSYS Graphics** window, as shown in Fig. 3.7.

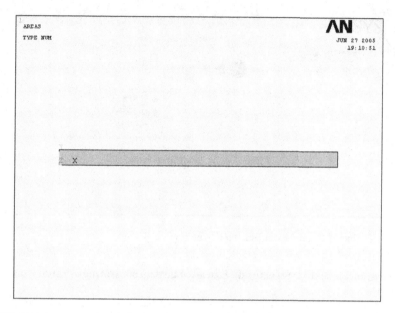

Fig. 3.7 2D beam created and displayed on the 'ANSYS Graphics' window.

How to correct the shape of the model

When correcting the model, delete the area first, and repeat procedures (1) and (2) above. In order to delete the area, execute the following commands:

[COMMAND]   ANSYS Main Menu → Preprocessor → Modeling → Delete → Area and Below

Then, the **Delete Area and Below** window opens and an upward arrow (↑) appears on the **ANSYS Graphics** window.

1. Move the arrow to the area to delete and click the left mouse button.
2. The colour of the area turns from light blue to pink.
3. Click the **OK** button and the area is deleted.

3.1.4.2 Input of the elastic properties of the beam material

Next, we specify elastic constants of the beam. In the case of isotropic material, the elastic constants are Young's modulus and Poisson's ratio. This procedure can be performed any time before the solution procedure—for instance, after setting boundary conditions. If this procedure is missed, we cannot perform the solution procedure.

[COMMAND]   ANSYS Main Menu → Preprocessor → Material Props → Material Models

Then the **Define Material Model Behavior** window opens, as shown in Fig. 3.8.

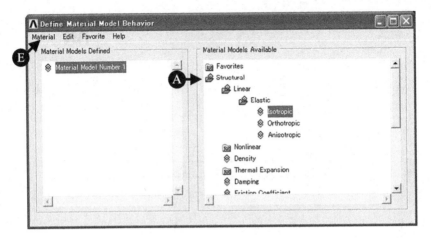

Fig. 3.8 'Define Material Model Behavior' window.

1. Double-click the [A] **Structural, Linear, Elastic** and **Isotropic** buttons one after another.
2. Input the value of Young's modulus, **2.1e11** (Pa), and that of Poisson's ratio, **0.3**, into [B] **EX** and [C] **PRXY** boxes, and click the [D] **OK** button of the **Linear Isotropic Properties for Materials Number 1** as shown in Fig. 3.9.

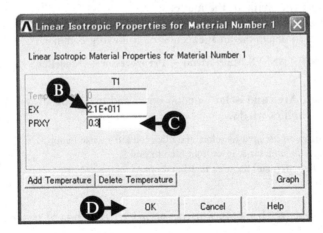

Fig. 3.9 Input of elastic constants through the 'Linear Isotropic Properties for Material Number 1' window.

3. Exit from the **Define Material Model Behavior** window by selecting **Exit** in [E] **Material** menu of the window (see Fig. 3.8).

3.1.4.3 Finite-element discretisation of the beam area

Here we shall divide the beam area into finite elements. The procedures for finite-element discretisation are firstly to select the element type, secondly to input the element thickness and finally to divide the beam area into elements.

(1) Selection of the element type

[COMMAND] **ANSYS Main Menu → Preprocessor → Element Type → Add/Edit/ Delete**

Then the **Element Types** window opens, as shown in Fig. 3.10.

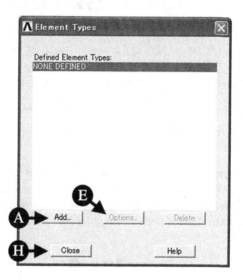

Fig. 3.10 'Element Types' window.

1. Click the [A] **Add** ... button to open the **Library of Element Types** window as shown in Fig. 3.11 and select the element type to use.

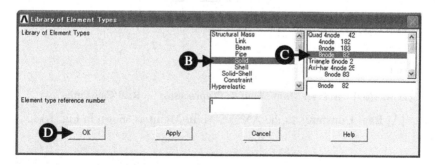

Fig. 3.11 'Library of Element Types' window.

2. To select the 8-node isoparametric element, select [B] **Structural Mass – Solid.**
3. Select [C] **Quad 8 node 82** and click the [D] **OK** button to choose the 8-node isoparametric element.
4. Click the [E] **Options ...** button in the **Element Types** window as shown in Fig. 3.10 to open the **PLANE82 element type options** window, as depicted in Fig. 3.12. Select the [F] **Plane strs w/thk** item in the **Element behavior** box and click the [G] **OK** button to return to the **Element Types** window. Click the [H] **Close** button to close the window.

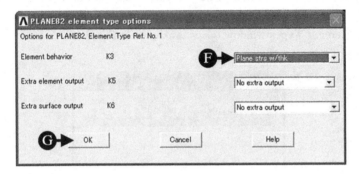

Fig. 3.12 'PLANE82 element type options' window.

The 8-node isoparametric element is a rectangular element which has four corner nodal points and four middle points, as shown in Fig. 3.13, and can realise the finite element analysis with higher accuracy than the 4-node linear rectangular element. The beam area is divided into these 8-node rectangular #82 finite elements.

Fig. 3.13 8-node isoparametric rectangular element.

(2) Input of the element thickness

    [COMMAND]   ANSYS Main Menu → Preprocessor → Real Constants

Select [A] **Real Constants** in the **ANSYS Main Menu** as shown in Fig. 3.14.

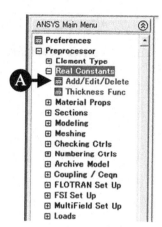

Fig. 3.14 Setting of the element thickness from the real constant command.

1. Click the [A] **Add/Edit/Delete** button to open the **Real Constants** window as shown in Fig. 3.15 and click the [B] **Add ...** button.

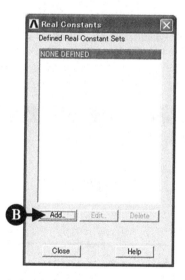

Fig. 3.15 'Real Constants' window before setting the element thickness.

2. Then the **Element Type for Real Constants** window opens (see Fig. 3.16). Click the [C] **OK** button.

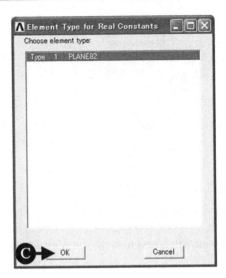

Fig. 3.16 'Element Type for Real Constants' window.

3. The **Element Type for Real Constants** window vanishes and the **Real Constants Set Number 1. for PLANE82** window appears instead, as shown in Fig. 3.17. Input a plate thickness of **0.01** (m) in the [D] **Thickness** box and click the [E] **OK** button.

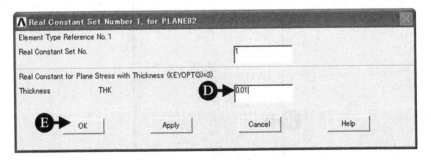

Fig. 3.17 'Real Constants Set Number 1. for PLANE82' window.

4. The **Real Constants** window returns with the display of the **Defined Real Constants Sets** box changed to **Set 1**, as shown in Fig. 3.18. Click the [F] **Close** button, which completes the operation of setting the plate thickness.

Fig. 3.18 'Real Constants' window after setting the element thickness.

(3) Sizing of the elements

[COMMAND] **ANSYS Main Menu → Preprocessor → Meshing → Size Cntrls → Manual Size → Global → Size**

The **Global Element Sizes** window opens, as shown in Fig. 3.19.

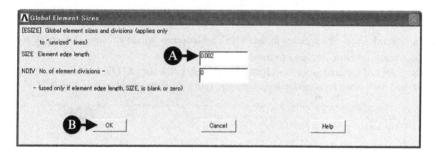

Fig. 3.19 'Global Element Sizes' window.

1. Input **0.002** (m) in the [A] **SIZE** box and click the [B] **OK** button.
   By the operations above, the element size of 0.002, i.e. 0.002 m or 2 mm, is specified and the beam of 5 mm by 90 mm is divided into rectangular finite elements with one side 2 mm and the other side 3 mm in length.
2. (4) Meshing

[COMMAND] **ANSYS Main Menu → Preprocessor → Meshing → Mesh → Areas → Free**

The **Mesh Areas** window opens, as shown in Fig. 3.20.

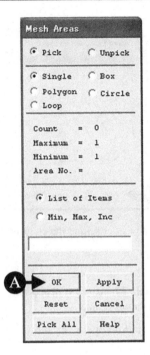

Fig. 3.20 'Mesh Areas' window.

1. An upward arrow (⬆) appears in the **ANSYS Graphics** window. Move this arrow to the beam area and click this area to mesh.
2. The colour of the area turns from light blue to pink. Click the [A] **OK** button to see the area meshed by 8-node rectangular isoparametric finite elements, as shown in Fig. 3.21.

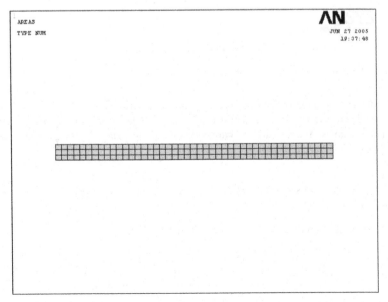

Fig. 3.21 Beam area subdivided into 8-node isoparametric rectangular elements.

How to modify meshing

In the case of modifying meshing, delete the elements, and repeat procedures (1)–(4) above. Repeat procedures (1), (2), or (3) to modify the element type, change the plate thickness without changing the element type, or change the element size only.

In order to delete the elements, execute the following commands:

[COMMAND]   ANSYS Main Menu → Preprocessor → Meshing → Clear → Areas

The **Clear Areas** window opens.

1. An upward arrow (⬆) appears in the **ANSYS Graphics** window. Move this arrow to the beam area and click this area.
2. The colour of the area turns from light blue to pink. Click the **OK** button to delete the elements from the beam area. After this operation, the area disappears from the display. Execute the following commands to replot the area.

[COMMAND]   ANSYS Utility Menu → Plot → Areas

3.1.4.4   Input of boundary conditions

Here we shall impose constraint and loading conditions on nodes of the beam model. Display the nodes first to define the constraint and loading conditions.

(1) Display of nodes

[COMMAND]   ANSYS Utility Menu → Plot → Nodes

The nodes are plotted in the **ANSYS Graphics** windows shown in Fig. 3.22.

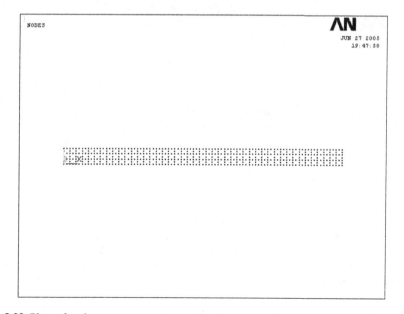

Fig. 3.22 Plots of nodes.

(2) Zoom in the node display

An enlarged view of the models is often convenient when imposing constraint and loading conditions on the nodes. In order to zoom in the node display, execute the following commands:

[COMMAND]   ANSYS Utility Menu → PlotCtrls → Pan Zoom Rotate ...

The **Pan-Zoom-Rotate** window opens, as shown in Fig. 3.23.

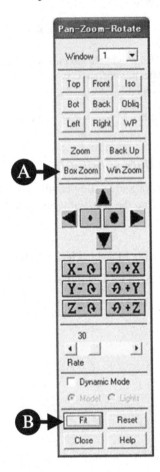

Fig. 3.23 'Pan-Zoom-Rotate' window.

1. Click [A] **Box Zoom** button.
2. The shape of the mouse cursor turns into a magnifying glass in the **ANSYS Graphics** window. Click the upper left point and then the lower right point, which enclose a portion of the beam area to enlarge, as shown in Fig. 3.24. Zoom in the left end of the beam.
3. In order to display the whole view of the beam, click the [B] **Fit** button in the **Pan-Zoom-Rotate** window.

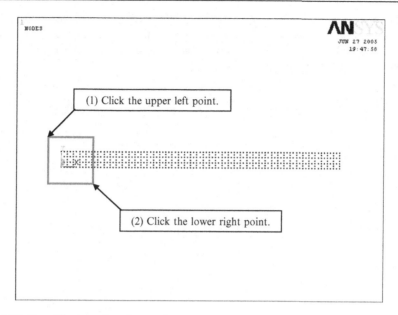

Fig. 3.24 Magnification of an observation area.

(3) Definition of constraint conditions

Selection of nodes

[COMMAND]   ANSYS Main Menu → Solution → Define
Loads → Apply →Structural → Displacement → On Nodes

The **Apply U. ROT on Nodes** window opens as shown in Fig. 3.25.

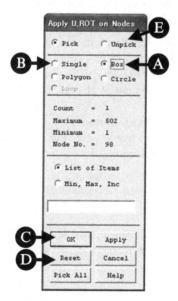

Fig. 3.25 'Apply U. ROT on Nodes' window.

1. Select [A] **Box** button and drag the mouse in the **ANSYS Graphics** window so as to enclose the nodes on the left edge of the beam area with the yellow rectangular frame, as shown in Fig. 3.26. The **Box** button is selected to pick multiple nodes at once, whereas the [B] **Single** button is chosen to pick a single node.

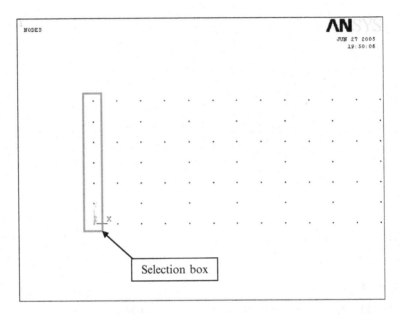

Fig. 3.26 Picking multiple nodes by box.

2. After confirming that only the nodes to impose constraints on are selected, i.e. the nodes on the left edge of the beam area, click the [C] **OK** button.

How to reselect nodes: Click [D] **Reset** button to clear the selection of the nodes before clicking the [C] **OK** button in procedure (2) above, and repeat procedures (1) and (2) above. The selection of nodes can be reset either by picking selected nodes after choosing the [E] **Unpick** button or clicking the right mouse button to turn the upward arrow upside down.

Imposing constraint conditions on nodes

The **Apply U. ROT on Nodes** window (see Fig. 3.27) opens after clicking the [C] **OK** button in procedure (2) in Section 3.1.

1. In case of selecting [A] **ALL DOF**, the nodes are to be clamped, i.e. the displacements are set to zero in the directions of the *x*- and *y*-axes. Similarly, the selection of **UX** makes the displacement in the *x*-direction equal to zero and the selection of **UY** makes the displacement in the *y*-direction equal to zero.
2. Click the [B] **OK** button and blue triangular symbols which denote the clamping conditions appear in the **ANSYS Graphics** window, as shown in Fig. 3.28. The upright triangles

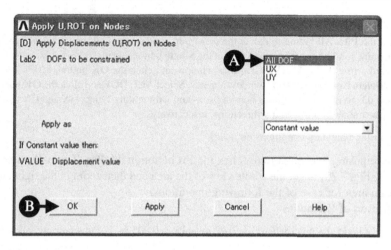

Fig. 3.27 'Apply U. ROT on Nodes' window.

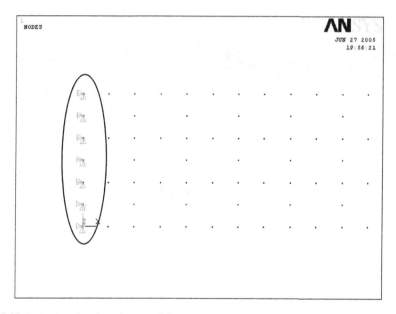

Fig. 3.28 Imposing the clamping conditions on nodes.

indicate that each node to which the triangular symbol is attached is fixed in the *x*-direction, whereas the tilted triangles indicate the fixed condition in the *y*-direction.

How to clear constraint conditions

[COMMAND]   **ANSYS Main Menu → Solution → Define Loads → Delete → Structural → Displacement → On Nodes**

The **Delete Node Constrai...** window opens.

1. Click the **Pick All** button to delete the constraint conditions of all the nodes on which the constraint conditions are imposed. Select the **Single** button and pick a particular node by the upward arrow in the **ANSYS Graphics** window and click the **OK** button.
2. The **Delete Node constraints** window appears. Select **ALL DOF** and click the **OK** button to delete the constraint conditions both in the x- and y-directions. Select **UX** and **UY** to delete the constraints in the x- and y-directions, respectively.

(4) Imposing boundary conditions on nodes

Before imposing load conditions, click the **Fit** button in the **Pan-Zoom-Rotate** window (see Fig. 3.23) to get the whole view of the area and then zoom in the right end of the beam area for ease of the following operations.

Selection of the nodes

1. Pick the node at a point where $x = 0.08$ m and $y = 0.005$ m. For this purpose, click

   [COMMAND]   ANSYS Utility Menu → PlotCtrls → Numbering ...

   consecutively to open the Plot Numbering Controls window, as shown in Fig. 3.29.

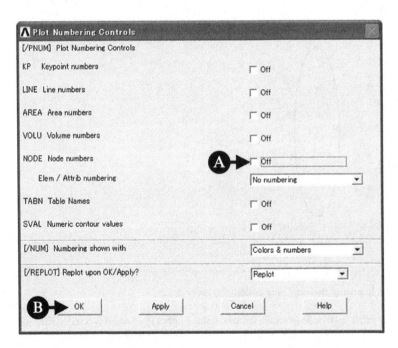

Fig. 3.29 'Plot Numbering Controls' window.

2. Click the [A] **NODE** ☐ **Off** box to change it to ☑ **On**.
3. Click the [B] **OK** button to display node numbers adjacent to corresponding nodes in the **ANSYS Graphics** window, as shown in Fig. 3.30.

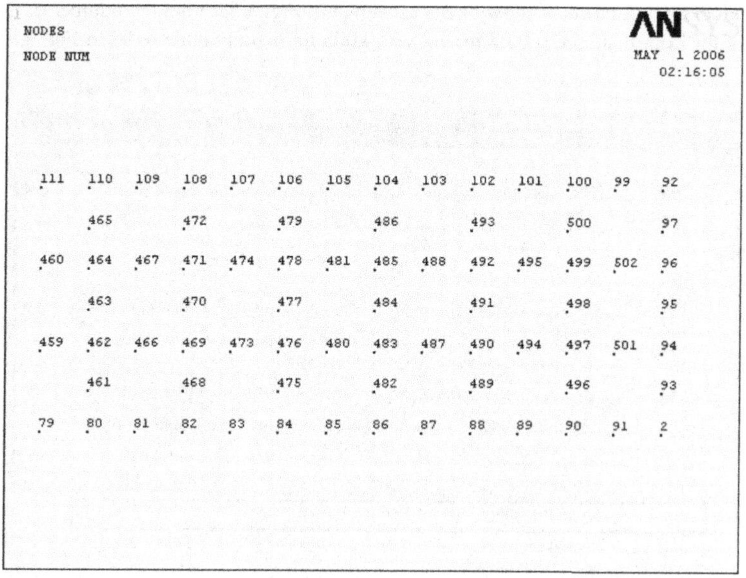

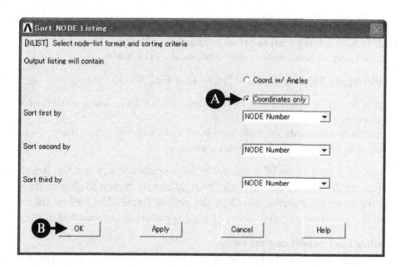

Fig. 3.30 Nodes and nodal numbers displayed on the results displaying window.

4. To delete node numbers, click the [A] **NODE** ☑ **On** box again to change it to ☐ **Off**.
5. Execute the following commands:

    [COMMAND]   **ANSYS Utility Menu → List → Nodes ...**

and the **Sort NODE Listing** window opens (see Fig. 3.31). Select the [A] **Coordinates only** button and then click the [B] **OK** button.

Fig. 3.31 'Sort NODE Listing' window.

6. The **NLIST Command** window opens as shown in Fig. 3.32. Find the number of the node having the coordinates $x = 0.08$ m and $y = 0.005$ m, namely node #108 in Fig. 3.32.

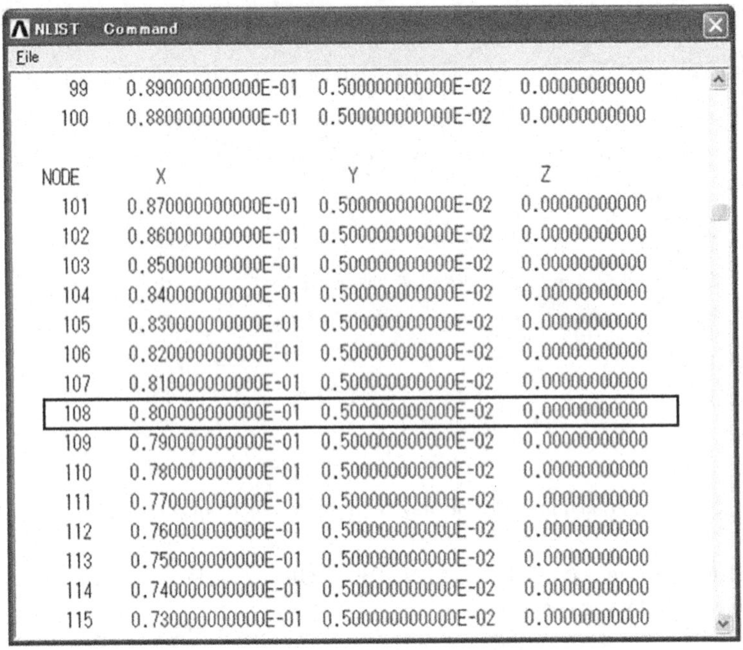

NODE	X	Y	Z
99	0.890000000000E-01	0.500000000000E-02	0.00000000000
100	0.880000000000E-01	0.500000000000E-02	0.00000000000
NODE	X	Y	Z
101	0.870000000000E-01	0.500000000000E-02	0.00000000000
102	0.860000000000E-01	0.500000000000E-02	0.00000000000
103	0.850000000000E-01	0.500000000000E-02	0.00000000000
104	0.840000000000E-01	0.500000000000E-02	0.00000000000
105	0.830000000000E-01	0.500000000000E-02	0.00000000000
106	0.820000000000E-01	0.500000000000E-02	0.00000000000
107	0.810000000000E-01	0.500000000000E-02	0.00000000000
108	0.800000000000E-01	0.500000000000E-02	0.00000000000
109	0.790000000000E-01	0.500000000000E-02	0.00000000000
110	0.780000000000E-01	0.500000000000E-02	0.00000000000
111	0.770000000000E-01	0.500000000000E-02	0.00000000000
112	0.760000000000E-01	0.500000000000E-02	0.00000000000
113	0.750000000000E-01	0.500000000000E-02	0.00000000000
114	0.740000000000E-01	0.500000000000E-02	0.00000000000
115	0.730000000000E-01	0.500000000000E-02	0.00000000000

Fig. 3.32 'NLIST Command' window showing the coordinates of the nodes; the framed portion indicates the coordinates of the node for load application.

7. Execute the following commands:

[COMMAND] ANSYS Main Menu → Solution → Define Loads → Apply → Structural → Force/Moment → On Nodes

to open the **Apply F/M on Nodes** window (see Fig. 3.33).

8. Pick only the #108 node having the coordinates $x = 0.08$ m and $y = 0.005$ m with the upward arrow, as shown in Fig. 3.34.
9. After confirming that only the #108 node is enclosed with the yellow frame, click the [A] **OK** button in the **Apply F/M on Nodes** window.

How to cancel the selection of the nodes of load application: Click the **Reset** button before clicking the **OK** button or click the right mouse button to change the upward arrow to the downward arrow and click the yellow frame. The yellow frame disappears and the selection of the node(s) of load application is cancelled.

Imposing load conditions on nodes
Click [A] **OK** in the **Apply F/M on Nodes** window to open another **Apply F/M on Nodes** window, as shown in Fig. 3.35.

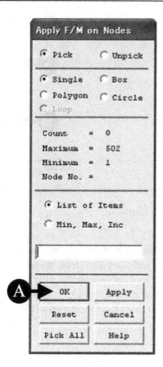

Fig. 3.33 'Apply F/M on Nodes' window.

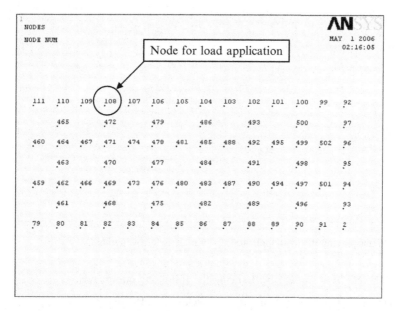

Fig. 3.34 Selection of a node for load application.

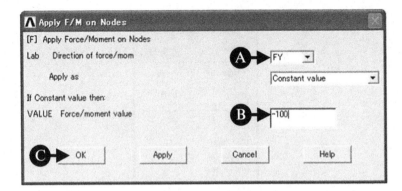

Fig. 3.35 'Apply F/M on Nodes' window.

1. Choose [A] **FY** in the **Lab Direction of force/mom** box and input [B] **-100** in the **VALUE** box. A positive value for load indicates load in the upward or rightward direction, whereas a negative value load in the downward or leftward direction.
2. Click the [C] **OK** button to display the downward arrow attached to the #108 node indicating the downward load applied to that point, as shown in Fig. 3.36.

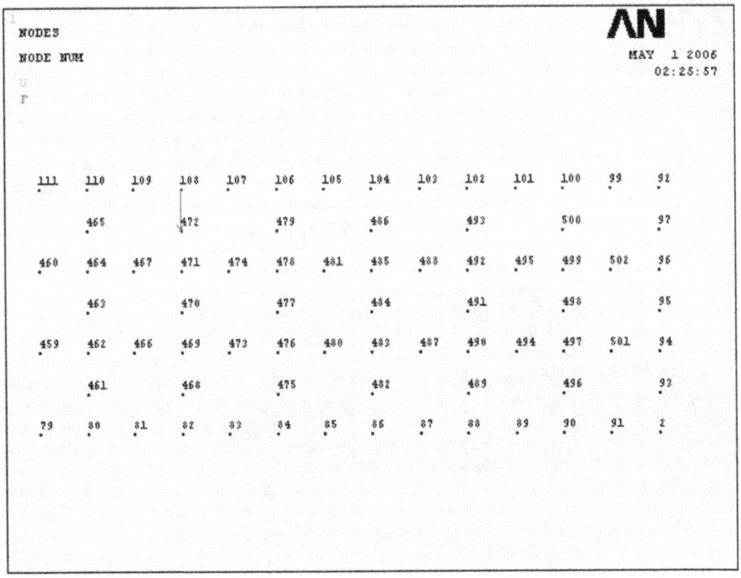

```
NODES                                                                    AN
NODE NUM
U                                                                  MAY  1 2006
F                                                                   02:25:57


   111    110   109   108   107   106   105   104   103   102   101   100   99   98
                                                                           .     .

          465         472         479         486         492         500        97
           .           .           .           .           .           .          .

   460    464   467   471   474   478   481   485   488   492   495   499   502   96
    .      .     .     .     .     .     .     .     .     .     .     .     .     .

          463         470         477         484         493         498        95
           .           .           .           .           .           .          .

   459    462   466   469   473   476   480   483   487   490   494   497   501   94
    .      .     .     .     .     .     .     .     .     .     .     .     .     .

          461         468         475         482         489         496        93
           .           .           .           .           .           .          .

    79     80    81    82    83    84    85    86    87    88    89    90    91    2
    .      .     .     .     .     .     .     .     .     .     .     .     .     .
```

Fig. 3.36 Displaying the load application on a node by arrow symbol.

How to delete load conditions: Execute the following commands:

[COMMAND]   ANSYS Main Menu → Solution → Define
Loads → Delete → Structural → Force/Moment → On Nodes

to open the **Delete F/M on Nodes** window (see Fig. 3.37). Choose [A] **FY** or **ALL** in the **Lab Force/moment to be deleted** and click the **OK** button to delete the downward load applied to the #108 node.

Delete F/M on Nodes
[FDELE] Delete Force/Moment on Nodes
Lab Force/moment to be deleted
A
ALL
B
OK
Apply
Cancel

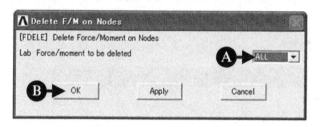

Fig. 3.37 'Delete F/M on Nodes' window.

3.1.4.5 Solution procedures

[COMMAND] ANSYS Main Menu → Solution → Solve → Current LS

The **Solve Current Load Step** and **/STATUS Command** windows appear as shown in Figs 3.38 and 3.39, respectively.

Solve Current Load Step
[SOLVE] Begin Solution of Current Load Step
Review the summary information in the lister window (entitled "/STATUS Command"), then press OK to start the solution.
A
OK
Cancel
Help

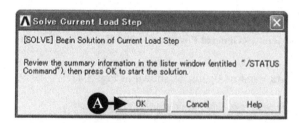

Fig. 3.38 'Solve Current Load Step' window.

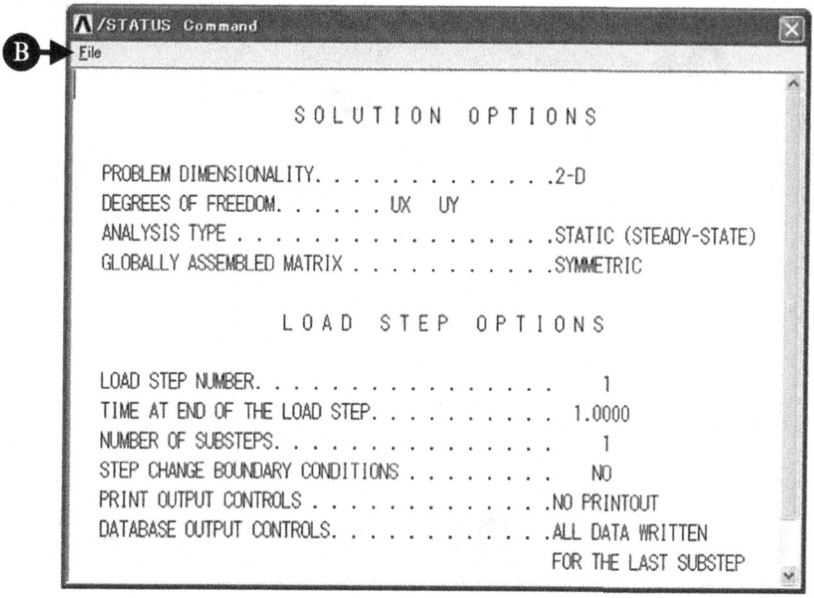

/STATUS Command
B
File
SOLUTION OPTIONS
PROBLEM DIMENSIONALITY. 2-D
DEGREES OF FREEDOM. . . . . . UX UY
ANALYSIS TYPE . . . . .STATIC (STEADY-STATE)
GLOBALLY ASSEMBLED MATRIX . . . . .SYMMETRIC
LOAD STEP OPTIONS
LOAD STEP NUMBER. 1
TIME AT END OF THE LOAD STEP. 1.0000
NUMBER OF SUBSTEPS. 1
STEP CHANGE BOUNDARY CONDITIONS . . . . NO
PRINT OUTPUT CONTROLS . . . . .NO PRINTOUT
DATABASE OUTPUT CONTROLS. . . . . .ALL DATA WRITTEN
FOR THE LAST SUBSTEP

Fig. 3.39 '/STATUS Command' window.

1. Click the [A] **OK** button in the **Solve Current Load Step** window, as shown in Fig. 3.38, to begin the solution of the current load step.
2. The **/STATUS Command** window displays information on solution and load step options. Select the [B] **File** button to open the submenu and select the **Close** button to close the **/STATUS Command** window.
3. When the solution is completed, the **Note** window (see Fig. 3.40) appears. Click the [C] **Close** button to close the window.

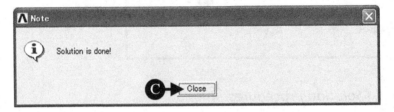

Fig. 3.40 'Note' window.

3.1.4.6   Graphical representation of the results

(1) Contour plot of displacements

   [COMMAND]   **ANSYS Main Menu → General Postproc → Plot Results → Contour Plot → Nodal Solution**

The **Contour Nodal Solution Data** window opens, as shown in Fig. 3.41.

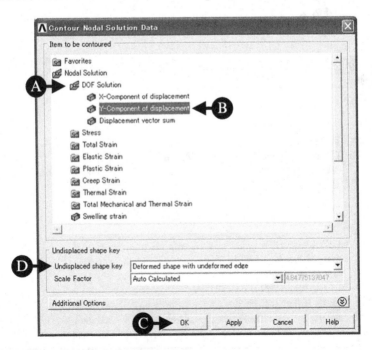

Fig. 3.41 'Contour Nodal Solution Data' window.

1. Select [A] **DOF Solution** and [B] **Y-Component of displacement**.
2. Click the [C] **OK** button to display the contour of the y-component of displacement, or the deflection of the beam in the **ANSYS Graphics** window (see Fig. 3.42). The **DMX** value shown in the Graphics window indicates the maximum deflection of the beam.

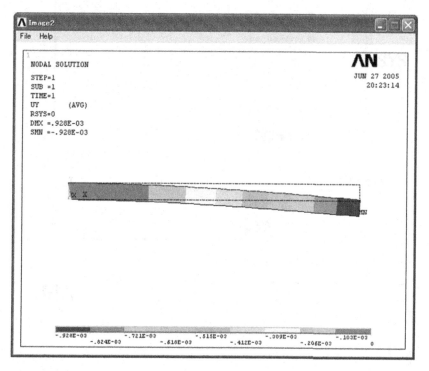

Fig. 3.42 Contour map representation of the distribution of displacement in the y- or vertical direction.

3. Select [D] **Deformed shape with undeformed edge** in the **Undisplaced shape key** box to compare the shapes of the beam before and after deformation.

(2) Contour plot of stresses

1. Select [A] **Stress** and [B] **X-Component of stress**, as shown in Fig. 3.43.
2. Click the [C] **OK** button to display the contour of the x-component of stress, or the bending stress in the beam in the **ANSYS Graphics** window (see Fig. 3.44). The **SMX** and **SMN** values shown in the Graphics window indicate the maximum and the minimum stresses in the beam, respectively.
3. Click the [D] **Additional Options** bar to open additional option items to choose. Select the [E] **All applicable** in the **Number of facets per element edge** box to calculate stresses and strains at middle points of the elements.

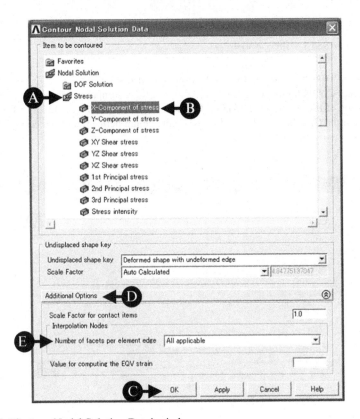

Fig. 3.43 'Contour Nodal Solution Data' window.

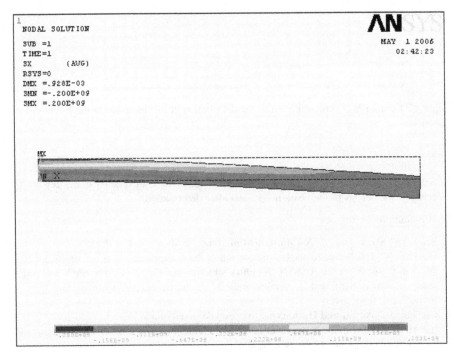

Fig. 3.44 Contour map representation of the distribution of normal stress in the *x*- or horizontal direction.

3.1.5 Comparison of FEM results with experimental ones

Fig. 3.45 compares longitudinal stress distributions obtained by ANSYS with those by experiments and by elementary beam theory. The results obtained by three different methods agree well with one another. As the applied load increases, however, errors among the three groups of results become larger, especially at the clamped end. This tendency arises from the fact that the clamped condition can be hardly realised in the strict sense.

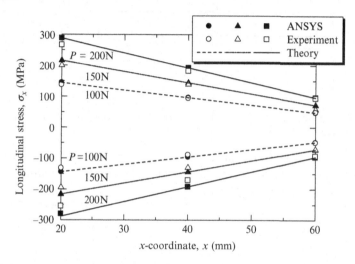

Fig. 3.45 Comparison of longitudinal stress distributions obtained by ANSYS with those by experiments and by the elementary beam theory.

3.1.6 Problems to solve

Problem 3.1. Change the point of load application and the intensity of the applied load in the cantilever beam model shown in Fig. 3.3, and calculate the maximum deflection.

Problem 3.2. Calculate the maximum deflection in a beam clamped at the both ends as shown in Fig. 3.46, where the thickness of the beam in the direction perpendicular to the page surface is 10 mm.

(Answer: 0.00337 mm)

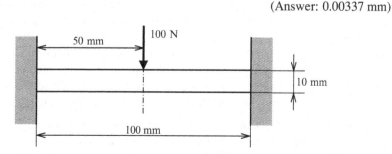

Fig. 3.46 A beam clamped at the both ends and subjected to a concentrated force of 100 N at the centre of the span.

Problem 3.3. Calculate the maximum deflection in a beam simply supported at the both ends, as shown in Fig. 3.47, where the thickness of the beam in the direction perpendicular to the page surface is 10 mm.

(Answer: 0.00645 mm)

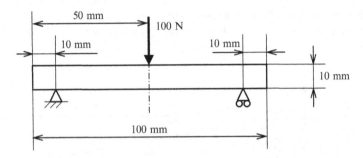

Fig. 3.47 A beam simply supported at the both ends and subjected to a concentrated force of 100 N at the centre of the span.

Problem 3.4. Calculate the maximum deflection in a beam shown in Fig. 3.48 where the thickness of the beam in the direction perpendicular to the page surface is 10 mm. Choose an element size of 1 mm.

(Answer: 0.00337 mm)

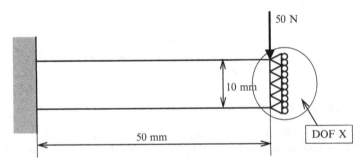

Fig. 3.48 A half model of the beam in Problem 3.2.

Note that the beam shown in Fig. 3.46 is bilaterally symmetric so that the x-component of the displacement (DOFX) is zero at the centre of the beam span. If the beam shown in Fig. 3.46 is cut at the centre of the span and the finite element calculation is made for only the left half of the beam by applying a half load of 50 N to its right end which is fixed in the x-direction but is deformed freely in the y-direction as shown in Fig. 3.48, the solution obtained is the same as that for the left half of the beam in Problem 3.2. This problem can be solved by its half model as shown in Fig. 3.48. A half model can achieve the efficiency of finite element calculations.

Problem 3.5. Calculate the maximum deflection in a beam shown in Fig. 3.49, where the thickness of the beam in the direction perpendicular to the page surface is 10 mm. This beam is the half model of the beam of Problem 3.3.

(Answer: 0.00645 mm)

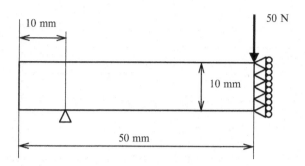

Fig. 3.49 A half model of the beam in Problem 3.3.

Problem 3.6. Calculate the maximum value of the von Mises stress in the stepped beam, as shown in Fig. 3.50, where Young's modulus $E = 210$ GPa, Poisson's ratio $\nu = 0.3$, the element size is 2 mm, and the beam thickness is 10 mm. Refer to the Appendix to create the stepped beam. The von Mises stress σ_{eq} is sometimes called the equivalent stress or the effective stress, and is expressed by the following formula:

$$\sigma_{eq} = \frac{1}{\sqrt{2}}\sqrt{\left(\sigma_x - \sigma_y\right)^2 + \left(\sigma_y - \sigma_z\right)^2 + \left(\sigma_z - \sigma_x\right)^2 + 6\left(\tau_{xy}^2 + \tau_{yz}^2 + \tau_{zx}^2\right)} \qquad (3.4)$$

in three-dimensional elasticity problems. It is often considered that a material yields when the value of the von Mises stress reaches the yield stress of the material σ_Y, which is determined by the uniaxial tensile tests of the material.

(Answer: 40.8 MPa)

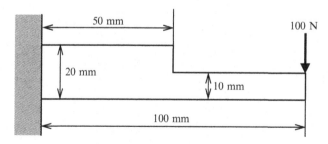

Fig. 3.50 A stepped cantilever beam subjected to a concentrated force of 100 N at the free end.

Problem 3.7. Calculate the maximum value of the von Mises stress in the stepped beam with a rounded fillet, as shown in Fig. 3.51, where Young's modulus $E = 210$ GPa, Poisson's ratio $\nu = 0.3$, the element size is 2 mm, and the beam thickness is 10 mm. Refer to the Appendix to create the stepped beam.

(Answer: 30.2 MPa)

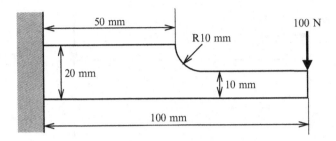

Fig. 3.51 A stepped cantilever beam with a rounded fillet subjected to a concentrated force of 100 N at the free end.

Appendix: Procedures for creating stepped beams

A.1 Creation of a stepped beam

A stepped beam as shown in Fig. 3.50 can be created by adding two rectangular areas of different sizes:

1. Create two rectangular areas of different sizes, say 50 mm by 20 mm and 60 mm by 10 mm, following operations described in Section 3.1.4.1.
2. Select the Boolean operation of adding areas as follows to open the **Add Areas** window (see Fig. 3.52):

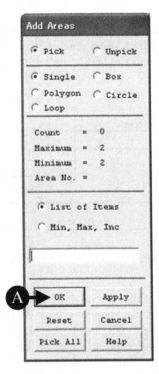

Fig. 3.52 'Add Areas' window.

[COMMAND] **ANSYS** **Main** **Menu** → **Preprocessor** → **Modeling** → **Operate** → **Booleans** → **Add** → **Areas**

3. Pick all the areas to add by the upward arrow.
4. The colour of the areas picked turns from light blue to pink (see Fig. 3.53). Click the [A] **OK** button to add the two rectangular areas to create a stepped beam area, as shown in Fig. 3.54.

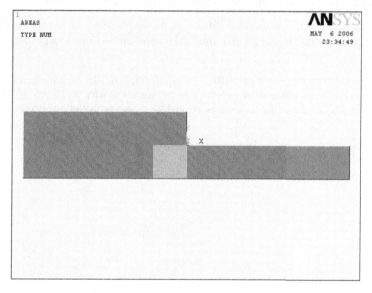

Fig. 3.53 Two rectangular areas of different sizes to be added to create a stepped beam area.

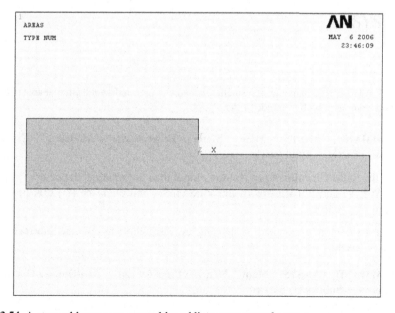

Fig. 3.54 A stepped beam area created by adding two rectangle areas.

A.1.1 How to cancel the selection of areas

Click the **Reset** button or click the right mouse button to change the upward arrow to the downward arrow and click the area(s) picked. The colour of the unpicked area(s) turns pink to light blue and the selection of the area(s) is cancelled.

A.2 Creation of a stepped beam with a rounded fillet

A stepped beam with a rounded fillet, as shown in Fig. 3.51, can be created by subtracting a smaller rectangular area and a solid circle from a larger rectangular area:

1. Create a larger rectangular area of 100 mm by 20 mm, a smaller rectangular area of, say 50 mm by 15 mm, and a solid circular area having a diameter of 10 mm as shown in Fig. 3.55. The solid circular area can be created by executing the following operation:

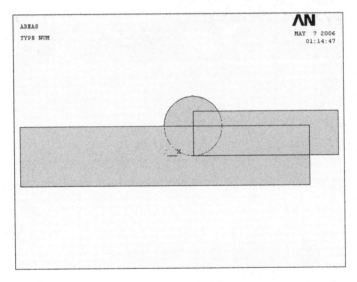

Fig. 3.55 A larger rectangular area, a smaller rectangular area and a solid circular area to create a stepped beam area with a rounded fillet.

　　[COMMAND] ANSYS Main Menu → **Preprocessor** → **Modeling** → **Create** → **Areas** → **Circle** → **Solid Circle**

to open the **Solid Circular Area** window. Input the coordinate of the centre ([A] **WP X**, [B] **WP Y**) and [C] **Radius** of the solid circle, and click the [D] **OK** button as shown in Fig. 3.56.

2. Select the Boolean operation of subtracting areas as follows to open the **Subtract Areas** window (see Fig. 3.57):

　　[COMMAND] ANSYS Main Menu → **Preprocessor** → **Modeling** → **Operate** → **Booleans** → **Subtract** → **Areas**

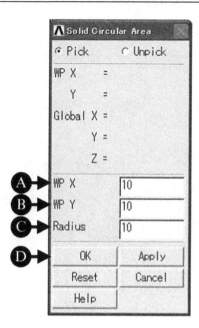

Fig. 3.56 'Solid Circular Area' window.

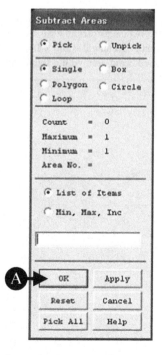

Fig. 3.57 'Subtract Areas' window.

3. Pick the larger rectangular area by the upward arrow as shown in Fig. 3.58 and the colour of the area picked turns from light blue to pink. Click the [A] **OK** button.

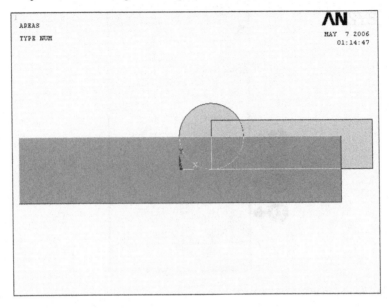

Fig. 3.58 A larger rectangular area picked.

4. Pick the smaller rectangular area and the solid circular area by the upward arrow as shown in Fig. 3.59 and the colour of the two areas picked turns from light blue to pink. Click the [A] **OK** button to subtract the smaller rectangular and solid circular areas from the larger rectangular area to create a stepped beam area with a rounded fillet, as shown in Fig. 3.60.

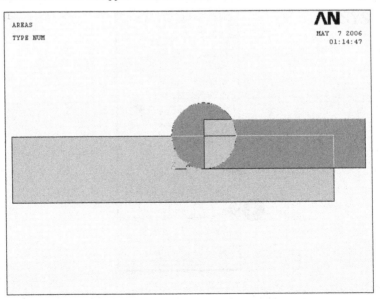

Fig. 3.59 A smaller rectangular area and a solid circular area picked to be subtracted from a larger rectangular area to create a stepped beam area with a rounded fillet.

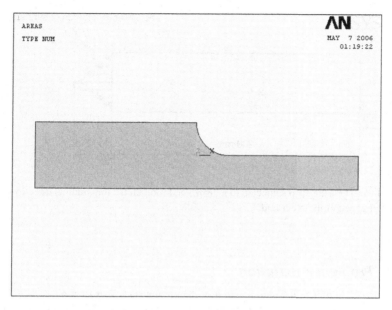

Fig. 3.60 A stepped cantilever beam area with a rounded fillet.

A.2.1 How to display area numbers

Area numbers can be displayed in the **ANSYS Graphics** window by the following procedure.

    [Commands] **ANSYS Utility Menu → PlotCtrls → Numbering …**

1. The **Plot Numbering Controls** window opens as shown in Fig. 3.29.
2. Click the **AREA** ☐ **Off** box to change it to ☑ **On**.
3. Click the **OK** button to display area numbers in corresponding areas in the **ANSYS Graphics** window.
4. To delete element numbers, click the **AREA** ☑ **On** box again to change it to ☐ **Off**.

3.2 The principle of St. Venant

3.2.1 Example problem

An elastic strip is subjected to distributed uniaxial tensile stress or negative pressure at one end and clamped at the other end.

    Perform an FEM analysis of a 2D elastic strip subjected to a distributed stress in the longitudinal direction at one end and clamped at the other end, as shown in Fig. 3.61, and calculate the stress distributions along the cross sections at different distances from the loaded end in the strip.

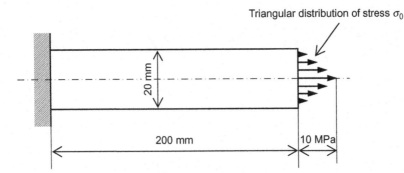

Fig. 3.61 A 2D elastic strip subjected to a distributed force in the longitudinal direction at one end and clamped at the other end.

3.2.2 Problem description

Geometry: length $l = 200$ mm, height $h = 20$ mm, thickness $b = 10$ mm.
Material: mild steel having Young's modulus $E = 210$ GPa and Poisson's ratio $\nu = 0.3$.
Boundary conditions: The elastic strip is subjected to a triangular distribution of stress in the longitudinal direction at the right end and clamped to a rigid wall at the left end.

3.2.3 Analytical procedures

3.2.3.1 Creation of an analytical model

[COMMAND] ANSYS Main Menu → **Preprocessor** → **Modeling** → **Create** → **Areas** → **Rectangle** → **By 2 Corners**

1. Input two **0**s into the **WP X** and **WP Y** boxes in the **Rectangle by 2 Corners** window to determine the lower left corner point of the elastic strip on the Cartesian coordinates of the working plane.
2. Input **200** and **20** (mm) into the **Width** and **Height** boxes, respectively, to determine the shape of the elastic strip model.
3. Click the **OK** button to create the rectangular area, or beam on the **ANSYS Graphics** window.

In the procedures above, the geometry of the strip is input in millimetres. You must decide what kind of units to use in finite element analyses. When you input the geometry of a model to analyse in millimetres, for example, you must input applied loads in N (Newton) and Young's modulus in MPa, since 1 MPa is equivalent to 1 N/mm^2. When you use metres and N as the units of length and load, respectively, you must input Young's modulus in Pa, since 1 Pa is equivalent to 1 N/m^2. You can choose any system of unit you would like to, but your unit system must be consistent throughout the analyses.

3.2.3.2   Input of the elastic properties of the strip material

[COMMAND]   ANSYS Main Menu → Preprocessor → Material Props → Material Models

1. The **Define Material Model Behavior** window opens.
2. Double-click the **Structural, Linear, Elastic,** and **Isotropic** buttons one after another.
3. Input the value of Young's modulus, **2.1e5** (MPa), and that of Poisson's ratio, **0.3**, into the **EX** and **PRXY** boxes, and click the **OK** button of the **Linear Isotropic Properties for Materials Number 1** window.
4. Exit from the **Define Material Model Behavior** window by selecting **Exit** in the **Material** menu of the window.

3.2.3.3   Finite-element discretisation of the strip area

(1) Selection of the element type

[COMMAND]   ANSYS Main Menu → Preprocessor → Element Type → Add/Edit/ Delete

1. The **Element Types** window opens.
2. Click the **Add ...** button in the **Element Types** window to open the **Library of Element Types** window and select the element type to use.
3. Select **Structural Mass – Solid** and **Quad 8 node 82**.
4. Click the **OK** button in the **Library of Element Types** window to use the 8-node isoparametric element.
5. Click the **Options ...** button in the **Element Types** window to open the **PLANE82 element type options** window. Select the **Plane strs w/thk** item in the **Element behavior** box and click the **OK** button to return to the **Element Types** window. Click the **Close** button in the **Element Types** window to close the window.

(2) Input of the element thickness

[COMMAND]   ANSYS Main Menu → Preprocessor → Real Constants → Add/Edit/ Delete

1. The **Real Constants** window opens.
2. Click [A] **Add/Edit/Delete** button to open the **Real Constants** window and click the **Add ...** button.
3. The **Element Type for Real Constants** window opens. Click the **OK** button.
4. The **Element Type for Real Constants** window vanishes and the **Real Constants Set Number 1. for PLANE82** window appears instead. Input a strip thickness of 10 mm in the **Thickness** box and click the **OK** button.
5. The **Real Constants** window returns with the display of the **Defined Real Constants Sets** box changed to **Set 1.** Click the **Close** button.

(3) Sizing of the elements

[COMMAND]   ANSYS Main Menu → Preprocessor → Meshing → Size Cntrls → Manual Size → Global → Size

1. The **Global Element Sizes** window opens.
2. Input **2** in the **SIZE** box and click the **OK** button.

(4) Dividing the right-end side of the strip area into two lines

Before proceeding to meshing, the right-end side of the strip area must be divided into two lines for imposing the triangular distribution of the applied stress or stress by executing the following commands.

[COMMAND] ANSYS Main Menu → Preprocessor → Modeling → Operate → Booleans → Divide → Lines w/ Options

1. The **Divide Multiple Lines** ... window opens as shown in Fig. 3.62.

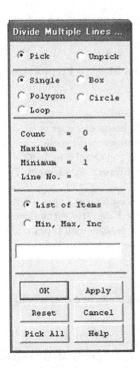

Fig. 3.62 'Divide Multiple Lines ...' window.

2. When the mouse cursor is moved to the **ANSYS Graphics** window, an upward arrow (⬆) appears.
3. Confirming that the **Pick** and **Single** buttons are selected, move the upward arrow onto the right-end side of the strip area and click the left mouse button.
4. Click the **OK** button in the **Divide Multiple Lines** ... window to display the **Divide Multiple Lines with Options** window, as shown in Fig. 3.63.
5. Input **2** in the [A] **NDIV** box and **0.5** in the [B] **RATIO** box, and select **Be modified** in the [C] **KEEP** box.
6. Click the [D] **OK** button.

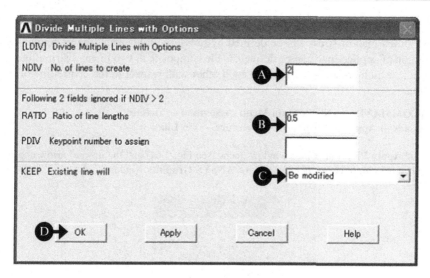

Fig. 3.63 'Divide Multiple Lines with Options' window.

(5) Meshing

[COMMAND] ANSYS Main Menu → Preprocessor → Meshing → Mesh → Areas → Free

1. The **Mesh Areas** window opens.
2. The upward arrow appears in the **ANSYS Graphics** window. Move this arrow to the elastic strip area and click this area.
3. The colour of the area turns from light blue to pink. Click the **OK** button to see the area meshed by 8-node rectangular isoparametric finite elements.

3.2.3.4 Input of boundary conditions

(1) Imposing constraint conditions on the left end of the strip

[COMMAND] ANSYS Main Menu → Solution → Define Loads → Apply → Structural → Displacement → On Lines

1. The **Apply U. ROT on Lines** window opens and the upward arrow appears when the mouse cursor is moved to the **ANSYS Graphics** window.
2. Confirming that the **Pick** and **Single** buttons are selected, move the upward arrow onto the left-end side of the strip area and click the left mouse button.
3. Click the **OK** button in the **Apply U. ROT on Lines** window to display another **Apply U. ROT on Lines** window.
4. Select **ALL DOF** in the **Lab2** box and click the **OK** button in the **Apply U. ROT on Lines** window.

(2) Imposing a triangular distribution of applied stress on the right end of the strip

Distributed load or stress can be defined by pressure on lines and the triangular distribution of applied load can be defined as the composite of two linear distributions of pressure which are antisymmetric to each other with respect to the centre line of the strip area.

[COMMAND]   ANSYS Main Menu → Solution → Define
Loads → Apply → Structural → Pressure → On Lines

1. The **Apply PRES on Lines** window opens (see Fig. 3.64) and the upward arrow appears when the mouse cursor is moved to the **ANSYS Graphics** window.

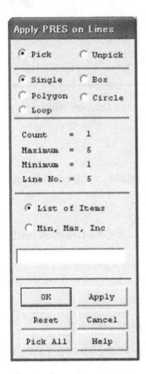

Fig. 3.64 'Apply PRES on Lines' window for picking the lines to which pressure is applied.

2. Confirming that the **Pick** and **Single** buttons are selected, move the upward arrow onto the upper line of the right-end side of the strip area and click the left mouse button. Remember that the right-end side of the strip area was divided into two lines in procedure (4) in Section 3.2.3.3.

3. Another **Apply PRES on Lines** window opens (see Fig. 3.65). Select a Constant value in the [A] **[SFL] Apply PRES on lines as a** box and input [B] **-10** (MPa) in the **VALUE Load PRES value** box and [C] **0** (MPa) in the **Value** box.

4. Click the [D] **OK** button in the window to define a linear distribution of pressure on the upper line which is zero at the upper right corner and − 10 MPa at the centre of the right-end side of the strip area (see Fig. 3.66).

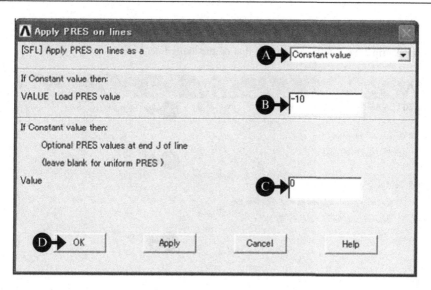

Fig. 3.65 'Apply PRES on Lines' window for applying linearly distributed pressure to the upper half of the right end of the elastic strip.

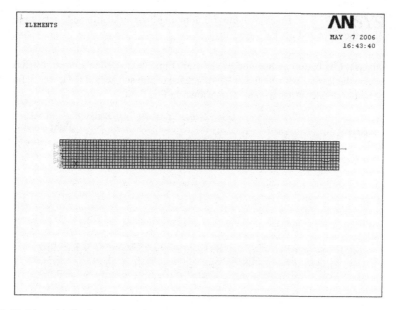

Fig. 3.66 Linearly distributed negative pressure applied to the upper half of the right-end side of the elastic strip.

5. For the lower line of the right-end side of the strip area, repeat the commands above and procedures (2)–(4).
6. Select a Constant value in the [A] **[SFL] Apply PRES on lines as a** box and input [B] **0** (MPa) in the **VALUE Load PRES value** box and [C] **-10** (MPa) in the **Value** box as shown

in Fig. 3.67. Note that the values to input in the lower two boxes in the **Apply PRES on Lines** window is interchanged, since the distributed pressure on the lower line of the right-end side of the strip area is antisymmetric to that on the upper line.

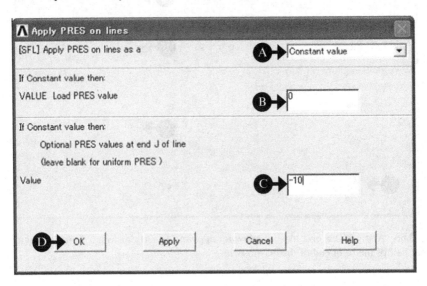

Fig. 3.67 'Apply PRES on Lines' window for applying linearly distributed pressure to the lower half of the right end of the elastic strip.

7. Click the [D] **OK** button in the window shown in Fig. 3.66 to define a linear distribution of pressure on the lower line, which is −10 MPa at the centre and zero at the lower right corner of the right-end side of the strip area, as shown in Fig. 3.68.

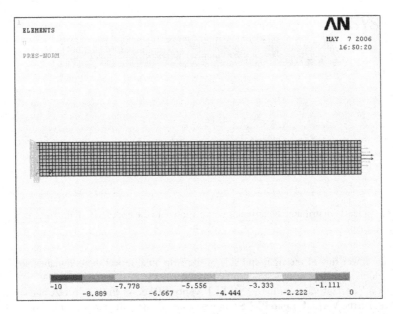

Fig. 3.68 Triangular distribution of pressure applied to the right end of the elastic strip.

3.2.3.5 Solution procedures

[COMMAND] ANSYS Main Menu → Solution → Solve → Current LS

1. The **Solve Current Load Step** and **/STATUS Command** windows appear.
2. Click the **OK** button in the **Solve Current Load Step** window to begin the solution of the current load step.
3. Select the **File** button in **/STATUS Command** window to open the submenu and select the **Close** button to close the **/STATUS Command** window.
4. When the solution is completed, the **Note** window appears. Click the **Close** button to close the **Note** window.

3.2.3.6 Contour plot of stress

[COMMAND] ANSYS Main Menu → General Postproc → Plot Results → Contour Plot → Nodal Solution

1. The **Contour Nodal Solution Data** window opens.
2. Select **Stress** and **X-Component of stress**.
3. Click the **OK** button to display the contour of the x-component of stress in the elastic strip in the **ANSYS Graphics** window, as shown in Fig. 3.69.

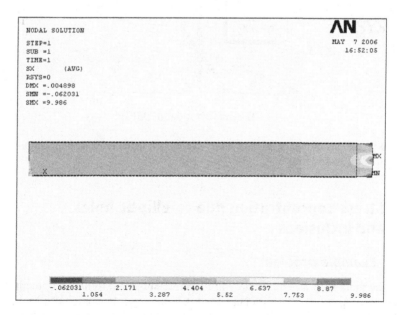

Fig. 3.69 Contour of the x-component of stress in the elastic strip showing uniform stress distribution at one width or larger distance from the right end of the elastic strip to which triangular distribution of pressure is applied.

3.2.4 Discussion

Fig. 3.70 shows the variations of the longitudinal stress distribution in the cross section with the x-position of the elastic strip. At the right end of the strip, or at $x = 200$ mm, the distribution of the applied longitudinal stress takes the triangular shape, which is zero at the upper and lower corners and 10 MPa at the centre of the strip. The longitudinal stress distribution varies as the distance of the cross section from the right end of the strip increases, and the distribution becomes almost uniform at $x = 180$ mm, i.e. at one width distance from the end of the stress application. The total amount of stress in any cross section is the same, i.e. 2 kN in the strip and at any cross section at one width or larger distance from the end of the stress application, stress is uniformly distributed and the magnitude of stress becomes 5 MPa.

The above result is known as the principle of St Venant and is very useful in practice, in the design of structural components. Even if the stress distribution is very complicated at the loading points due to the complicated shape of load transfer equipment, one can assume a uniform stress distribution in the main parts of structural components or machine elements at some distance from the load transfer equipment.

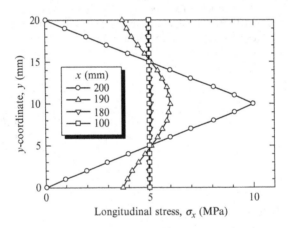

Fig. 3.70 Variations of the longitudinal stress distribution in the cross section with the x-position of the elastic strip.

3.3 Stress concentration due to elliptic holes and inclusions

3.3.1 Example problem

An elastic plate with an elliptic hole in its centre subjected to uniform longitudinal tensile stress σ_0 at one end and clamped at the other end. Perform an FEM analysis of a 2D elastic plate with an elliptic hole in its centre subjected to a uniform tensile stress σ_0 in the longitudinal direction at one end and clamped at the other end, as shown in Fig. 3.71, and calculate the maximum longitudinal stress σ_{max} in the plate. Calculate the stress concentration factor $\alpha = \sigma_{max}/\sigma_0$ and observe the variation of the longitudinal stress distribution in the ligament between the foot of the hole and the edge of the plate.

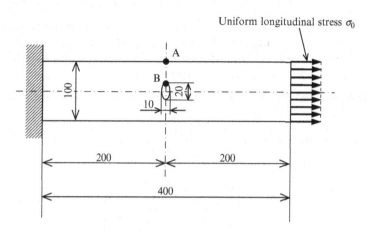

Fig. 3.71 A 2D elastic plate with an elliptic hole in its centre subjected to a uniform longitudinal stress at one end and clamped at the other end.

3.3.2 Problem description

Plate geometry: length $l = 400$ mm, height $h = 100$ mm, thickness $b = 10$ mm.
Material: mild steel having Young's modulus $E = 210$ GPa and Poisson's ratio $\nu = 0.3$.
Elliptic hole: An elliptic hole has a minor radius of 5 mm in the longitudinal direction and a major radius of 10 mm in the transversal direction.
Boundary conditions: The elastic plate is subjected to a uniform tensile stress in the longitudinal direction at the right end and clamped to a rigid wall at the left end.

3.3.3 Analytical procedures

3.3.3.1 Creation of an analytical model

Let us use a quarter model of the elastic plate with an elliptic hole as illustrated in Fig. 3.71, since the plate is symmetric about the horizontal and vertical centre lines. The quarter model can be created by a slender rectangular area from which an elliptic area is subtracted by using the Boolean operation described in Section A.1.

First, create the rectangular area by the following operation:

[COMMAND] ANSYS Main Menu → Preprocessor → Modeling → Create → Areas → Rectangle → By 2 Corners

1. Input two **0**s into the **WP X** and **WP Y** boxes in the **Rectangle by 2 Corners** window to determine the lower left corner point of the elastic plate on the Cartesian coordinates of the working plane.
2. Input **200** and **50** (mm) into the **Width** and **Height** boxes, respectively, to determine the shape of the elastic plate model.
3. Click the **OK** button to create the elastic plate on the **ANSYS Graphics** window.

Then, create a circular area with a diameter of 10 mm, and reduce its diameter in the longitudinal direction to half of the original value to obtain the elliptic area. The following commands create a circular area by designating the coordinates (**UX, UY**) of the centre and the radius of the circular area:

[COMMAND] ANSYS Main Menu → Preprocessor → Modeling → Create → Areas → Circle → Solid Circle

1. The **Solid Circular Area** window opens as shown in Fig. 3.72.

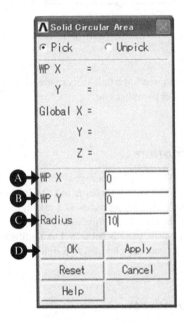

Fig. 3.72 'Solid Circular Area' window.

2. Input two **0**s into the [A] **WP X** and [B] **WP Y** boxes to determine the centre position of the circular area.
3. Input [C] **10** (mm) in the **Radius** box to determine the radius of the circular area.
4. Click the [D] **OK** button to create the circular area superimposed on the rectangular area in the **ANSYS Graphics** window, as shown in Fig. 3.73.

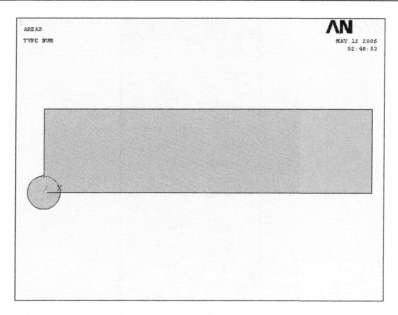

Fig. 3.73 Circular area superimposed on the rectangular area.

In order to reduce the diameter of the circular area in the longitudinal direction to half of the original value, use the following **Scale → Areas** operation:

[COMMAND] ANSYS Main Menu → **Preprocessor → Modeling → Operate → Scale → Areas**

1. The **Scale Areas** window opens, as shown in Fig. 3.74.
2. The upward arrow appears in the **ANSYS Graphics** window. Move the arrow to the circular area and select it by clicking the left mouse button. The colour of the circular area turns from light blue to pink. Click the [A] **OK** button.
3. The colour of the circular area turns to light blue and another **Scale Areas** window opens, as shown in Fig. 3.75.
4. Input [A] **0.5** in the **RX** box, select [B] **Areas only** in the **NOELEM** box and select the [C] **Moved** in **IMOVE** box.
5. Click the [D] **OK** button. An elliptic area appears and the circular area still remains. The circular area is an afterimage and does not exist in reality. To erase this afterimage, perform the following commands:

[COMMAND] ANSYS Utility Menu → **Plot → Replot**

The circular area vanishes. Subtract the elliptic area from the rectangular area in a similar manner as described in Section A.1, i.e.

[COMMAND] ANSYS Main Menu → **Preprocessor → Modeling → Operate → Booleans → Subtract → Areas**

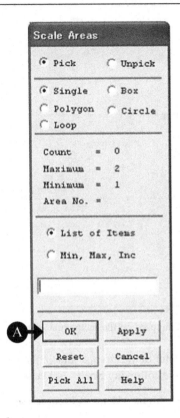

Fig. 3.74 'Scale Areas' window.

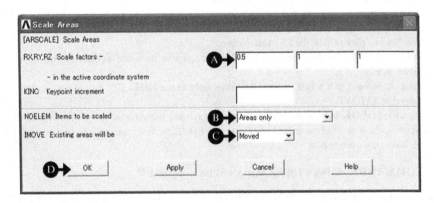

Fig. 3.75 'Scale Areas' window.

1. Pick the rectangular area by the upward arrow and confirm that the colour of the area picked turns from light blue to pink. Click the **OK** button.
2. Pick the elliptic area by the upward arrow and confirm that the colour of the elliptic area picked turns from light blue to pink. Click the **OK** button to subtract the elliptic area from

the rectangular area to get a quarter model of a plate with an elliptic hole in its centre, as shown in Fig. 3.76.

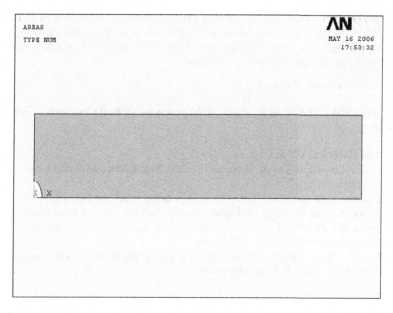

Fig. 3.76 Quarter model of a plate with an elliptic hole in its centre created by subtracting an elliptic area from a rectangular area.

3.3.3.2 Input of the elastic properties of the plate material

[COMMAND] ANSYS Main Menu → Preprocessor → Material Props → Material Models

1. The **Define Material Model Behavior** window opens.
2. Double-click the **Structural**, **Linear**, **Elastic**, and **Isotropic** buttons one after another.
3. Input the value of Young's modulus, **2.1e5** (MPa), and that of Poisson's ratio, **0.3**, into the **EX** and **PRXY** boxes, and click the **OK** button of the **Linear Isotropic Properties for Materials Number 1** window.
4. Exit from the **Define Material Model Behavior** window by selecting **Exit** in the **Material** menu of the window.

3.3.3.3 Finite-element discretisation of the quarter plate area

(1) Selection of the element type

[COMMAND] ANSYS Main Menu → Preprocessor → Element Type → Add/Edit/Delete

1. The **Element Types** window opens.
2. Click the **Add ...** button in the **Element Types** window to open the **Library of Element Types** window and select the element type to use.

3. Select **Structural Mass – Solid** and **Quad 8 node 82**.
4. Click the **OK** button in the **Library of Element Types** window to use the 8-node isoparametric element.
5. Click the **Options** ... button in the **Element Types** window to open the **PLANE82 element type options** window. Select the **Plane strs w/thk** item in the **Element behavior** box and click the **OK** button to return to the **Element Types** window. Click the **Close** button in the **Element Types** window to close the window.

(2) Input of the element thickness

   [COMMAND]   ANSYS Main Menu → Preprocessor → Real Constants → Add/Edit/ Delete

1. The **Real Constants** window opens.
2. Click the **[A] Add/Edit/Delete** button to open the **Real Constants** window and click the **Add** ... button.
3. The **Element Type for Real Constants** window opens. Click the **OK** button.
4. The **Element Type for Real Constants** window vanishes and the **Real Constants Set Number 1. for PLANE82** window appears instead. Input a strip thickness of 10 mm in the **Thickness** box and click the **OK** button.
5. The **Real Constants** window returns with the display of the **Defined Real Constants Sets** box changed to **Set 1**. Click the **Close** button.

(3) Sizing of the elements

   [COMMAND]   ANSYS   Main   Menu → Preprocessor → Meshing → Size Cntrls → Manual Size → Global → Size

1. The **Global Element Sizes** window opens.
2. Input **1.5** in the **SIZE** box and click the **OK** button.

(4) Meshing

   [COMMAND]   ANSYS   Main   Menu → Preprocessor → Meshing → Mesh → Areas → Free

1. The **Mesh Areas** window opens.
2. The upward arrow appears in the **ANSYS Graphics** window. Move this arrow to the quarter plate area and click this area.
3. The colour of the area turns from light blue to pink. Click the **OK** button to see the area meshed by 8-node isoparametric finite elements, as shown in Fig. 3.77.

3.3.3.4   Input of boundary conditions

(1) Imposing constraint conditions on the left end and the bottom side of the quarter plate model

Due to the symmetry, the constraint conditions of the quarter plate model are **UX**-fixed condition on the left end and **UY**-fixed condition on the bottom side of the quarter plate model. Apply the constraint conditions onto the corresponding lines using the following commands:

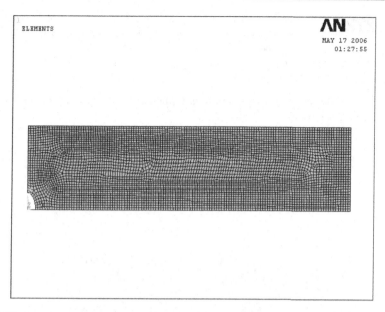

Fig. 3.77 Quarter model of a plate with an elliptic hole meshed by 8-node isoparametric finite elements.

[COMMAND]   ANSYS Main Menu → Solution → Define
Loads → Apply → Structural → Displacement → On Lines

1. The **Apply U. ROT on Lines** window opens and the upward arrow appears when the mouse cursor is moved to the **ANSYS Graphics** window.
2. Confirming that the **Pick** and **Single** buttons are selected, move the upward arrow onto the left-end side of the quarter plate area and click the left mouse button.
3. Click the **OK** button in the **Apply U. ROT on Lines** window to display another **Apply U. ROT on Lines** window.
4. Select **UX** in the **Lab2** box and click the **OK** button in the **Apply U. ROT on Lines** window.

Repeat the commands and operations (1)–(3) above for the bottom side of the model. Then, select **UY** in the **Lab2** box and click the **OK** button in the **Apply U. ROT on Lines** window.

(2) Imposing a uniform longitudinal stress on the right end of the quarter plate model

A uniform longitudinal stress can be defined by pressure on the right-end side of the plate model as follows:

[COMMAND]   ANSYS Main Menu → Solution → Define
Loads → Apply → Structural → Pressure → On Lines

1. The **Apply PRES on Lines** window opens and the upward arrow appears when the mouse cursor is moved to the **ANSYS Graphics** window.
2. Confirming that the **Pick** and **Single** buttons are selected, move the upward arrow onto the right-end side of the quarter plate area and click the left mouse button.

3. Another **Apply PRES on Lines** window opens. Select Constant value in the **[SFL] Apply PRES on lines as a** box and input **-10** (MPa) in the **VALUE Load PRES value** box and leave a blank in the **Value** box.
4. Click the **OK** button in the window to define a uniform tensile stress of 10 MPa applied to the right end of the quarter plate model.

Fig. 3.78 illustrates the boundary conditions applied to the centre notched plate model by the above operations.

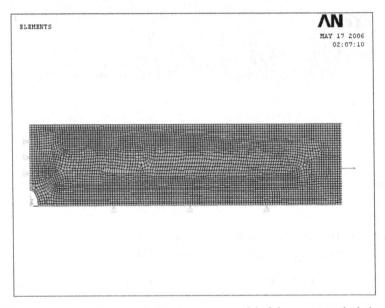

Fig. 3.78 Boundary conditions applied to the quarter model of the centre notched plate.

3.3.3.5 Solution procedures

[COMMAND]   ANSYS Main Menu → Solution → Solve → Current LS

1. The **Solve Current Load Step** and **/STATUS Command** windows appear.
2. Click the **OK** button in the **Solve Current Load Step** window to begin the solution of the current load step.
3. Select the **File** button in the **/STATUS Command** window to open the submenu and select the **Close** button to close the **/STATUS Command** window.
4. When the solution is completed, the **Note** window appears. Click the **Close** button to close the **Note** window.

3.3.3.6 Contour plot of stress

[COMMAND]   ANSYS Main Menu → General Postproc → Plot Results → Contour Plot → Nodal Solution

1. The **Contour Nodal Solution Data** window opens.
2. Select **Stress** and **X-Component of stress**.

3. Click the **OK** button to display the contour of the *x*-component of stress in the quarter model of the centre notched plate in the **ANSYS Graphics** window, as shown in Fig. 3.79.

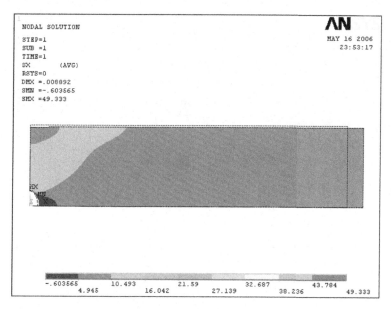

Fig. 3.79 Contour of the *x*-component of stress in the quarter model of the centre notched plate.

3.3.3.7 Observation of the variation of the longitudinal stress distribution in the ligament region

In order to observe the variation of the longitudinal stress distribution in the ligament between the foot of the hole and the edge of the plate, carry out the following commands:

 [COMMAND] **ANSYS Utility Menu → PlotCtrls → Symbols ...**

The **Symbols** window opens, as shown in Fig. 3.80.

1. Select [A] **All Reactions** in the **[/PBC] Boundary condition symbol** buttons, [B] **Pressures** in the **[/PSF] Surface Load Symbols** box, and [C] **Arrows** in the **[/PSF] Show pres and convect as** box.
2. The distributions of the reaction force and the longitudinal stress in the ligament region (see [A] in Fig. 3.81) as well as that of the applied stress on the right end of the plate area are superimposed on the contour *x*-component of stress in the plate in the **ANSYS Graphics** window. The reaction force is indicated by the leftward red arrows, and the longitudinal stress are shown by the rightward red arrows in the ligament region.

The longitudinal stress reaches its maximum value at the foot of the hole and is decreased approaching to a constant value almost equal to the applied stress σ_0 at some distance, say, about one major diameter distance, from the foot of the hole. This tendency can be explained by the principle of St Venant, as discussed in the previous section.

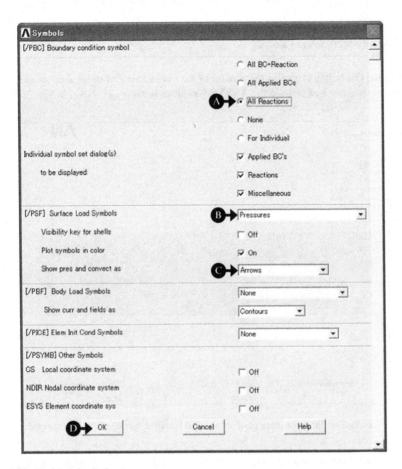

Fig. 3.80 'Symbols' window.

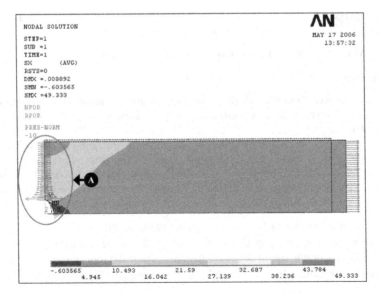

Fig. 3.81 Applied stress on the right-end side of the plate and resultant reaction force and longitudinal stress in its ligament region.

3.3.4 Discussion

If an elliptical hole is placed in an infinite body and subjected to a uniform stress in a remote uniform stress of σ_0, the maximum stress σ_{max} occurs at the foot of the hole, i.e. Point B in Fig. 3.71, and is expressed by the following formula:

$$\sigma_{max} = \left(1 + 2\frac{a}{b}\right)\sigma_0 = \left(1 + \sqrt{\frac{a}{\rho}}\right)\sigma_0 = \alpha_0\sigma_0 \qquad (3.5)$$

where a is the major radius, b is the minor radius, ρ is the local radius of curvature at the foot of the elliptic hole, and α_0 is the stress concentration factor, which is defined as

$$\alpha_0 \equiv \frac{\sigma_{max}}{\sigma_0} = \left(1 + 2\frac{a}{b}\right) = \left(1 + \sqrt{\frac{a}{\rho}}\right) \qquad (3.6)$$

The stress concentration factor α_0 varies inversely proportional to the aspect ratio of elliptic hole b/a, namely the smaller the value of the aspect ratio b/a or the radius of curvature ρ becomes, the larger the value of the stress concentration factor α_0 becomes.

In a finite plate, the maximum stress at the foot of the hole is increased due to the finite boundary of the plate. Fig. 3.82 shows the variation of the stress concentration factor α for elliptic holes having different aspect ratios with normalised major radius $2a/h$, indicating that the stress concentration factor in a plate with finite width α is increased dramatically as the ligament between the foot of the hole and the plate edge becomes smaller.

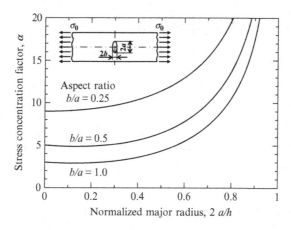

Fig. 3.82 Variation of the stress concentration factor α for elliptic holes having different aspect ratios with normalised major radius $2a/h$.

From Fig. 3.82, the value of the stress intensity factor for the present elliptic hole is approximately 5.16, whereas Fig. 3.81 shows that the maximum value of the longitudinal stress obtained by the present FEM calculation is approximately 49.3 MPa, i.e.

the value of the stress concentration factor is approximately $49.3/10 = 4.93$. Hence, the relative error of the present calculation is approximately $(4.93 - 5.16)/5.16 \approx -0.0446 = 4.46\%$ which may be reasonably small.

3.3.5 Problems to solve

Problem 3.8. Calculate the value of stress concentration factor for the elliptic hole shown in Fig. 3.71 by using the whole model of the plate, and compare the result with that obtained and discussed in Section 3.3.4.

Problem 3.9. Calculate the values of stress concentration factor α for circular holes for different values of the normalised major radius $2a/h$ and plot the results as the α vs $2a/h$ diagram, as shown in Fig. 3.80.

Problem 3.10. Calculate the values of stress concentration factor α for elliptic holes having different aspect ratios b/a for different values of the normalised major radius $2a/h$, and plot the results as the α vs $2a/h$ diagram.

Problem 3.11. For smaller values of $2a/h$, the disturbance of stress in the ligament between the foot of a hole and the plate edge due to the existence of the hole is decreased and stress in the ligament approaches to a constant value equal to the remote stress σ_0 at some distance from the foot of the hole (remember the principle of St Venant in the previous section). Find at which distance from the foot of the hole the stress in the ligament region can be considered to be almost equal to the value of the remote stress σ_0.

3.4 Stress singularity problem

3.4.1 Example problem

An elastic plate with a crack of length $2a$ in its centre is subjected to uniform longitudinal tensile stress σ_0 at one end and clamped at the other end, as illustrated in Fig. 3.83. Perform an FEM analysis of the 2D elastic plate having a crack of length $2a$ in its centre subjected to a uniform tensile stress σ_0 in the longitudinal direction at one end and clamped at the other end, and calculate the value of the mode I (crack-opening mode) stress intensity factor.

3.4.2 Problem description

Specimen geometry: length $l = 400$ mm, height $h = 100$ mm.
Material: mild steel having Young's modulus $E = 210$ GPa and Poisson's ratio $\nu = 0.3$.
Crack: A crack is placed perpendicular to the loading direction in the centre of the plate and has a length of 20 mm. The centre-cracked tension plate is assumed to be in the plane strain condition in the present analysis.
Boundary conditions: The elastic plate is subjected to a uniform tensile stress in the longitudinal direction at the right end and clamped to a rigid wall at the left end.

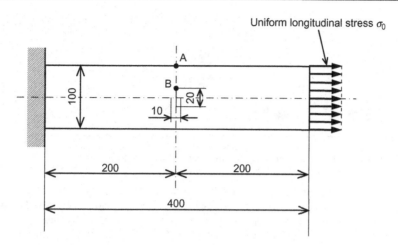

Fig. 3.83 Two dimensional elastic plate with a crack of length $2a$ in its centre subjected to a uniform tensile stress σ_0 in the longitudinal direction at one end and clamped at the other end.

3.4.3 Analytical procedures

3.4.3.1 Creation of an analytical model

Let us use a quarter model of the centre-cracked tension plate as illustrated in Fig. 3.83, since the plate is symmetric about the horizontal and vertical centre lines.

Here we use the singular element or the quarter point element, which can interpolate the stress distribution in the vicinity of the crack tip at which stress has the $1/\sqrt{r}$ singularity where r is the distance from the crack tip ($r/a \ll 1$). An ordinary isoparametric element which is familiar as **Quad 8 node 82** has nodes at corners and also at the midpoint on each side of the element. A singular element, however, has the midpoint moved one quarter side distance from the original midpoint position to the node which is placed at the crack tip position. This is why the singular element is often called the quarter point element instead. ANSYS is equipped with a 2D triangular singular element only, not with 2D rectangular or 3D singular elements. Around the node at the crack tip, a circular area is created and is divided into a designated number of triangular singular elements. Each of these elements has its vertex placed at the crack tip position and has the quarter points on the two sides joining the vertex and the other two nodes.

In order to create the singular elements, the plate area must be created via key points set at the four corner points and at the crack tip position on the left-end side of the quarter plate area.

[COMMAND] ANSYS   Main   Menu → **Preprocessor** → **Modeling** → **Create** → **Keypoints** → **In Active CS**

1. The **Create Keypoints in Active Coordinate System** window opens as shown in Fig. 3.84.

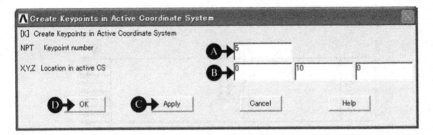

Fig. 3.84 'Create Keypoints in Active Coordinate System' window.

2. Input the [A] **key point number** in the **NPT** box and the [B] *x-*, *y-* and *z-***coordinates** of the key point in the three **X**, **Y**, and **Z** boxes, respectively. Fig. 3.84 shows the case of Key point #5 which is placed at the crack tip having the coordinates (0, 10, 0). In the present model, let us create Key point #1 to #5 at the coordinates (0, 0, 0), (200, 0, 0), (200, 50, 0), (0, 50, 0), and (0, 10, 0), respectively. Note that the *z*-coordinate is always 0 in 2D elasticity problems.
3. Click the [C] **Apply** button and create Key point2 #1 to #4 not to exit from the window and finally click the [D] **OK** button to create Key point #5 at the crack tip position and exit from the window (see Fig. 3.85).

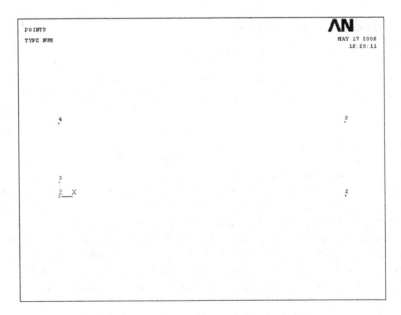

Fig. 3.85 Five key points created in the '**ANSYS Graphics**' window.

Then create the plate area via the five key points created by the procedures above, using the following commands:

    [COMMAND] ANSYS   Main   Menu → **Preprocessor** → **Modeling** → **Create** → **Areas** → **Arbitrary** → **Through KPs**

1. The **Create Area thru KPs** window opens.
2. The upward arrow appears in the **ANSYS Graphics** window. Move this arrow to Key point #1 and click this point. Click Key points #1–#5 one by one counter-clockwise (see Fig. 3.86).

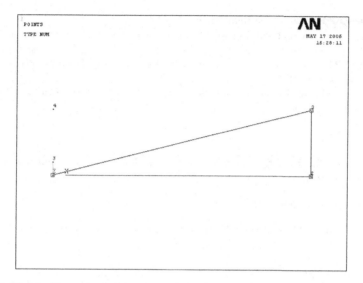

Fig. 3.86 Clicking Key points #1 through #5 one by one counter-clockwise to create the plate area.

3. Click the **OK** button to create the plate area, as shown in Fig. 3.87.

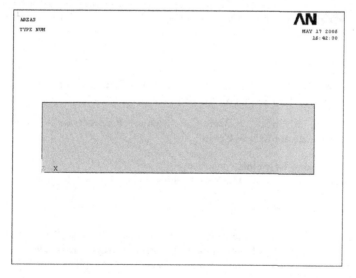

Fig. 3.87 Quarter model of the centre cracked tension plate.

3.4.3.2   Input of the elastic properties of the plate material

[COMMAND]   ANSYS Main Menu → Preprocessor → Material Props → Material
Models

1. The **Define Material Model Behavior** window opens.
2. Double-click the **Structural, Linear, Elastic,** and **Isotropic** buttons one after another.
3. Input the value of Young's modulus, **2.1e5** (MPa), and that of Poisson's ratio, **0.3,** into the
   EX and **PRXY** boxes, and click the **OK** button of the **Linear Isotropic Properties for
   Materials Number 1** window.
4. Exit from the **Define Material Model Behavior** window by selecting **Exit** in the **Material**
   menu of the window.

3.4.3.3   Finite-element discretisation of the centre-cracked tension
plate area

(1) Selection of the element type

[COMMAND]   ANSYS Main Menu → Preprocessor → Element Type → Add/Edit/
Delete

1. The **Element Types** window opens.
2. Click the **Add ...** button in the **Element Types** window to open the **Library of Element
   Types** window and select the element type to use.
3. Select **Structural Mass – Solid** and **Quad 8 node 82**.
4. Click the **OK** button in the **Library of Element Types** window to use the 8-node
   isoparametric element.
5. Click the **Options ...** button in the **Element Types** window to open the **PLANE82 element
   type options** window. Select the **Plane strain** item in the **Element behavior** box and click
   the **OK** button to return to the **Element Types** window. Click the **Close** button in the **Element Types** window to close the window.

(2) Sizing of the elements

Before meshing, the crack tip point around which the triangular singular elements will
be created must be specified using the following commands:

[COMMAND] ANSYS   Main   Menu → Preprocessor → Meshing → Size
Cntrls → Concentrate KPs → Create

1. The **Concentration Keypoints** window opens.
2. Display the key points in the **ANSYS Graphics** window by the following:

[COMMAND]   ANSYS Utility Menu → Plot → Keypoints → Keypoints.

3. Pick Key point #5 by placing the upward arrow on Key point #5 and clicking the left mouse
   button. Then click the **OK** button in the **Concentration Keypoints** window.
4. Another **Concentration Keypoints** window opens, as shown in Fig. 3.88.

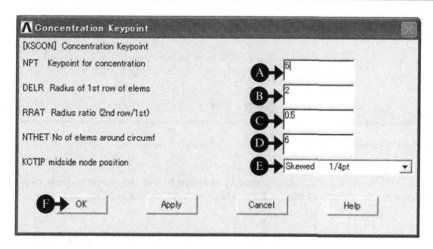

Fig. 3.88 'Concentration Keypoints' window.

5. Confirming that [A] **5**, i.e. the key point number of the crack tip position is input in the **NPT** box, input [B] **2** in the **DELR** box, [C] **0.5** in the **RRAT** box, and [D] **6** in the **NTHET** box, then select [E] **Skewed 1/4pt** in the **KCTIP** box in the window. Refer to the explanations of the numerical data described after the names of the respective boxes on the window. **Skewed 1/4pt** in the last box means that the mid-nodes of the sides of the elements which contain Key point #5 are the quarter points of the elements.
6. Click the [F] **OK** button in the **Concentration Keypoints** window.

The size of the meshes other than the singular elements and the elements adjacent to them can be controlled by the same procedures, as has been described in the previous sections of this chapter.

[COMMAND] ANSYS Main Menu → Preprocessor → Meshing → Size Cntrls → Manual Size → Global → Size

1. The **Global Element Sizes** window opens.
2. Input **1.5** in the **SIZE** box and click the **OK** button.

(3) Meshing

The meshing procedures are also the same as before.

[Commands] ANSYS Main Menu → Preprocessor → Meshing → Mesh → Areas → Free

1. The **Mesh Areas** window opens.
2. The upward arrow appears in the **ANSYS Graphics** window. Move this arrow to the quarter plate area and click this area.
3. The colour of the area turns from light blue to pink. Click the **OK** button.
4. The **Warning** window appears as shown in Fig. 3.89 due to the existence of six singular elements. Click the [A] **Close** button and proceed to the next operation below.

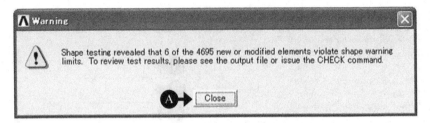

Fig. 3.89 'Warning' window.

5. Fig. 3.90 shows the plate area meshed by ordinary 8-node isoparametric finite elements except for the vicinity of the crack tip, where we have six singular elements.

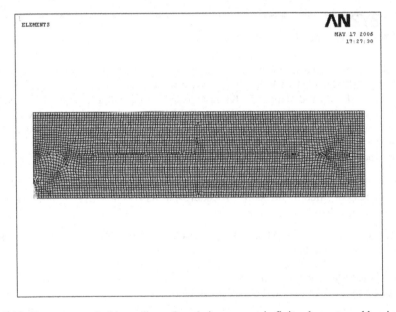

Fig. 3.90 Plate area meshed by ordinary 8-node isoparametric finite elements and by singular elements.

Fig. 3.91 shows an enlarged view of the singular elements around the [A] crack tip, showing that six triangular elements are placed in a radial manner and that the size of the second row of elements is half the radius of the first row of elements, i.e. triangular singular elements.

3.4.3.4 Input of boundary conditions

(1) Imposing constraint conditions on the ligament region of the left end and the bottom side of the quarter plate model

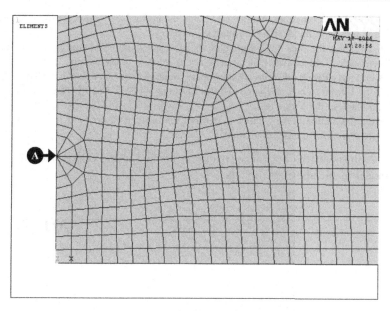

Fig. 3.91 Enlarged view of the singular elements around the crack tip.

Due to the symmetry, the constraint conditions of the quarter plate model are **UX**-fixed condition on the ligament region of the left end, i.e. the line between Key points #4 and #5, and **UY**-fixed condition on the bottom side of the quarter plate model. Apply these constraint conditions onto the corresponding lines by the following commands:

[COMMAND]   ANSYS Main Menu → Solution → Define Loads → Apply → Structural → Displacement → On Lines

1. The **Apply U. ROT on Lines** window opens and the upward arrow appears when the mouse cursor is moved to the **ANSYS Graphics** window.
2. Confirming that the **Pick** and **Single** buttons are selected, move the upward arrow onto the line between Key points #4 and #5, and click the left mouse button.
3. Click the **OK** button in the **Apply U. ROT on Lines** window to display another **Apply U. ROT on Lines** window.
4. Select **UX** in the **Lab2** box and click the **OK** button in the **Apply U. ROT on Lines** window.

Repeat the commands and operations (1)–(3) above for the bottom side of the model. Then, select **UY** in the **Lab2** box and click the **OK** button in the **Apply U. ROT on Lines** window.

(2) Imposing a uniform longitudinal stress on the right end of the quarter plate model

A uniform longitudinal stress can be defined by pressure on the right-end side of the plate model, as described below:

[COMMAND]   ANSYS Main Menu → Solution → Define Loads → Apply → Structural → Pressure → On Lines

1. The **Apply PRES on Lines** window opens and the upward arrow appears when the mouse cursor is moved to the **ANSYS Graphics** window.
2. Confirming that the **Pick** and **Single** buttons are selected, move the upward arrow onto the right-end side of the quarter plate area and click the left mouse button.
3. Another **Apply PRES on Lines** window opens. Select a Constant value in the **[SFL] Apply PRES on lines as a** box and input **-10** (MPa) in the **VALUE Load PRES value** box, and leave a blank in the **Value** box.
4. Click the **OK** button in the window to define a uniform tensile stress of 10 MPa applied to the right end of the quarter plate model.

Fig. 3.92 illustrates the boundary conditions applied to the centre-cracked tension plate model by the above operations.

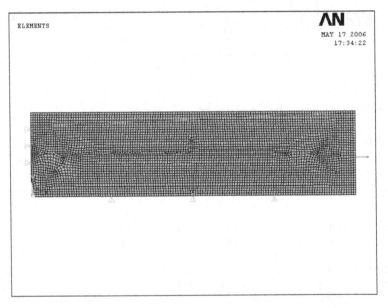

Fig. 3.92 Boundary conditions applied to the centre-cracked tension plate model.

3.4.3.5 Solution procedures

The solution procedures are also the same as usual.

   [COMMAND]   ANSYS Main Menu → **Solution** → **Solve** → **Current LS**

1. The **Solve Current Load Step** and **/STATUS Command** windows appear.
2. Click the **OK** button in the **Solve Current Load Step** window to begin the solution of the current load step.
3. The **Verify** window opens, as shown in Fig. 3.93. Proceed to the next operation below by clicking the [A] **Yes** button in the window.

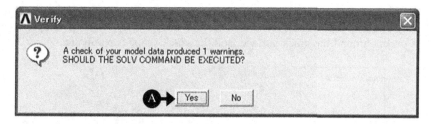

Fig. 3.93 'Verify' window.

4. Select the **File** button in the **/STATUS Command** window to open the submenu and select the **Close** button to close the **/STATUS Command** window.
5. When the solution is completed, the **Note** window appears. Click the **Close** button to close this window.

3.4.3.6   Contour plot of stress

[COMMAND]   ANSYS Main Menu → General Postproc → Plot Results → Contour Plot → Nodal Solution

1. The **Contour Nodal Solution Data** window opens.
2. Select **Stress** and **X-Component of stress**.
3. Click the **OK** button to display the contour of the x-component of stress, or longitudinal stress in the centre-cracked tension plate in the **ANSYS Graphics** window, as shown in Fig. 3.94.

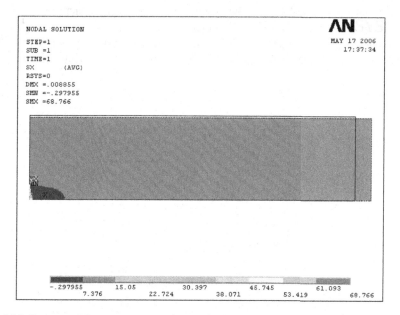

Fig. 3.94 Contour of the x-component of stress in the centre-cracked tension plate.

Fig. 3.95 is an enlarged view of the longitudinal stress distribution around the crack tip showing that very high tensile stress is induced at the crack tip, whereas compressive stress occurs around the crack surface, and the crack shape is parabolic.

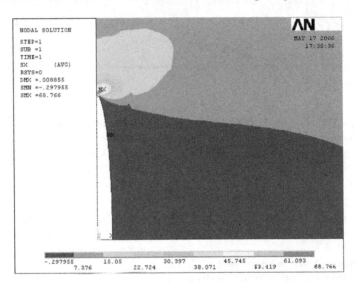

Fig. 3.95 Enlarged view of the longitudinal stress distribution around the crack tip.

3.4.4 Discussion

Fig. 3.96 shows extrapolation of the values of the correction factor, or the non-dimensional mode I (the crack-opening mode) stress intensity factor $F_I = K_I/\sigma\sqrt{\pi a}$ to the point where $r = 0$, i.e. the crack tip position by the hybrid extrapolation method [1]. The plots in the right region of the figure are obtained by the following formula:

$$K_I = \lim_{r \to 0} \sqrt{2\pi r}\sigma_x(\theta = 0) \qquad\qquad (3.7)$$

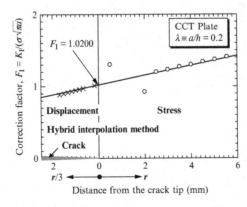

Fig. 3.96 Extrapolation of the values of the correction factor $F_I = K_I/\sigma\sqrt{\pi a}$ to the crack tip point where $r = 0$.

or

$$F_I(\lambda) = \frac{K_I}{\sigma\sqrt{\pi a}} = \lim_{r \to 0} \frac{\sqrt{2\pi r}\sigma_x(\theta = 0)}{\sigma\sqrt{\pi a}} \quad \text{where } \lambda = 2a/\lambda \tag{3.7'}$$

whereas those in the left region are obtained by the following formula:

$$K_I = \lim_{r \to 0} \sqrt{\frac{2\pi}{r}} \frac{E}{(1+\nu)(\kappa+1)} u_x(\theta = \pi) \tag{3.8}$$

or

$$F_I(\lambda) = \frac{K_I}{\sigma\sqrt{\pi a}} = \lim_{r \to 0} \sqrt{\frac{2\pi}{r}} \frac{E}{(1+\nu)(\kappa+1)\sigma\sqrt{\pi a}} u_x(\theta = \pi) \tag{3.8'}$$

where r is the distance from the crack tip, σ_x is the x-component of stress, σ is the applied uniform stress in the direction of the x-axis, a is the half crack length, E is Young's modulus, $u_x(\theta = \pi)$, $\kappa = 3 - 4\nu$ for plane strain condition and $\kappa = (3 - \nu)/(1 + \nu)$ for plane stress condition, and θ is the angle around the crack tip. The correction factor F_I accounts for the effect of the finite boundary, or the edge effect of the plate. The value of F_I can be evaluated either by the stress extrapolation method (the right region) or by the displacement extrapolation method (the left region). The hybrid extrapolation method [1] is a blend of the two types of the extrapolation methods above and can bring about better solutions. As illustrated in Fig. 3.95, the extrapolation lines obtained by the two methods agree well with each other when the gradient of the extrapolation line for the displacement method is rearranged by putting $r' = r/3$. The value of F_I obtained for the present CCT plate by the hybrid method is 1.0200.

Table 3.1 lists the values of F_I calculated by the following four equations [2–6] for various values of nondimensional crack length $\lambda = 2a/h$:

$$F_I(\lambda) = 1 + \sum_{k=1}^{35} A_k \lambda^{2k} \tag{3.9}$$

Table 3.1 Correction factor $F_I(\lambda)$ as a function of nondimensional crack length $\lambda = 2a/h$

$\lambda = 2a/h$	Isida Eq. (3.7)	Feddersen Eq. (3.8)	Koiter Eq. (3.9)	Tada Eq. (3.10)
0.1	1.0060	1.0062	1.0048	1.0060
0.2	1.0246	1.0254	1.0208	1.0245
0.3	1.0577	1.0594	1.0510	1.0574
0.4	1.1094	1.1118	1.1000	1.1090
0.5	1.1867	1.1892	1.1757	1.1862
0.6	1.3033	1.3043	1.2921	1.3027
0.7	1.4882	1.4841	1.4779	1.4873
0.8	1.8160	1.7989	1.8075	1.8143
0.9	2.5776	–	2.5730	2.5767

$$F_I(\lambda) = \sqrt{\sec\left(\pi\lambda/2\right)} \quad (\lambda \leq 0.8) \tag{3.10}$$

$$F_I(\lambda) = \frac{1 - 0.5\lambda + 0.326\lambda^2}{\sqrt{1-\lambda}} \tag{3.11}$$

$$F_I(\lambda) = \left(1 - 0.025\lambda^2 + 0.06\lambda^4\right)\sqrt{\sec\left(\pi\lambda/2\right)} \tag{3.12}$$

Among these, Isida's solution is considered to be the best. The value of $F_I(0.2)$ by Isida is 1.0246. The relative error of the present result is $(1.0200 - 1.0246)/1.0246 = -0.45\%$, which is a reasonably good result.

3.4.5 Problems to solve

Problem 3.12. Calculate the values of the correction factor for the mode I stress intensity factor for the centre-cracked tension plate having normalised crack lengths of $\lambda = 2a/h = 0.1$–0.9. Compare the results with the values of the correction factor $F(\lambda)$ listed in Table 3.1.

Problem 3.13. Calculate the values of the correction factor for the mode I stress intensity factor for a double edge cracked tension plate having normalised crack lengths of $\lambda = 2a/h = 0.1$–0.9 (see Fig. 3.97). Compare the results with the values of the correction factor $F(\lambda)$ calculated by Eq. (3.13) [7,8]. Note that the quarter model described in the present section can be used for the FEM calculations of this problem simply by changing the boundary conditions.

$$F_I(\lambda) = \frac{1.122 - 0.561\lambda - 0.205\lambda^2 + 0.471\lambda^3 - 0.190\lambda^4}{\sqrt{1-\lambda}} \quad \text{where } \lambda = 2a/h \tag{3.13}$$

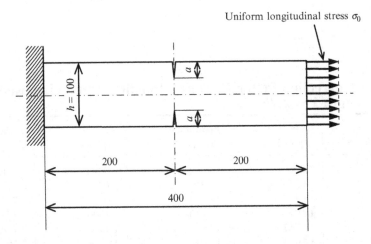

Fig. 3.97 Double edge cracked tension plate subjected to a uniform longitudinal stress at one end and clamped at the other end.

Problem 3.14. Calculate the values of the correction factor for the mode I stress intensity factor for a single edge cracked tension plate having normalised crack lengths of $\lambda = 2a/h = 0.1$–0.9 (see Fig. 3.98). Compare the results with the values of the correction factor $F(\lambda)$ calculated by Eq. (3.14) [7,9]. In this case, a half plate model is clamped along the ligament and is loaded at the right end of the plate. Note that the whole plate model with an edge crack needs a somewhat complicated procedure. A quarter model described in the present section gives good solutions with an error of a few percent or less, and is effective in practice.

$$F_I(\lambda) = \frac{0.752 + 2.02\lambda + 0.37[1 - \sin(\pi\lambda/2)]^3}{\cos(\pi\lambda/2)} \sqrt{\frac{2}{\pi\lambda} \tan\left(\frac{\pi\lambda}{2}\right)} \quad \text{where } \lambda$$
$$= a/h \tag{3.14}$$

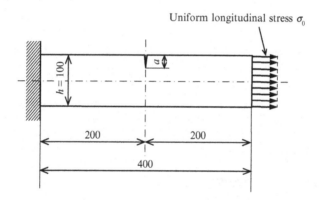

Fig. 3.98 Single edge cracked tension plate subjected to a uniform longitudinal stress at one end and clamped at the other end.

3.5 Two-dimensional contact stress

3.5.1 Example problem

An elastic cylinder with a radius of length a is pressed against a flat surface of a linearly elastic medium by a force P'. Perform an FEM analysis of an elastic cylinder of mild steel with a radius of length a pressed against an elastic flat plate of the same steel by a force P' illustrated in Fig. 3.99 and calculate the resulting contact pressure induced in the flat plate.

3.5.2 Problem description

Geometry of the cylinder: radius $a = 500$ mm.
Geometry of the flat plate: width $W = 500$ mm, height $h = 500$ mm.

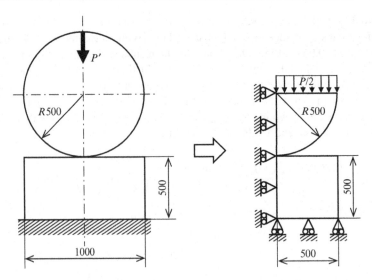

Fig. 3.99 Elastic cylinder of mild steel with a radius of length *a* pressed against an elastic flat plate of the same steel by a force *P'* and its FEM model.

Material: mild steel having Young's modulus $E = 210$ GPa and Poisson's ratio $\nu = 0.3$. Boundary conditions: The elastic cylinder is pressed against an elastic flat plate of the same steel by a force of $P' = 1$ kN/mm and the plate is clamped at the bottom.

3.5.3 Analytical procedures

3.5.3.1 Creation of an analytical model

Let us use a half model of the indentation problem illustrated in Fig. 3.99, since the problem is symmetric about the vertical centre line.

(1) Creation of an elastic flat plate

> **[COMMAND]** ANSYS   Main   Menu → **Preprocessor** → **Modeling** → **Create** → **Areas** → **Rectangle** → **By 2 Corners**

1. Input two **0**s into the **WP X** and **WP Y** boxes in the **Rectangle by 2 Corners** window to determine the upper left corner point of the elastic flat plate on the Cartesian coordinates of the working plane.
2. Input **500** and **-500** (mm) into the **Width** and **Height** boxes, respectively, to determine the shape of the elastic flat plate model.
3. Click the **OK** button to create the plate on the **ANSYS Graphics** window.

(2) Creation of an elastic cylinder

> **[COMMAND]** ANSYS   Main   Menu → **Preprocessor** → **Modeling** → **Create** → **Areas** → **Circle** → **Partial Annulus**

1. The **Part Annular Circ Area** window opens, as shown in Fig. 3.100.

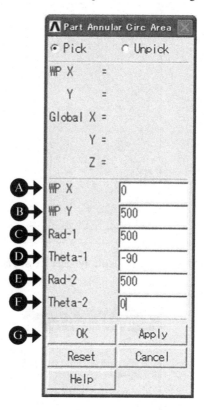

Fig. 3.100 'Part Annular Circ Area' window.

2. Input **0** and **500** into the [A] **WP X** and [B] **WP Y** boxes, respectively, in the **Part Annular Circ Area** window to determine the centre of the elastic cylinder on the Cartesian coordinates of the working plane.
3. Input two **500**s (mm) into the [C] **Rad-1** and [E] **Rad-2** boxes to designate the radius of the elastic cylinder model.
4. Input **-90** and **0** (mm) into the [D] **Theta-1** and [F] **Theta -2** boxes, respectively, to designate the starting and ending angles of the quarter elastic cylinder model.
5. Click the [G] **OK** button to create the quarter cylinder model on the **ANSYS Graphics** window, as shown in Fig. 3.101.

(3) Combining the elastic cylinder and the flat plate models

The elastic cylinder and the flat plate models must be combined into one model so as not to allow the rigid-body motions of the two bodies.

[COMMAND] ANSYS Main Menu → **Preprocessor** → **Modeling** → **Operate** → **Booleans** → **Glue** → **Areas**

1. The **Glue Areas** window opens, as shown in Fig. 3.102.

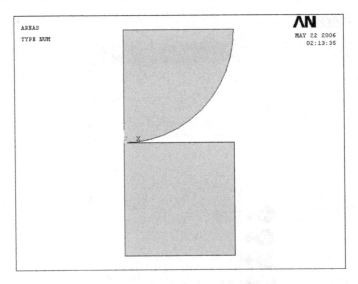

Fig. 3.101 Areas of the FEM model for an elastic cylinder pressed against an elastic flat plate.

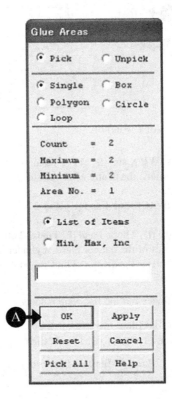

Fig. 3.102 'Glue Areas' window.

2. The upward arrow appears in the **ANSYS Graphics** window. Move this arrow to the quarter cylinder area and click this area, and then move the arrow on to the half plate area and click this area to combine these two areas.
3. Click the [A] **OK** button to create the contact model on the **ANSYS Graphics** window, as shown in Fig. 3.102.

3.5.3.2   Input of the elastic properties of the materials for the cylinder and the flat plate

[COMMAND]   ANSYS Main Menu → Preprocessor → Material Props → Material Models

1. The **Define Material Model Behavior** window opens.
2. Double-click the **Structural, Linear, Elastic,** and **Isotropic** buttons one after another.
3. Input the value of Young's modulus, **2.1e5** (MPa), and that of Poisson's ratio, **0.3**, into the **EX** and **PRXY** boxes, and click the **OK** button of the **Linear Isotropic Properties for Materials Number 1** window.
4. Exit from the **Define Material Model Behavior** window by selecting **Exit** in the **Material** menu of the window.

3.5.3.3   Finite-element discretisation of the cylinder and the flat plate areas

(1) Selection of the element types

Select the 8-node isoparametric elements as usual.

[COMMAND]   ANSYS Main Menu → Preprocessor → Element Type → Add/Edit/ Delete

1. The **Element Types** window opens.
2. Click the **Add** ... button in the **Element Types** window to open the **Library of Element Types** window and select the element type to use.
3. Select **Structural Mass – Solid** and **Quad 8 node 82**.
4. Click the **OK** button in the **Library of Element Types** window to use the 8-node isoparametric element. Note that this element is defined as a Type 1 element, as indicated in the **Element type reference number** box.
5. Click the **Options** ... button in the **Element Types** window to open the **PLANE82 element type options** window. Select the **Plane strain** item in the **Element behavior** box and click the **OK** button to return to the **Element Types** window. Click the **Close** button in the **Element Types** window to close the window.

In contact problems, two mating surfaces come in contact with each other exerting great force on each other. Contact elements must be used in contact problems to prevent penetration of one object into the other.

6. Repeat-click the **Add** ... button in the **Element Types** window to open the **Library of Element Types** window and select **Contact** and **2D target 169**.

7. Click the **OK** button in the **Library of Element Types** window to select the target element. Note that this target element is defined as a Type 2 element, as indicated in the **Element type reference number** box.

8. Click the **Add** ... button again in the **Element Types** window to open the **Library of Element Types** window and select **Contact** and **3 nd surf 172**.

9. Click the **OK** button in the **Library of Element Types** window to select the surface effect element. Note that this target element is defined as a Type 3 element, as indicated in the **Element type reference number** box.

(2) Sizing of the elements

Here let us designate the element size by specifying number of divisions of lines picked. For this purpose, carry out the following commands:

[COMMAND]   ANSYS Main Menu → Preprocessor → Meshing → Size Cntrls → Manual Size → Lines → Picked Lines

1. The **Element Size on Pick** ... window opens, as shown in Fig. 3.103.

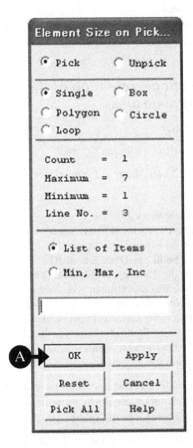

Fig. 3.103 'Element Size on Pick ...' window.

2. Move the upward arrow to the upper and the right sides of the flat plate area. Click the [A] **OK** button and the **Element Sizes on Picked Lines** window opens, as shown in Fig. 3.104.

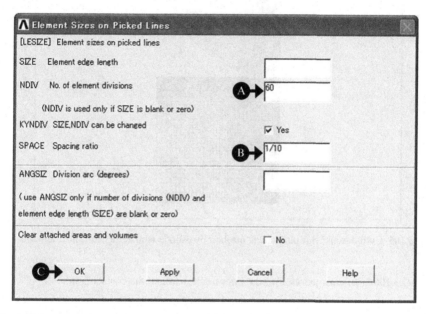

Fig. 3.104 'Element Sizes on Picked Lines' window.

3. Input [A] **60** in the **NDIV** box and [B] **1/10** in the **SPACE** box to divide the upper and right side of the plate area into 60 subdivisions. The size of the subdivisions decreases in geometric progression as they approach the left side of the plate area. It would be preferable to have smaller elements around the contact point where high stress concentration occurs. Input **60** in the **NDIV** box and **10** in the **SPACE** box for the left and the bottom sides of the plate area, **40** and **10** for the left side and the circumference of the quarter circular area, and **40** and **1/10** for the upper side of the quarter circular area in the respective boxes.

(3) Meshing

The meshing procedures are the same as usual.

[COMMAND] ANSYS   Main   Menu → Preprocessor → Meshing → Mesh → Areas → Free

1. The **Mesh Areas** window opens.
2. The upward arrow appears in the **ANSYS Graphics** window. Move this arrow to the quarter cylinder and the half plate areas and click these areas.
3. The colour of the areas turns from light blue to pink. Click the **OK** button to see the areas meshed by 8-node isoparametric finite elements, as shown in Fig. 3.105.

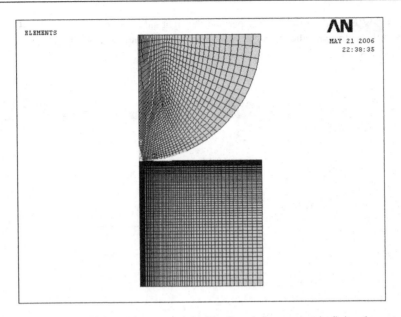

Fig. 3.105 Cylinder and flat plate areas meshed by 8-node isoparametric finite elements.

4. Fig. 3.106 is an enlarged view of the finer meshes around the contact point.

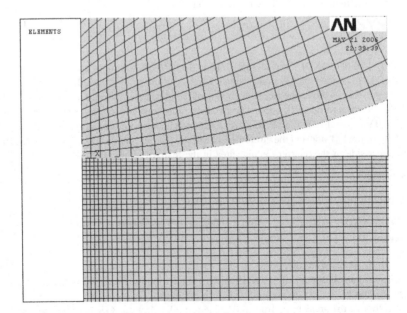

Fig. 3.106 Enlarged view of the finer meshes around the contact point.

(4) Creation of target and contact elements

First select the lower surface of the quarter cylinder to which the contact elements are attached.

    [Commands] **ANSYS Utility Menu → Select → Entities ...**

1. The **Select Entities** window opens, as shown in Fig. 3.107.

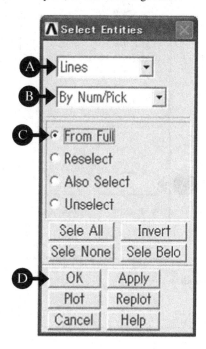

Fig. 3.107 'Select Entities' window.

2. Select [A] **Lines**, [B] **By Num/Pick**, and [C] **From Full** in the **Select Entities** window.

3. Click the [D] **OK** button in the **Select Entities** window and the **Select lines** window opens, as shown in Fig. 3.108.

4. The upward arrow appears in the **ANSYS Graphics** window. Move this arrow to the quarter circumference of the cylinder and click this line. The lower surface of the cylinder is ' selected, as shown in Fig. 3.109.

Repeat the **Select → Entities ...** commands to select nodes on the lower surface of the cylinder.

    [COMMAND] **ANSYS Utility Menu → Select → Entities ...**

1. The **Select Entities** window opens, as shown in Fig. 3.110.

2. Select [A] **Nodes**, [B] **Attached to**, [C] **Lines, all**, and [D] **Reselect** in the **Select Entities** window.

3. Click the [E] **OK** button and perform the following commands:

    [COMMAND] **ANSYS Utility Menu → Plot → Nodes**

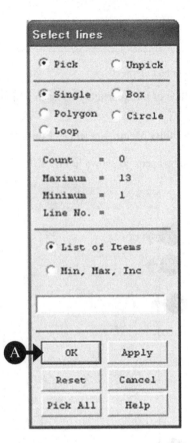

Fig. 3.108 'Select lines' window.

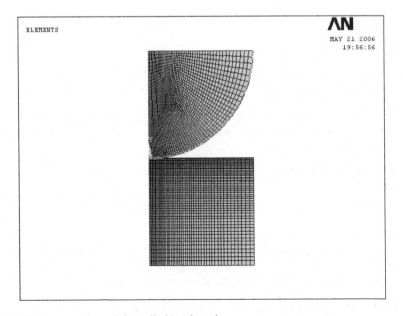

Fig. 3.109 Lower surface of the cylinder selected.

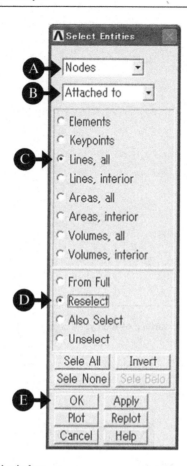

Fig. 3.110 'Select Entities' window.

Only nodes on the lower surface of the cylinder are plotted in the **ANSYS Graphics** window, as shown in Fig. 3.111.

Repeat the **Select → Entities ...** commands again to select a smaller number of the nodes on the portion of the lower surface of the cylinder in the vicinity of the contact point.

   [COMMAND]   ANSYS Utility Menu → **Select** → **Entities** ...

1. The **Select Entities** window opens, as shown in Fig. 3.112.
2. Select [A] **Nodes**, [B] **By Num/Pick**, and [C] **Reselect** in the **Select Entities** window.
3. Click the [D] **OK** button and the **Select nodes** window opens, as shown in Fig. 3.113.
4. The upward arrow appears in the **ANSYS Graphics** window. Select [A] **Box** instead of **Single**, then move the upward arrow to the leftmost end of the array of nodes and select nodes by enclosing them by a rectangle formed by dragging the mouse, as shown in Fig. 3.114. Click the [B] **OK** button to select the nodes on the bottom surface of the cylinder.

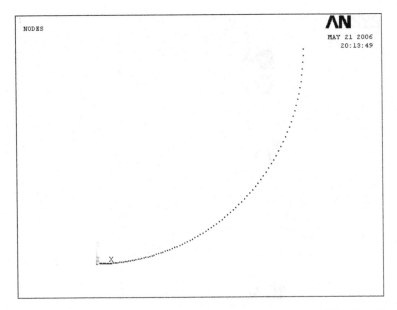

Fig. 3.111 Nodes on the lower surface of the cylinder plotted in the 'ANSYS Graphics' window.

Fig. 3.112 'Select Entities' window.

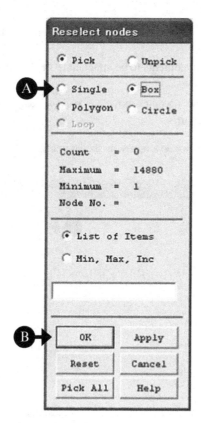

Fig. 3.113 'Select nodes' window.

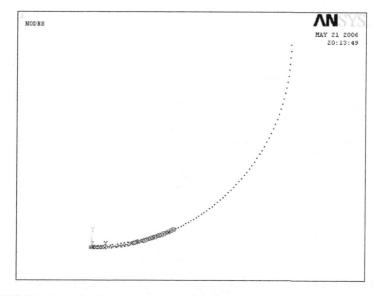

Fig. 3.114 Selection of nodes on the portion of the bottom surface of the cylinder.

Next, let us define contact elements to be attached to the lower surface of the cylinder by the commands as follows:

[COMMAND] ANSYS Main Menu → Preprocessor → Modeling → Create → Elements → Elem Attributes

1. The **Element Attributes** window opens, as shown in Fig. 3.115.

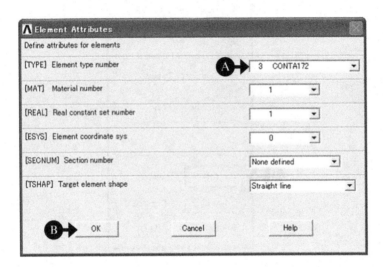

Fig. 3.115 'Element Attributes' window.

2. Select [A] **3 CONTA172** in the **TYPE** box and click the [B] **OK** button in the **Element Attributes** window.

Then, perform the following commands to attach the contact elements to the lower surface of the cylinder.

[COMMAND] ANSYS Main Menu → Preprocessor → Modeling → Create → Elements → Surf / Contact → Surf to Surf

1. The **Mesh Free Surfaces** window opens, as shown in Fig. 3.116.

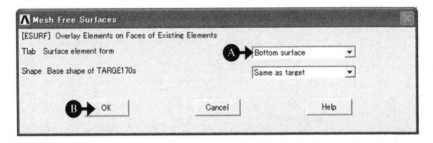

Fig. 3.116 'Mesh Free Surfaces' window.

2. Select [A] **Bottom surface** in the **Tlab** box and click the [B] **OK** button in the window.
3. **CONTA172** elements are created on the lower surface of the cylinder, as shown in Fig. 3.117.

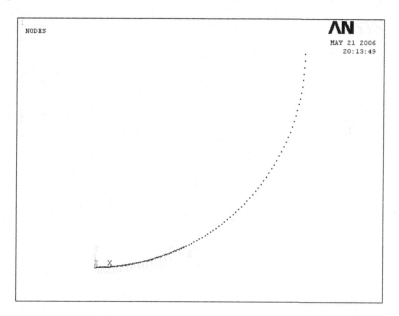

Fig. 3.117 'CONTA172' elements created on the lower surface of the cylinder.

Repeat operations similar to the above in order to create **TARGET169** elements on the top surface of the flat plate. Namely, perform the **Select → Entities** ... commands to select only nodes on the portion of the top surface of the flat plate in the vicinity of the contact point. Then, define the target elements to attach to the top surface of the flat plate using the following commands:

[COMMAND] ANSYS Main Menu → Preprocessor → Modeling → Create → Elements → Elem Attributes

1. The **Element Attributes** window opens, as shown in Fig. 3.118.
2. Select [A] **2 TARGET169** in the **TYPE** box and click the [B] **OK** button in the **Element Attributes** window.

Then, perform the following commands to attach the target elements to the top surface of the flat plate.

[COMMAND] ANSYS Main Menu → Preprocessor → Modeling → Create → Elements → Surf / Contact → Surf to Surf

3. The **Mesh Free Surfaces** window opens, as shown in Fig. 3.119.
4. Select [A] **Top surface** in the **Tlab** box and click the [B] **OK** button in the window.
5. **TARGET169** elements are created on the top surface of the flat plate, as shown in Fig. 3.120.

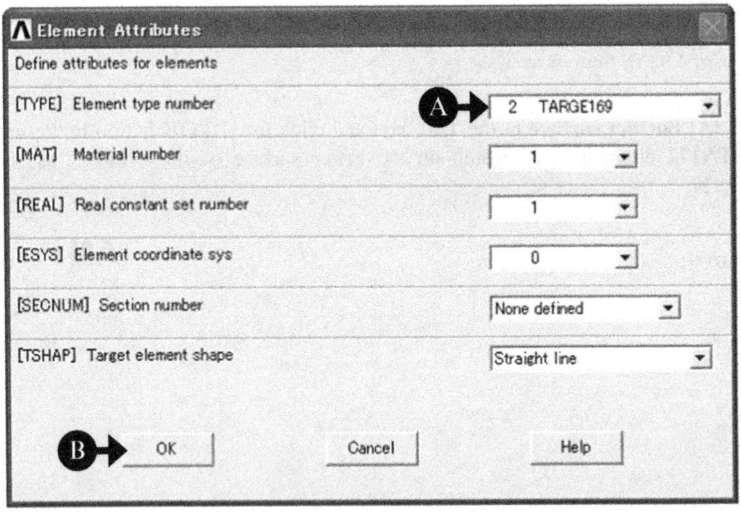

Fig. 3.118 'Element Attributes' window.

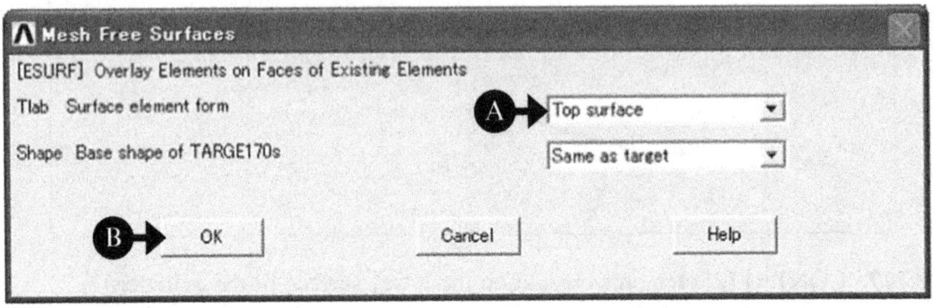

Fig. 3.119 'Mesh Free Surfaces' window.

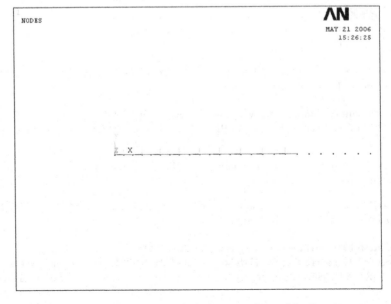

Fig. 3.120 'TARGET169' elements created on the top surface of the flat plate.

3.5.3.4 Input of boundary conditions

(1) Imposing constraint conditions on the left sides of the quarter cylinder and the half flat plate models

Due to the symmetry, the constraint conditions of the present model are in a UX-fixed condition on the left sides of the quarter cylinder and the half flat plate models, and a UY-fixed condition on the bottom side of the half flat plate model. Apply these constraint conditions to the corresponding lines by the following commands:

[COMMAND]   ANSYS Main Menu → Solution → Define
Loads → Apply → Structural → Displacement → On Lines

1. The **Apply U. ROT on Lines** window opens and the upward arrow appears when the mouse cursor is moved to the **ANSYS Graphics** window.
2. Confirming that the **Pick** and **Single** buttons are selected, move the upward arrow onto the left sides of the quarter cylinder and the half flat plate models, and click the left mouse button on each model.
3. Click the **OK** button in the **Apply U. ROT on Lines** window to display another **Apply U. ROT on Lines** window.
4. Select **UX** in the **Lab2** box and click the **OK** button in the **Apply U. ROT on Lines** window.

Repeat the commands and operations (1)–(3) above for the bottom side of the flat plate model. Then, select **UY** in the **Lab2** box and click the **OK** button in the **Apply U. ROT on Lines** window.

(2) Imposing a uniform pressure on the upper side of the quarter cylinder model

[COMMAND]   ANSYS Main Menu → Solution → Define
Loads → Apply → Structural → Pressure → On Lines

1. The **Apply PRES on Lines** window opens and the upward arrow appears when the mouse cursor is moved to the **ANSYS Graphics** window.
2. Confirming that the **Pick** and **Single** buttons are selected, move the upward arrow onto the upper side of the quarter cylinder model and click the left mouse button.
3. Another **Apply PRES on Lines** window opens. Select a Constant value in the **[SFL] Apply PRES on lines as a** box and input **1** (MPa) in the **VALUE Load PRES value** box, and leave a blank in the **Value** box.
4. Click the **OK** button in the window to define a downward force of $P' = 1$ kN/mm applied to the elastic cylinder model.

Fig. 3.121 illustrates the boundary conditions applied to the FE model of the present contact problem by the above operations.

3.5.3.5 Solution procedures

[COMMAND]   ANSYS Main Menu → Solution → Solve → Current LS

1. The **Solve Current Load Step** and **/STATUS Commands** windows appear.

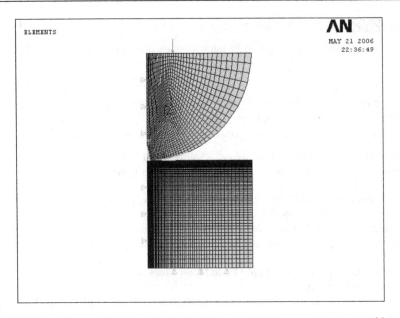

Fig. 3.121 Boundary conditions applied to the FE model of the present contact problem.

2. Click the **OK** button in the **Solve Current Load Step** window to begin the solution of the current load step.
3. Select the **File** button in **/STATUS Commands** window to open the submenu and select the **Close** button to close this window.
4. When the solution is completed, the **Note** window appears. Click the **Close** button to close this window.

3.5.3.6   Contour plot of stress

[COMMAND]   ANSYS Main Menu → **General Postproc** → **Plot Results** → **Contour Plot** → **Nodal Solution**

1. The **Contour Nodal Solution Data** window opens.
2. Select **Stress** and **Y-Component of stress**.
3. Click the **OK** button to display the contour of the y-component of stress in the present model of the contact problem in the **ANSYS Graphics** window, as shown in Fig. 3.122.

Fig. 3.123 is an enlarged contour of the y-component of stress around the contact point, showing that very high compressive stress is induced in the vicinity of the contact point; this is known as Hertz contact stress.

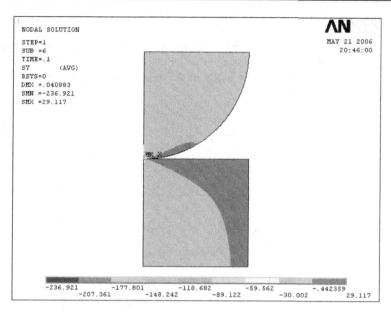

Fig. 3.122 Contour of the y-component of stress in the present model of the contact problem.

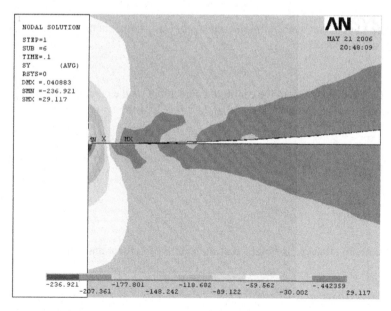

Fig. 3.123 Enlarged contour of the y-component of stress around the contact point.

3.5.4 Discussion

The open circular symbols in Fig. 3.124 show the plots of the y-component stress, or vertical stress, along the upper surface of the flat plate obtained by the present FE calculation. The solid line in the figure indicates the theoretical curve $p(x)$ expressed by a parabola [10,11], i.e.

$$p(x) = k\sqrt{b^2 - x^2} = 115.4\sqrt{2.349^2 - x^2} \tag{3.15}$$

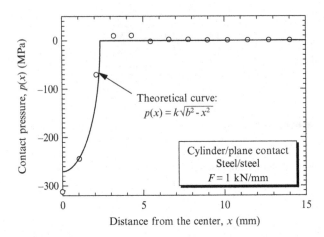

Fig. 3.124 Plots of the y-component stress along the upper surface of the flat plate obtained by the present FE calculation.

where x is the distance from the centre of contact, $p(x)$ is the contact stress at a point x, k is the coefficient given by Eq. (3.16), and b is half the width of contact given by Eq. (3.17).

$$k = \frac{2P'}{\pi b^2} = 115.4\,\text{N/mm}^2 \tag{3.16}$$

$$b = 1.522\sqrt{\frac{P'R}{E}} = 2.348\,\text{mm} \tag{3.17}$$

The maximum stress p_0 is obtained at the centre of contact and expressed as

$$p_0 = 0.418\sqrt{\frac{P'E}{R}} = 270.9\,\text{MPa} \tag{3.18}$$

The length of contact is as small as 4.7 mm, so that the maximum value of the contact stress is as high as 271 MPa. The number of nodes that fall within the contact region is only 3, although the total number of nodes is approximately 15,000 and meshes are finer in the vicinity of the contact point. The three plots of the present results agree reasonably well with the theoretical curve, as shown in Fig. 3.124.

3.5.5 Problems to solve

Problem 3.15. Calculate the contact stress distribution between a cylinder having a radius of curvature $R_1 = 500$ mm pressed against another cylinder having $R_2 = 1000$ mm by a force $P' = 1$ kN/mm, as shown in Fig. 3.125. The two cylinders are made of the same steel having Young's modulus $= 210$ GPa and Poisson's ratio $\nu = 0.3$. The theoretical expression for the contact stress distribution $p(x)$ is given by Eq. (3.19a), with the parameters given by Eqs (3.19b)–(3.19d).

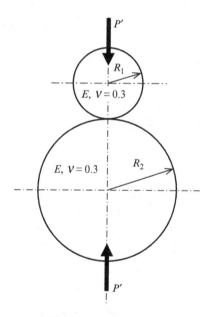

Fig. 3.125 A cylinder having a radius of curvature $R_1 = 500$ mm pressed against another cylinder having $R_2 = 1000$ mm by a force $P' = 1$ kN/mm.

$$p(x) = k\sqrt{b^2 - x^2} \tag{3.19a}$$

$$k = \frac{2P'}{\pi b^2} \tag{3.19b}$$

$$b = 1.522\sqrt{\frac{P'}{E} \cdot \frac{R_1 R_2}{R_1 + R_2}} \tag{3.19c}$$

$$p_0 = 0.418\sqrt{\frac{P'E(R_1 + R_2)}{R_1 R_2}} \tag{3.19d}$$

Problem 3.16. Calculate the contact stress distribution between a cylinder having a radius of curvature $R = 500$ mm pressed against a flat plate of 1000 mm width by 500 mm height by a force of $P' = 1$ kN/mm, as shown in Fig. 3.126. The cylinder is made of fine ceramics having Young's modulus $E_1 = 320$ GPa and Poisson's ratio $\nu_1 = 0.3$, and the flat plate of stainless steel having $E_2 = 192$ GPa and $\nu_2 = 0.3$. The theoretical curve of the contact stress $p(x)$ is given by Eq. (3.20a), with the parameters expressed by Eqs (3.20b)–(3.20d). Compare the result obtained by the FEM calculation with the theoretical distribution given by Eqs (3.20a)–(3.20d).

$$p(x) = k\sqrt{b^2 - x^2} \tag{3.20a}$$

$$k = \frac{2P'}{\pi b^2} \tag{3.20b}$$

$$b = \frac{2}{\sqrt{\pi}} \sqrt{P'R_1 \left(\frac{1 - \nu_1^2}{E_1} + \frac{1 - \nu_2^2}{E_2} \right)} \tag{3.20c}$$

$$p_0 = \frac{1}{\sqrt{\pi}} \sqrt{\frac{P'}{R_1} \cdot \frac{1}{\left(\dfrac{1 - \nu_1^2}{E_1} + \dfrac{1 - \nu_2^2}{E_2} \right)}} \tag{3.20d}$$

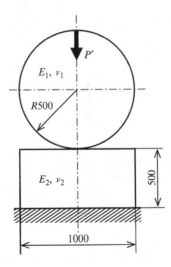

Fig. 3.126 A cylinder having a radius of curvature $R = 500$ mm pressed against a flat plate.

Problem 3.17. Calculate the contact stress distribution between a cylinder having a radius of curvature $R_1 = 500$ mm pressed against another cylinder having $R_2 = 1000$ mm by a force $P' = 1$ kN/mm, as shown in Fig. 3.125. The upper cylinder is made of fine ceramics having Young's modulus $E_1 = 320$ GPa and Poisson's ratio $v_1 = 0.3$, and the lower cylinder of stainless steel having $E_2 = 192$ GPa and $v_2 = 0.3$. The theoretical curve of the contact stress $p(x)$ is given by Eq. (3.21a), with the parameters expressed by Eqs (3.21b)–(3.21d). Compare the result obtained by the FEM calculation with the theoretical distribution given by Eqs (3.21a)–(3.21d).

$$p(x) = k\sqrt{b^2 - x^2} \tag{3.21a}$$

$$k = \frac{2P'}{\pi b^2} \tag{3.21b}$$

$$b = \frac{2}{\sqrt{\pi}} \sqrt{\frac{P'R_1R_2}{R_1 + R_2} \left(\frac{1 - v_1^2}{E_1} + \frac{1 - v_2^2}{E_2} \right)} \tag{3.21c}$$

$$p_0 = \frac{1}{\sqrt{\pi}} \sqrt{\frac{P'(R_1 + R_2)}{R_1 R_2} \cdot \frac{1}{\left(\dfrac{1 - v_1^2}{E_1} + \dfrac{1 - v_2^2}{E_2} \right)}} \tag{3.21d}$$

Problem 3.18. Calculate the contact stress distribution between a cylinder of fine ceramics having a radius of curvature $R_1 = 500$ mm pressed against a cylindrical seat of stainless steel having $R_2 = 1000$ mm by a force $P' = 1$ kN/mm, as shown in Fig. 3.127. Young's moduli of fine ceramics and stainless steel are $E_1 = 320$ GPa and $E_2 = 192$ GPa, respectively, and Poisson's ratio of each material is the same, i.e. $v_1 = v_2 = 0.3$. The theoretical curve of the contact stress $p(x)$ is given by Eq. (3.22a) with the parameters expressed by Eqs (3.22b)–(3.22d). Compare the result obtained by the FEM calculation with the theoretical distribution given by Eqs (3.22a)–(3.22d).

$$p(x) = k\sqrt{b^2 - x^2} \tag{3.22a}$$

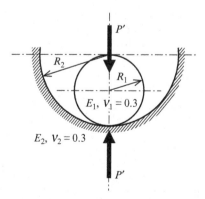

Fig. 3.127 A cylinder having a radius of curvature $R_1 = 500$ mm pressed against a cylindrical seat having $R_2 = 1000$ mm by a force $P' = 1$ kN/mm.

$$k = \frac{2P'}{\pi b^2} \tag{3.22b}$$

$$b = \frac{2}{\sqrt{\pi}} \sqrt{\frac{P'R_1 R_2}{R_2 - R_1}\left(\frac{1-\nu_1^2}{E_1} + \frac{1-\nu_2^2}{E_2}\right)} \tag{3.22c}$$

$$p_0 = \frac{1}{\sqrt{\pi}} \sqrt{\frac{P'(R_2 - R_1)}{R_1 R_2} \cdot \frac{1}{\left(\dfrac{1-\nu_1^2}{E_1} + \dfrac{1-\nu_2^2}{E_2}\right)}} \tag{3.22d}$$

Note that Eqs (3.22a)–(3.22d) cover Eqs (3.19a)–(3.19d), (3.20a)–(3.20d), and (3.21a)–(3.21d).

3.6 Three-dimensional stress analysis: Stress concentration in a stepped round bar subjected to torsion

Stepped round bars are also important fundamental structural and/or machine elements; they are found in automobile axels, pump shafts, shafts in other rotary machines, etc. Such shafts are subjected to torsion.

Torsion problems of circular shafts are one of the most important problems in engineering.

3.6.1 Example problem: A stepped round bar subjected to torsion

Perform an FEM analysis of a 3D stepped round bar subjected to an applied torque at the free end and clamped to the rigid wall at the other end, as shown in Fig. 3.128. Calculate the stress concentration at the foot of the fillet of the bar and the shear stress distribution on a cross section of the bar.

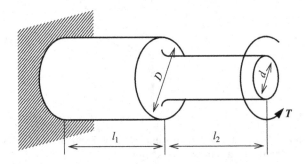

Fig. 3.128 Stepped round bar subjected to the applied torque T.

3.6.2 Problem description

Geometry: lengths $l_1 = 50$ mm, $l_2 = 50$ mm, diameters $D = 40$ mm, $d = 20$ mm, fillet radius $\rho = 3$ mm.
Material: mild steel having Young's modulus $E = 210$ GPa and Poisson's ratio $\nu = 0.3$.
Boundary conditions: A stepped bar is rigidly clamped to a wall at the left end and twisted at the right free end by a torque (twisting moment) of $T = 100$ N mm applied in a plane perpendicular to the central axis of the bar, as shown in Fig. 3.128.

3.6.3 Review of the solutions obtained by the elementary mechanics of materials

Before proceeding to the FEM analysis of the stepped round bar, let us review the solution to the torsion problem of a straight cylindrical bar obtained by the elementary mechanics of materials, as shown in Fig. 3.129.

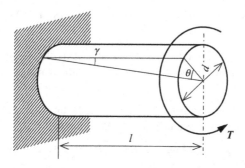

Fig. 3.129 A circular cylindrical bar subjected to the applied torque T.

In a circular cylindrical bar subjected to the applied torque, shear strain γ is induced and varies linearly from the central axis reaching its maximum value $\gamma_{\max}$ at the periphery of the bar, as given by Eq. (3.23).

$$\gamma_{\max} = \gamma|_{r=d/2} = \frac{d\theta}{2l} \tag{3.23}$$

where d is the diameter of the bar, l is its length, and θ is the angle of twist. The twist angle θ is obtained by the following equation:

$$\theta = \frac{32T}{\pi G d^4} = \frac{T}{GI_\mathrm{p}} \quad \text{where } I_\mathrm{p} = \frac{\pi d^4}{32} \tag{3.24}$$

where G is the shear modulus and I_p is the polar moment of inertia of area of the bar.
The maximum shearing stress τ_0 is thus obtained on the periphery of the bar at the right end where the torque T is applied, and is given by the following equation:

$$\tau_0 = \tau_{\theta z}|_{r=d/2} = \frac{16T}{\pi d^3} \tag{3.25}$$

3.6.4 Analytical procedures

3.6.4.1 Creation of an analytical model

A 3D stepped round bar can be created by rotating the 2D stepped beam described in Section A.1.1 around the longitudinal axis, or the *x*-axis. At first, create a 2D stepped beam having a rounded fillet referring to the operations described in Section A.1.1.

[COMMAND] ANSYS Main Menu → Preprocessor → Modeling → Create → Areas → Rectangle → By 2 Corners

1. Input two **0**s into the **WP X** and **WP Y** boxes in the **Rectangle by 2 Corners** window to determine the lower left corner point of the thicker portion of the bar on the Cartesian coordinates of the working plane.
2. Input **100** and **20** (mm) into the **Width and Height** boxes, respectively, to determine the shape of the length and the radius of the thicker portion of the bar.

[COMMAND] ANSYS Main Menu → Preprocessor → Modeling → Create → Areas → Circle → Solid Circle

1. Input **53**, **13**, and **3** (mm) into the **WP X**, **WP Y**, and **Radius** boxes, respectively, in the **Solid Circular Area** window to create the rounded fillet having the curvature radius of **3** (mm).

Two more rectangles are needed to create a stepped beam with a rounded fillet. Create two rectangles by carrying out the following commands:

[COMMAND] ANSYS Main Menu → Preprocessor → Modeling → Create → Areas → Rectangle → By 2 Corners

1. Input **53** and **10** (mm) into the **WP X** and **WP Y** boxes in the **Rectangle by 2 Corners** window, and **50** and **15** (mm) into the **Width and Height** boxes, respectively. Press the **Apply** button.
2. Input **50** and **13** (mm) into the **WP X** and **WP Y** boxes in the **Rectangle by 2 Corners** window, and **10** and **15** (mm) into the **Width and Height** boxes, respectively. Press the **OK** button.

The stepped beam with a rounded fillet can be obtained by subtracting the solid circle and the two small rectangles from the **100** by **20** (mm) rectangle created by the operations mentioned above.

[COMMAND] ANSYS Main Menu → Preprocessor → Modeling → Operate → Booleans → Subtract > Areas

1. Pick the largest rectangular area by the upward arrow to turn the colour of the rectangular area to pink. Click the **OK** button.
2. Pick the solid circular area and the two smaller rectangular areas by the upward arrow to turn the colour of the areas to pink. Click the **OK** button to get the stepped beam with a rounded fillet, as shown in Fig. 3.130.

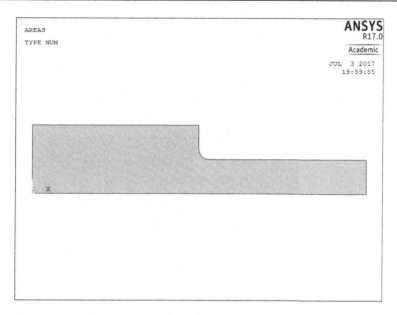

Fig. 3.130 2D stepped beam with a rounded fillet.

A stepped round bar with a rounded fillet can be obtained by rotating the 2D stepped beam around the x-axis.

> **[COMMAND]** ANSYS Main Menu → **Preprocessor** → **Modeling** → **Operate** → **Extrude** → **Areas** → **About Axis**

1. The **Sweep Areas about Axis** window appears, as shown in Fig. 3.131. Pick the stepped beam area by the upward arrow to turn the colour of the rectangular area to pink. Click the **OK** button of the window.
2. Pick the two corner points of the bottom side of the stepped beam area by the upward arrow and click the **OK** button of the window.
3. Another **Sweep Areas about Axis** window appears, as shown in Fig. 3.132. Input **360** (degrees) into the **ARC** box to rotate the 2D stepped beam at **360** degrees about the x-axis to get the 3D stepped round bar with circumferential shoulder fillet, as shown in Fig. 3.133.

3.6.4.2 Input of the elastic properties of the stepped round bar material

> **[COMMAND]** ANSYS Main Menu → **Preprocessor** → **Material Props** → **Material Models**

1. The **Define Material Model Behavior** window opens.
2. Double-click the **Structural**, **Linear**, **Elastic**, and **Isotropic** buttons one after another.

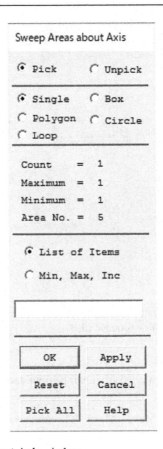

Fig. 3.131 Sweep Areas about Axis window.

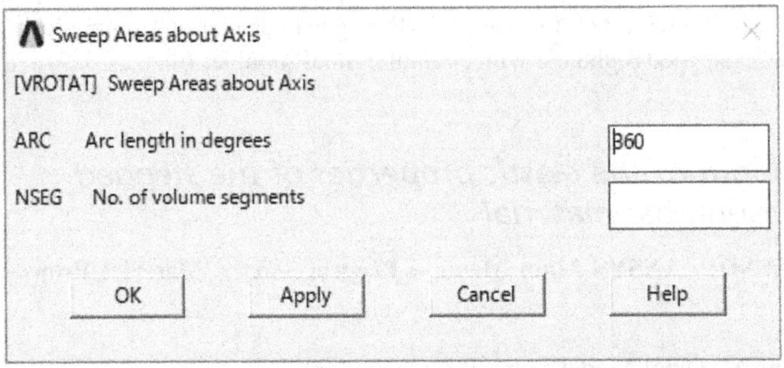

Fig. 3.132 Another **Sweep Areas about Axis** window appearing after the window shown in Fig. 3.131.

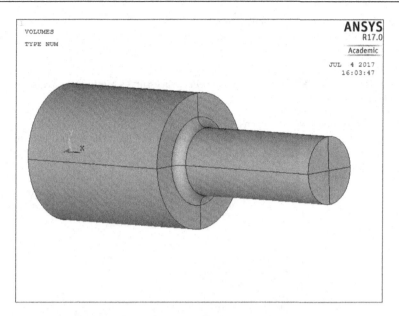

Fig. 3.133 3D stepped round bar with circumferential shoulder fillet created.

3. Input the value of Young's modulus, **2.1e5** (MPa), and that of Poisson's ratio, **0.3**, into the **EX** and the **PRXY** boxes, respectively, and click the **OK** button of the **Linear Isotropic Properties for Materials Number 1** window.
4. Exit from the **Define Material Model Behavior** window by selecting **Exit** in the **Material** menu of the window.

3.6.4.3 Finite-element discretisation of the round bar volume

(1) Selection of the element type

   [COMMAND]   ANSYS Main Menu → Preprocessor → Element Type → Add/Edit/ Delete

1. The **Element Types** window opens.
2. Click the **Add …** button in the **Element Types** window to open the **Library of Element Types** window and select the element type to use.
3. Select **Structural Mass – Solid** and **Tet 10 node 187.**
4. Click the **OK** button in the **Library of Element Types** window to use the 10-node isoparametric tetrahedral element.

(2) Sizing of the elements

   [COMMAND]   ANSYS Main Menu → Preprocessor → Meshing → Size Cntrls → Manual Size → Global → Size

1. The **Global Element Sizes** window opens.
2. Input **2** in the **SIZE** box and click the **OK** button.

(3) Meshing

   [COMMAND]   ANSYS Main Menu → Preprocessor → Meshing → Mesh →
   Volumes → Free

1. The **Mesh Volumes** window opens.
2. The upward arrow appears in the **ANSYS Graphics** window. The stepped bar volume is quadrisected, as shown in Fig. 3.134. Point the arrow at each quarter volume and click one by one.

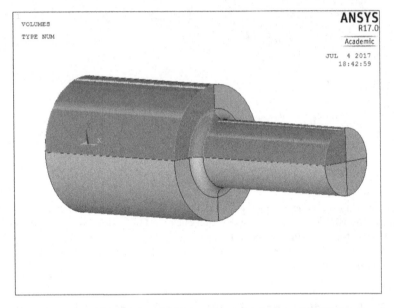

Fig. 3.134 A quadrisected volume of the stepped bar model selected for finite-element discretisation.

3. After the colour of the whole volume turns from light blue to pink, click the **OK** button to see the whole stepped round bar model meshed by 10-node tetrahedral isoparametric elements, as shown in Fig. 3.135.

3.6.4.4 Input of boundary conditions

(1) Imposing constraint conditions on the left-end face of the stepped round bar

Click the [A] **Dynamic Model Mode** button and drag the mouse while pressing its right button to rotate the round bar model to display the left end face, as shown in Fig. 3.136.

   [COMMAND]   ANSYS Main Menu → Preprocessor → Loads → Define
   Loads → Apply → Structural → Displacement → On Areas

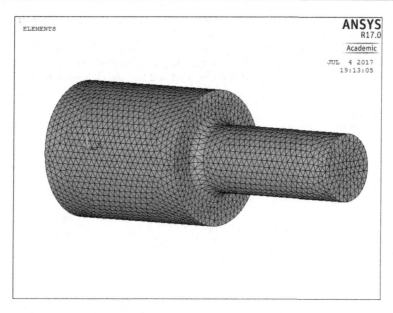

Fig. 3.135 A whole view of the present stepped bar model discretised by 10-node tetrahedral isoparametric elements.

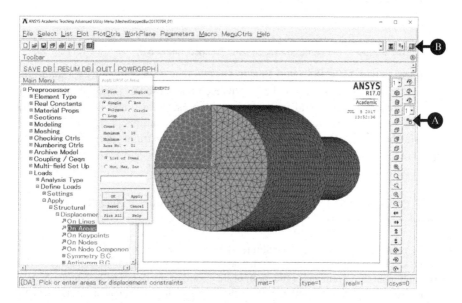

Fig. 3.136 Selected one of the four quadrants of the left-end face of the stepped bar model.

1. The **Apply U, ROT on Areas** window opens and the upward arrow appears when the mouse cursor is moved to the **ANSYS Graphics** window.
2. Click and select the four quadrants into which the left face is divided. Click the **OK** button in the **Apply U, ROT on Areas** window. Another **Apply U, ROT on Areas** window opens, as shown in Fig. 3.137. Select [A] **All DOF** in the **Lab2** box and click the **OK** button.

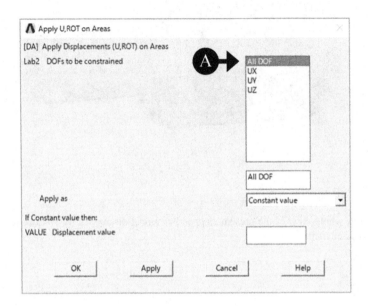

Fig. 3.137 Apply U, ROT on Areas window.

(2) Applying torque to the right-end face of the stepped round bar

An element called **MPC184 rigid link/beam** element is used to apply torque to the right-end face of the stepped round bar. This element is used as a rigid component to transmit forces, moments, and torques as in the present case, and is well suited for linear, large rotation, and/or large strain nonlinear applications [12].

Create the **MPC184** element, or the multipoint constraint element, and apply torque via this element to the stepped round bar around its centre axis, or the x-axis. For this purpose, the pilot node will be defined first.

[COMMAND] ANSYS Main Menu → **Preprocessor** → **Modeling** → **Create** → **Nodes** → **In Active CS**

1. The **Create Nodes in Active Coordinate System** window opens as shown in Fig. 3.138.
2. Input, say **150000**, the node number which does not exist in the present model, in the **NODE** box and input the coordinates (**100, 0, 0**) of the centre of the right-end face in the **X, Y, Z** window, then click the **OK** button. Fig. 3.139 shows the pilot node number 150000 depicted on the right-end face of the stepped bar model.

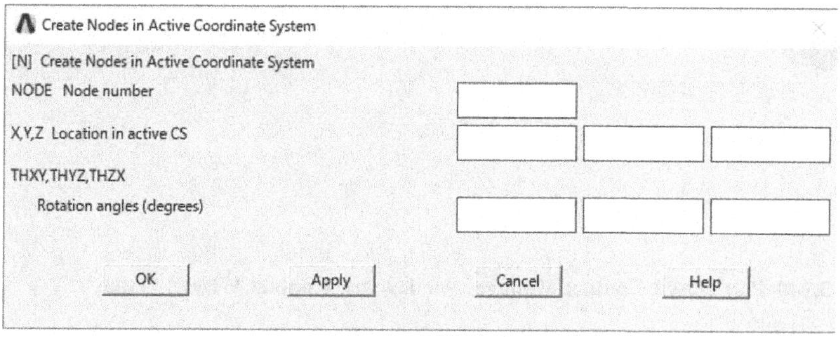

Fig. 3.138 Create Nodes in Active Coordinate System window.

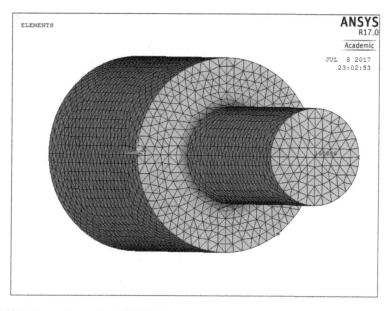

Fig. 3.139 Pilot node number 150000 depicted on the right-end face of the stepped bar model.

Click the **Pair Based Contact Manager** button shown by [B] in Fig. 3.136, and the **Pair Based Contact Manager** window opens. Click the [A] **Contact Wizard** button shown in Fig. 3.140.

The **Contact Wizard** window opens. Choose **Areas** as the **Target Surface** and **Pilot Node Only** as the **Target Type**. Click the **Next >** button. Input N_PILOT in the **Pilot name** box, choose **Pick existing node** ..., and click the **Pick Entity ...** button. The **Select Node for Pilot Node** window opens as shown in Fig. 3.141. Input the pilot node number, i.e. **150000**, in the **List of Items** box and click the **OK** button to return to the **Contact Wizard** window. Click the **Next >** button in the window.

Fig. 3.140 **Pair Based Contact Manager** window and **Contact Wizard** button.

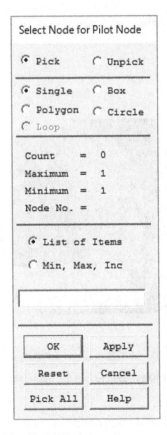

Fig. 3.141 **Select Node for Pilot Node** window.

The next screen of the **Contact Wizard** window appears. Choose **Areas** as the **Contact Surface** and **Surface-to-Surface** as the **Contact Element Type**. Click the **Pick Contact** ... button. The **Select Areas for Contact** window opens, as shown in Fig. 3.142. Pick the four quadrants of the right-end face of the stepped bar to which

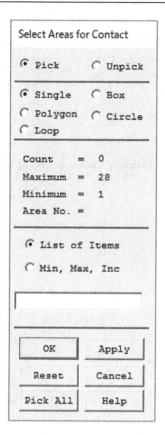

Fig. 3.142 Select Areas for Contact window.

the torque is to be applied to in the **ANSYS Graphics** window, and click the OK button to come back to the **Contact Wizard** window. Click the **Next >** button.

The next screen of the **Contact Wizard** window appears. Choose **Rigid constraint** as the **Constraint Surface Type** and **User specified** as the **Boundary conditions on target**. Click the **Create >** button. The message 'The contact pair has been created. To interact with the contact pair use real set ID 3' appears in the **Contact Wizard** window. Click the **Finish** button.

The **Pair Based Contact Manager** window opens, as shown in Fig. 3.143, describing the specifications of the contact pairs created. Fig. 3.144 shows in red the right-end face of the stepped bar to which the torque is applied.

[COMMAND]   ANSYS Utility Menu → Plot → Elements

The finite element mesh of the model is then recovered.

[COMMAND]   ANSYS   Main   Menu → Preprocessor → Loads → Define Loads → Apply → Structural → Force/Element → On Nodes

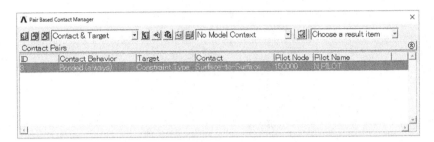

Fig. 3.143 Pair Based Contact Manager window describing the specifications of the contact pairs created.

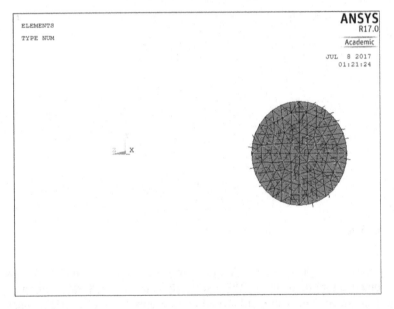

Fig. 3.144 Highlighted right-end face of the stepped bar to which the torque is applied.

1. The **Apply F/M on Nodes** window opens as shown in Fig. 3.145. Input the pilot node number, i.e. **150000**, in the **List of Items** box and click the **OK** button.
2. Another **Apply F/M on Nodes** window opens as shown in Fig. 3.146. Choose **MX**, or torsional moment around the *x*-axis, from among the items listed in the **Lab** box and **Constant Value** in the **Apply as** box. Input 100 (N mm) in the VALUE box, then click the **OK** button.

3.6.4.5 Solution procedures

Torsion analysis often needs the large rotation/displacement analysis. Without the large rotation/displacement option, the result of the torsion analysis brings about

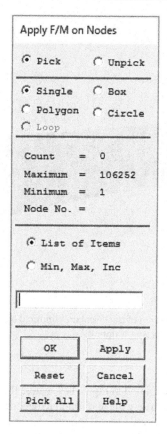

Fig. 3.145 **Apply F/M on Nodes** window.

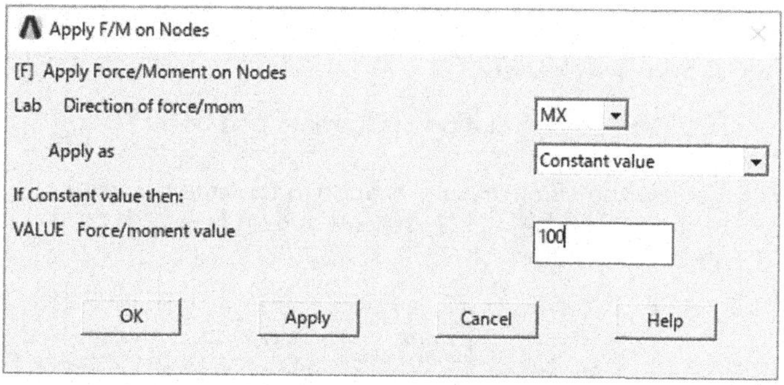

Fig. 3.146 Next scene of the **Apply F/M on Nodes** window.

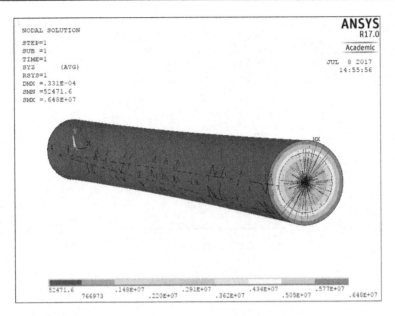

Fig. 3.147 Anomalous deformation of the circular cylindrical bar as a result of torsion analysis without the large rotation/displacement option.

anomalous deformation as depicted in Fig. 3.147, where the deformation is enlarged to approximately 302 times the true scale and found to become larger for approaching the torque-bearing right-end face.

In order to set up the large deformation analysis, the following commands should be carried out:

[COMMAND] ANSYS Main Menu → Solution → Analysis Type → Sol'n Controls

1. The **Solution Controls** window opens, as shown in Fig. 3.148.

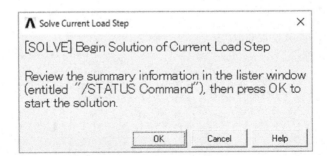

Fig. 3.148 Solution Controls window.

2. Choose **Large Displacement Static** from among the items listed in the **Analysis Options** box and choose **On** in the **Automatic time stepping** box in the **Time Control** pane.
3. When the **Automatic time stepping** option is chosen to be **On**, the number, the maximum and the minimum numbers of substeps shall be specified. Specify 10, 100, and 5 as the three numbers, respectively, for the present analysis.
4. Click the **OK** button.

[COMMAND]   ANSYS Main Menu → Solution → Solve → Current LS

(1) The **Solve Current Load Step** window opens.
(2) Click the **OK** button.

The ANSYS Process Status window appears, describing 'Nonlinear Solution' during the solution process. When the solution is complete, click the **Close** button in the **Note** window and close the **/STATUS Command** window by clicking the cross mark ('x') of the upper right corner of the window. The solution convergence diagram appears in the **ANSYS Graphics** window, as shown in Fig. 3.149.

[COMMAND]   ANSYS Utility Menu → Plot → Volumes

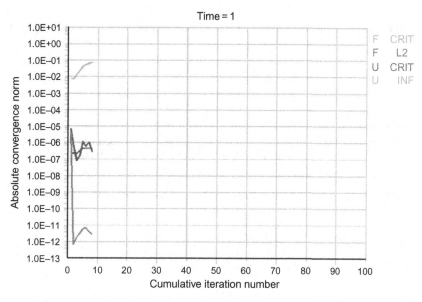

Fig. 3.149 Solution convergence diagram displayed in the **ANSYS Graphics** window after the end of nonlinear analysis.

1. The 3D model of the stepped bar is displayed as shown in Fig. 3.150, where both the global coordinate system (X, Y, Z) and the local one (WX, WY, WZ) are depicted.

3.6.4.6 Contour plot of stress

(1) Changing the coordinate system (CS): from the global Cartesian to the global cylindrical CS

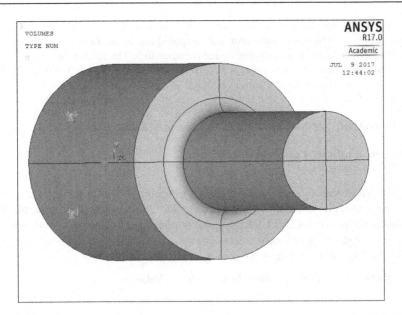

Fig. 3.150 Global coordinate system (X, Y, Z) and local coordinate (WX, WY, WZ) depicted on the present 3D model of the stepped bar.

It is preferable to present stress components induced in the present stepped round bar in the circular cylindrical coordinate system (r, θ, z) rather than by the usual Cartesian coordinate system (x, y, z).

Before changing the coordinate systems, the Cartesian coordinate system will be rotated around the y-axis by 90 degrees so that the new z-axis becomes the longitudinal axis of the stepped bar. On ANSYS, the x-, y-, and z-axes should be transformed to the radial axis r, the azimuthal axis θ, and the longitudinal axis z, respectively.

    [COMMAND]   **ANSYS Utility Menu → Work Plane → Offset WP by Increments …**

1. The **Offset WP** window opens, as shown in Fig. 3.151.
2. Move the [A] slide bar for the coordinate rotation to 90 degrees.
3. Click the [B] **+Y** rotation button and then the **OK** button. The local coordinate system is rotated around the y-axis by 90 degrees.

The coordinate system can now be changed by the following commands:

    [COMMAND]   **ANSYS Utility Menu → Work Plane → Local Coordinate Systems → Create CS → At WP Origin**

1. The **Create Local CS at WP Origin** window opens, as shown in Fig. 3.152.
2. Input any integer larger than 10 in the [A] **KCN** box and choose [B] **Cylindrical 1** in the **KCS** box.
3. Click the **OK** button.

    [COMMAND]   **ANSYS Main Menu → General Postprocessor → Options for Output**

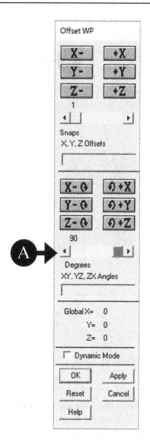

Fig. 3.151 Offset WP window.

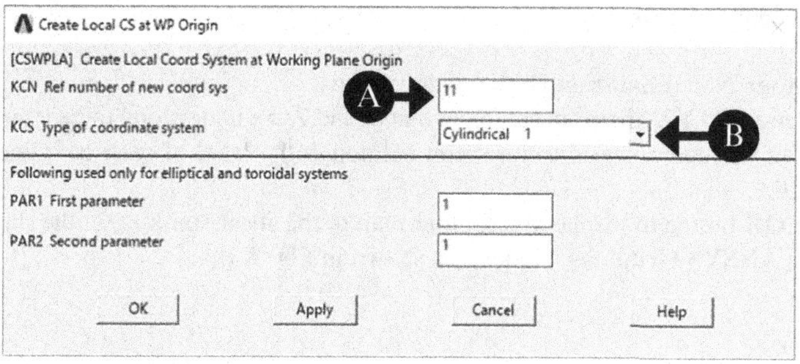

Fig. 3.152 Create Local CS at WP Origin window.

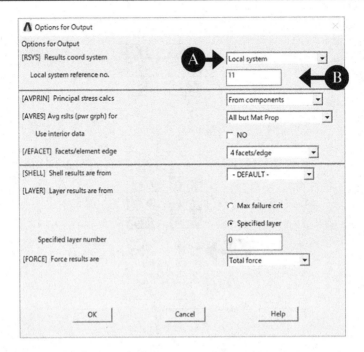

Fig. 3.153 Options for Output window.

1. The **Options for Output** window opens, as shown in Fig. 3.153.
2. Choose [A] **Local system** from among the items listed in the **RSYS** box, and input the [B] integer 11 in the **Local system reference no.** box.
3. Click the **OK** button to change the coordinate systems; i.e. from the Cartesian coordinate system to the local cylindrical one.

(2) Contour plot of the shear stress $\tau_{r\theta}$ component in the global cylindrical coordinate system

   [COMMAND] ANSYS Main Menu > General Postproc → Plot Results → Contour
   Plot → Nodal Solution

1. The **Contour Nodal Solution Data** window opens.
2. Select **Stress** and **YZ Shear stress**. Note that Y and Z are understood to be θ and z, respectively, after the coordinate transformation, although the label of each coordinate remains unchanged.
3. Click the **OK** button to display the contour map of the shear stress $\tau_{r\theta}$ in the stepped round bar in the **ANSYS Graphics** window, as shown in Fig. 3.154.

3.6.5 Discussion

Fig. 3.155 shows the relationship between stress concentration factor α and radius of curvature $2\rho/d$ at the foot of shoulder fillet for torsion of a stepped round bar with the shoulder fillet [13]. The relationship can be approximated by Eq. (3.27) [12].

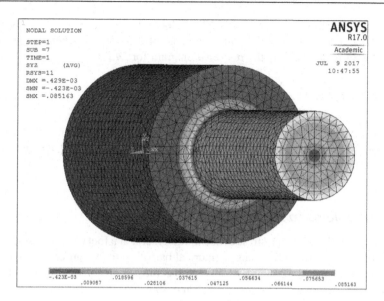

Fig. 3.154 Contour map of the shear stress $\tau_{r\theta}$ in the stepped round bar showing stress concentration at the foot of the circumferential shoulder fillet.

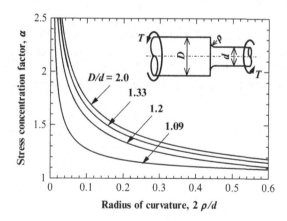

Fig. 3.155 Relationship between stress concentration factor α and radius of curvature $2\rho/d$ at the shoulder fillet for torsion of a stepped round bar with a shoulder fillet [12].

$$\alpha = \tau_{\max}/\tau_0 \tag{3.26}$$

$$\alpha = 1 + K\left(\frac{d}{\rho}\right)^{0.65} \tag{3.27}$$

The maximum stress obtained by ANSYS for the present stepped bar is about 0.0852 MPa, as shown in Fig. 3.154. The value of the stress concentration factor α of about 1.34 is obtained by the $\alpha - 2\rho/d$ diagram for $D/d = 2$; i.e.

$$\tau_{max} = \alpha\tau_0 \approx 1.34 \times \frac{16 \times 100}{\pi \times 20^3} \approx 0.0853 \text{ MPa} \tag{3.28}$$

The value of the maximum shear stress $\tau_{\theta z}|_{max}$ is a little bit smaller than that obtained by the tensile stress concentration vs radius of curvature diagram [13], and the relative difference the two values is -0.16%.

3.6.6   Problems to solve

Problem 3.19. Calculate the stress concentration factor at a foot of the circumferential rounded fillet in a stepped circular cylindrical bar subjected to uniform tensile stress σ_0 on the right-end face of the bar. The bar has the same geometry and mechanical properties as described in Section 3.6.2 and shown in Fig. 3.128, and is rigidly clamped to the rigid wall at the other end. Note that the large displacement option is not necessary to be imposed in this case.

Answer: The tensile stress concentration factor is obtained as $\alpha = \sigma_{max}/\sigma_0 \approx 1.78$ by the tensile stress concentration vs radius of curvature diagram [13].

Problem 3.20. Calculate the stress concentration factor at a foot of the circumferential rounded fillet in a stepped circular cylindrical bar subjected to point load P perpendicular to the longitudinal axis of the bar at the periphery of the right-end face of the bar. The bar has the same geometry and mechanical properties as described in Section 3.6.2 and shown in Fig. 3.128, and is rigidly clamped to the rigid wall at the other end. Note that the large displacement option is not necessary to be imposed in this case, either.

Answer: The bending stress concentration factor is obtained as $\alpha = \sigma_{max}/\sigma_0 \approx 1.55$ where $\sigma_0 = 32Pl_2/(\pi d^3)$ by the bending stress concentration vs radius of curvature diagram [13].

References

[1] R. Yuuki, H. Kisu, Elastic Analysis With the Boundary Element Method, Baifukan Co., Ltd., Tokyo, 1987 (in Japanese).

[2] M. Isida, Elastic Analysis of Cracks and Their Stress Intensity Factors, Baifukan Co., Ltd., Tokyo, 1976 (in Japanese).

[3] M. Isida, Analysis of stress intensity factors for the tension of a centrally cracked strip with stiffened edges, Eng. Fract. Mech. 5 (3) (1973) 647–665.

[4] R.E. Feddersen, Discussion, ASTM STP 410, 1967, pp. 77–79.

[5] W.T. Koiter, Delft Technological University, Department of Mechanical Engineering, Report, No. 314, 1965.

[6] H. Tada, A note on the finite width corrections to the stress intensity factor, Eng. Fract. Mech. 3 (2) (1971) 345–347.

[7] H. Okamura, Introduction to the Linear Elastic Fracture Mechanics, Baifukan Co., Ltd., Tokyo, 1976 (in Japanese).

[8] H.L. Ewalds, R.J.H. Wanhill, Fracture Mechanics, Edward Arnold, Ltd., London, 1984.

[9] A. Saxena, Nonlinear Fracture Mechanics for Engineers, CRC Press, Florida, 1997.

[10] I. Nakahara, Strength of Materials, vol. 2, Yokendo Co., Ltd., Tokyo, 1966 (in Japanese).

[11] S.P. Timoshenko, J.N. Goodier, Theory of Elasticity, third ed., McGraw-Hill Kogakusha, Ltd., Tokyo, 1971.

[12] ANSYS help documents, SAS IP, Inc., 2015.

[13] M. Nishida, Stress Concentration, Morikita Pub. Co., Ltd., Tokyo, 1967 (in Japanese).

Mode analysis

4

4.1 Introduction

When a steel bar is hit by a hammer, a clear sound can be heard because the steel bar vibrates at its resonant frequency. If the bar is oscillated at this resonant frequency, it will be found that the vibration amplitude of the bar becomes very large. Therefore, when a machine is designed, it is important to know the resonant frequency of the machine. The analysis to obtain the resonant frequency and the vibration mode of an elastic body is called 'mode analysis'.

It is said that there are two methods for mode analysis. One is the theoretical analysis and the other is the finite element method (FEM). Theoretical analysis is usually used for a simple shape of an elastic body, such as a flat plate and a straight bar but theoretical analysis cannot give us the vibration mode for the complex shape of an elastic body. FEM analysis can obtain the vibration mode for it.

In this chapter, three examples are presented, which use beam elements, shell elements and solid elements, respectively, as shown in Fig. 4.1.

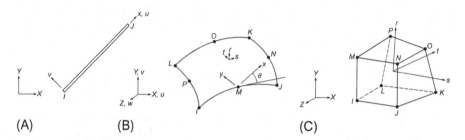

Fig. 4.1 FEM element types (A) beam element, (B) shell element, and (C) solid element.

4.2 Mode analysis of a straight bar

4.2.1 Problem description

Obtain the lowest three vibration modes and resonant frequencies in the y direction of the straight steel bar shown in Fig. 4.2 when the bar can move only in the y direction.

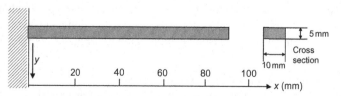

Fig. 4.2 Cantilever beam for mode analysis.

Engineering Analysis with ANSYS Software. https://doi.org/10.1016/B978-0-08-102164-4.00004-2

Thickness of the bar is 0.005 m; width of the bar is 0.01 m; length of the bar is 0.09 m. Material of the bar is steel with Young's modulus, $E = 206$ GPa, and Poisson's ratio $\nu = 0.3$. Density $\rho = 7.8 \times 10^3$ kg/m^3.

Boundary condition: All freedoms are constrained at the left end.

4.2.2 Analytical solution

Before mode analysis is attempted using ANSYS, an analytical solution for resonant frequencies will be obtained to confirm the validity of ANSYS solution. The analytical solution of resonant frequencies for a cantilever beam in y direction is given by,

$$f_i = \frac{\lambda_i^2}{2\pi L^2} \sqrt{\frac{EI}{M}} \quad (i = 1, 2, 3, \ldots) \tag{4.1}$$

where length of the cantilever beam, $L = 0.09$ m, cross section area of the cantilever beam, $A = 5 \times 10^{-5}$ m^2,

Young's modulus $E = 206$ GPa.

The area moment of inertia of the cross section of the beam is

$$I = bt^3/12 = \left(0.01 \times 0.005^3\right)/12 = 1.042 \times 10^{-10} \text{m}^4$$

Mass per unit width $M = \rho AL/L = \rho A = 7.8 \times 10^3$ kg/m$^3 \times 5 \times 10^{-5}$ m$^2 = 0.39$ kg/m. $\lambda_1 = 1.875$, $\lambda_2 = 4.694$, $\lambda_3 = 7.855$.

For that set of data the following solutions is obtained: $f_1 = 512.5$ Hz; $f_2 = 3212$ Hz; $f_3 = 8994$ Hz.

Fig. 4.3 shows the vibration modes and the positions of nodes obtained by Eq. (4.1).

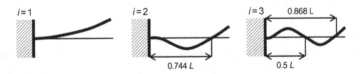

Fig. 4.3 Analytical vibration mode and the node position.

4.2.3 Model for FE analysis

4.2.3.1 Element type selection

In FEM analysis, it is very important to select a proper element type which influences the accuracy of solution, working time for model construction and CPU time. In this example, the two-dimensional elastic beam, as shown in Fig. 4.4, is selected for the following reasons:

(a) Vibration mode is constrained in the two-dimensional plane.
(b) Number of elements can be reduced; the time for model construction and CPU time are both shortened.

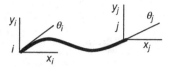

Fig. 4.4 Two-dimensional beam element.

A two-dimensional elastic beam has three degrees of freedom at each node (i, j), which are translatory deformations in the x and y directions and rotational deformation around the z axis. This beam can be subject to extension or compression bending due to its length and the magnitude of the area moment of inertia of its cross section.

From the ANSYS Main Menu, select **Preprocessor → Element Type → Add/Edit/Delete**.

Then the **Element Types** window, as shown in Fig. 4.5, is opened.

1. Click [A] **Add**. Then the **Library of Element Types** window as shown in Fig. 4.6 opens.
2. Select [B] **Beam** in the **Library of Element Types** table, and then select [C] **2 node 188**.
3. Ensure **Element type reference number** is set to **1** and click the [D] **OK** button. Then the **Library of Element Types** window is closed.
4. Click the [E] **Close** button in the window of Fig. 4.7.

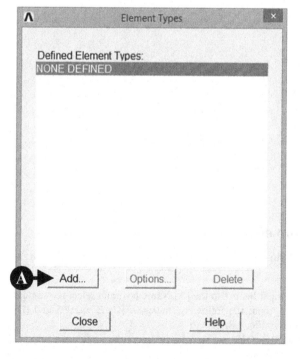

Fig. 4.5 Element Types window.

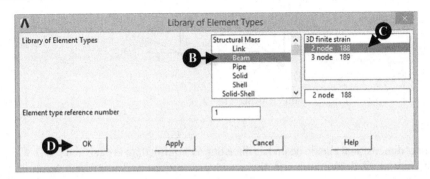

Fig. 4.6 **Library of Element Types** window.

Fig. 4.7 **Library of Element Types** window.

4.2.3.2 Sections for beam element

From the ANSYS Main Menu, select **Preprocessor → Sections → Beam → Common Sections**.

1. The **Beam Tool** window opens. Input **beam1** to the [A] **Name** box, and select rectangular shape [B] in the **Sub-Type** box. Input the following values of [C] B = 0.005 and [D] H = 0.01, respectively, as shown in Fig. 4.8.
2. After inputting these values, click [E] **OK**.

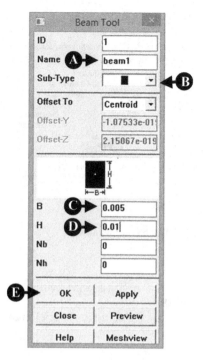

Fig. 4.8 Beam Tool window.

4.2.3.3 Material properties

This section describes the procedure of defining the material properties of the beam element.

From the ANSYS Main Menu, select **Preprocessor** → **Material Props** → **Material Models**.

1. Click the above buttons in specified order and the **Define Material Model Behavior** window as shown in Fig. 4.9 opens.
2. Click the following terms in the window: [A] **Structural** → **Linear** → **Elastic** → **Isotropic**. As a result, the window of **Linear Isotropic Properties for Material Number 1** as shown in Fig. 4.10 opens.
3. Input Young's modulus of **206e9** to the [B] **EX** box and Poisson ratio of **0.3** to the [C] **PRXY** box. Then click the [D] **OK** button.

Next, define the value of density of material.

1. Click the term [E] **Density** in Fig. 4.9 and the window **Density for Material Number 1**, Fig. 4.11, opens.
2. Input the value of density, **7800**, to the [F] **DENS** box and click the [G] **OK** button. Finally, close the window **Define Material Model Behavior** by clicking the [H] X mark at the upper right corner).

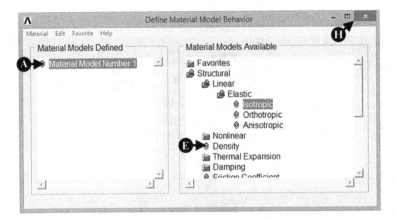

Fig. 4.9 Define Material Model Behavior window.

Fig. 4.10 Linear Isotropic Properties for Material Number 1 window.

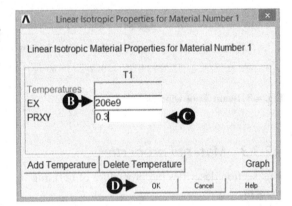

Fig. 4.11 Density for Material Number 1 window.

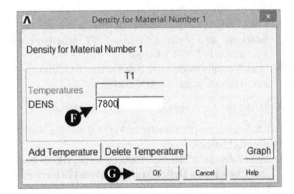

4.2.3.4   Create keypoints

To draw a cantilever beam for analysis, the method of using keypoints is described.

From the ANSYS Main Menu, select **Preprocessor** → **Modeling** → **Create** → **Keypoints** → **In Active CS**.

The **Create Keypoints in Active Coordinate System** window, Fig. 4.12, opens.

1. Input **1** to the [A] **NPT Key Point number** box, **0, 0, 0** to the [B] **X,Y,Z Location in active CS** box, and then click the [C] **Apply** button. Do not click the **OK** button at this stage. If you do, the window will be closed. In this case, open the **Create Keypoints in Active Coordinate System** window as shown in Fig. 4.13 and then proceed to step 2.
2. In the same window, input **2** to the [D] **NPT Key Point number** box, **0.09, 0, 0** to the [E] **X, Y,Z Location in active CS** box, and then click the [F] **OK** button.
3. After finishing the above steps, two keypoints appear in the window as shown in Fig. 4.14.

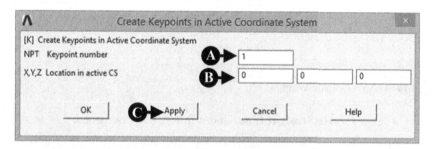

Fig. 4.12 **Create Keypoints in Active Coordinate System** window.

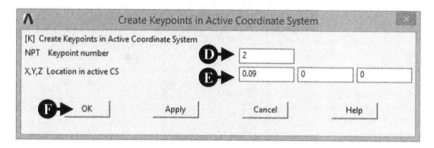

Fig. 4.13 **Create Keypoints in Active Coordinate System** window.

4.2.3.5   Create a line for beam element

By implementing the following steps, a line between two keypoints is created.

From the ANSYS Main Menu, select **Preprocessor** → **Modeling** → **Create** → **Lines** → **Lines** → **Straight Line**.

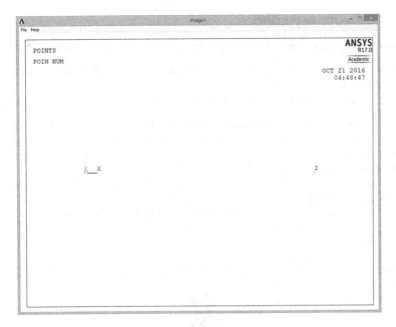

Fig. 4.14 ANSYS Graphic Window.

The window **Create Straight Line**, shown in Fig. 4.15, is opened.

1. Pick the keypoints, [A] **1** and [B] **2**, shown in Fig. 4.16, and click the [C] **OK** button in the window **Create Straight Line** shown in Fig. 4.15. A line is created as shown in Fig. 4.16.

Fig. 4.15 Create Straight Line window.

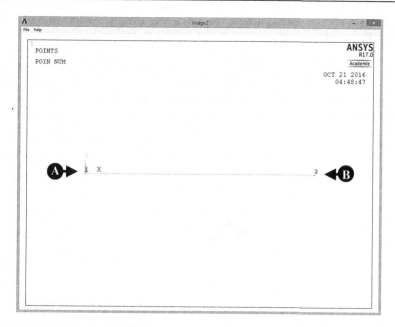

Fig. 4.16 ANSYS Graphic Window.

4.2.3.6 Create mesh in a line

From the ANSYS Main Menu, select **Preprocessor → Meshing → Size Cntrls →
Manual Size → Lines → All Lines**.

The window **Element Sizes on All Selected Lines**, shown in Fig. 4.17, is opened.

1. Input [A] the number of **20** to the **NDIV** box. This means that a line is divided into 20 elements.
2. Click the [B] **OK** button and close the window.

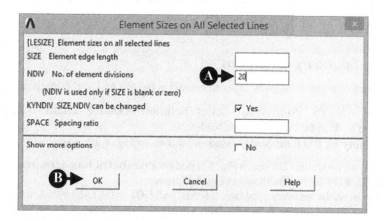

Fig. 4.17 Element Sizes on All Selected Lines window.

From the **ANSYS Graphic Window**, the preview of the divided line is available, as shown in Fig. 4.18, but the line is not really divided at this stage.

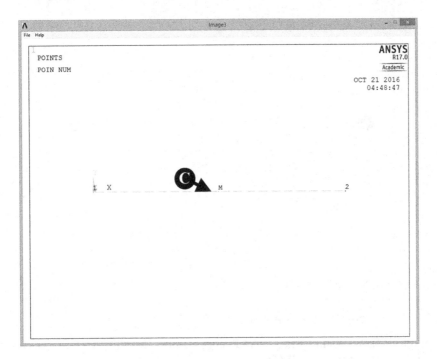

Fig. 4.18 Preview of the divided line.

From the ANSYS Main Menu, select **Preprocessor** → **Meshing** → **Mesh** → **Lines**. The **Mesh Lines** window, shown in Fig. 4.19, opens.

1. Click [C] the line shown in the **ANSYS Graphic Window**, as seen in Fig. 4.18, and then the [D] **OK** button to finish dividing the line.

4.2.3.7   Boundary conditions

The left end of nodes is fixed in order to constrain the left end of the cantilever beam.

From the ANSYS Main Menu, select **Solution** → **Define Loads** → **Apply** → **Structural** → **Displacement** → **On Nodes**.
   The **Apply U, ROT on Nodes** window, shown in Fig. 4.20, opens.

1. Pick [A] the node at the left end in Fig. 4.21 and click the [B] **OK** button. Then the window **Apply U, ROT on Nodes**, shown in Fig. 4.22, opens.
2. In order to set the boundary condition, select [C] **All DOF** in the **Lab2** box. In the **VALUE** box, input **0**, and then click the [E] **OK** button. After these steps, the **ANSYS Graphic Window** is changed as shown in Fig. 4.23.

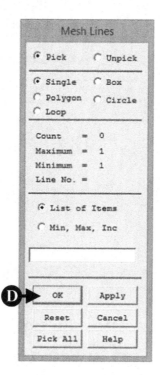

Fig. 4.19 Mesh Lines window.

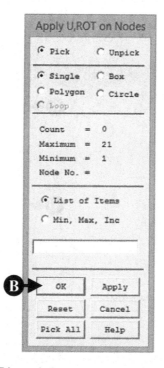

Fig. 4.20 Apply U, ROT on Lines window.

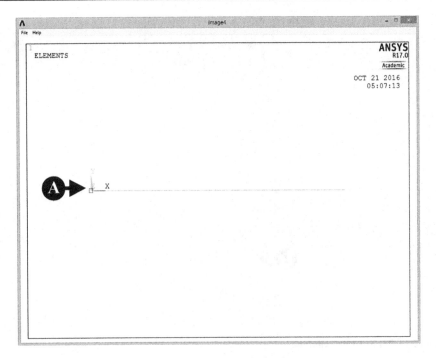

Fig. 4.21 ANSYS Graphic Window.

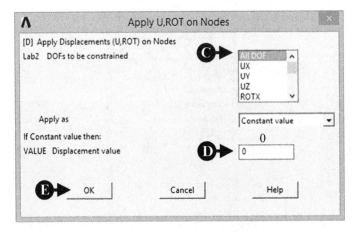

Fig. 4.22 Apply U, ROT on Lines window.

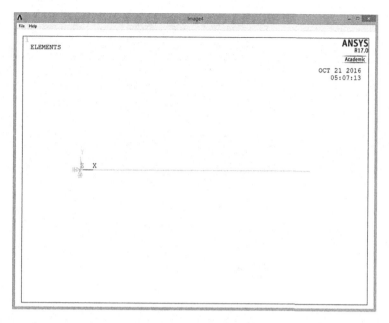

Fig. 4.23 Window after the boundary condition was set.

In order to obtain the vibration modes only in the *y* direction, the following boundary conditions are set.

From the ANSYS Main Menu, select **Solution** → **Define Loads** → **Apply** → **Structural** → **Displacement** → **On Lines**.
The window **Apply U, ROT on Nodes**, shown in Fig. 4.24, opens.

3. Pick [A] the line in Fig. 4.25 and click the [B] **OK** button. Then the window **Apply U, ROT on Lines**, shown in Fig. 4.26, opens. Select [C] **UZ** in the **Lab2** box. In the **VALUE** box, input **0**, and then click the [E] **Apply** button. After these steps, the **ANSYS Graphic Window** is changed as shown in Fig. 4.27.
4. Pick [A] the line again as indicated in Fig. 4.25, and click the [B] **OK** button as shown in Fig. 4.28. Then the window **Apply U, ROT on Lines**, shown in Fig. 4.29, opens again. Select [C] **ROTX** in the **Lab2** box. In the **VALUE** box input **0**, and then click the [E] **OK** button. After these steps, the **ANSYS Graphic Window** is changed as shown in Fig. 4.30.

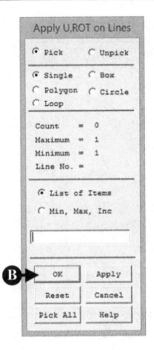

Fig. 4.24 Apply U, ROT on Lines window.

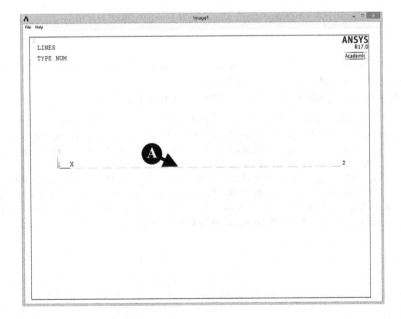

Fig. 4.25 ANSYS Graphic Window.

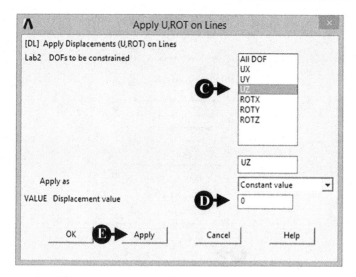

Fig. 4.26 Apply U, ROT on Lines window.

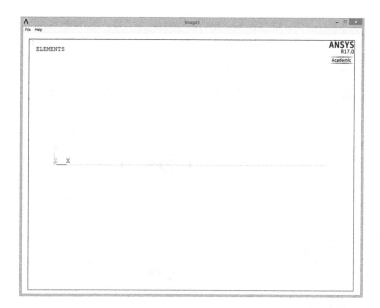

Fig. 4.27 Window after the boundary condition was set.

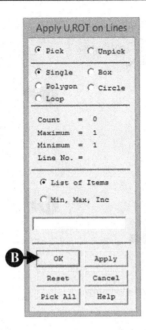

Fig. 4.28 **Apply U, ROT on Lines** window.

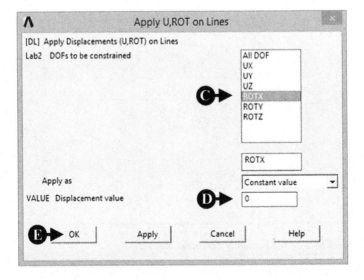

Fig. 4.29 **Apply U, ROT on Lines** window.

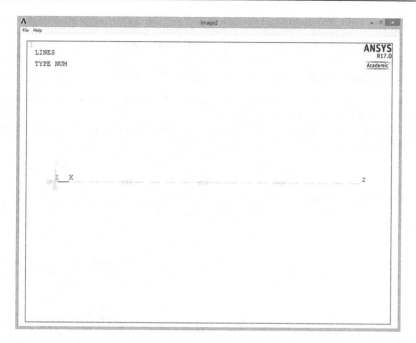

Fig. 4.30 Window after the boundary condition was set.

4.2.4 Execution of the analysis

4.2.4.1 Definition of the type of analysis

The following steps are used to define the type of analysis.

From the ANSYS Main Menu, select **Solution → Analysis Type → New Analysis**. The **New Analysis** window, shown in Fig. 4.31, opens.

1. Check [A] **Modal**, and then click the [B] **OK** button.

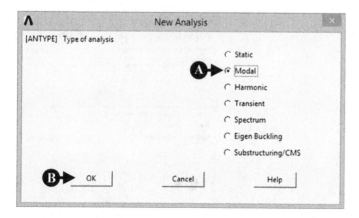

Fig. 4.31 New Analysis window.

In order to define the number of modes to extract, the following procedure is used.

From the ANSYS Main Menu, select **Solution → Analysis Type → Analysis Options**.

The **Modal Analysis** window, shown in Fig. 4.32, opens.

1. Check [C] **Supernode** of **MODOPT** and input [D] **3** in the **No. of modes to extract** box, then click the [E] **OK** button.
2. Then, the window **Subspace Modal Analysis**, shown in Fig. 4.33 opens. Input [F] **10000** in the **FREQE** box and click the [G] **OK** button.

Fig. 4.32 **Modal Analysis** window.

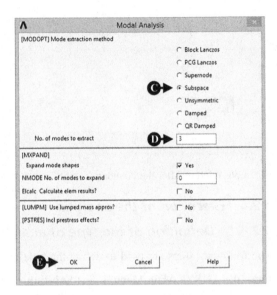

Fig. 4.33 **Subspace Modal Analysis** window.

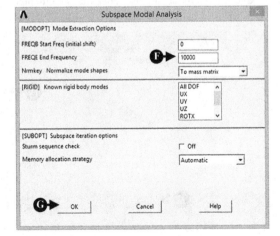

4.2.4.2   Execute calculation

From the ANSYS Main Menu, select **Solution** → **Solve** → **Current LS**.
The **Solve Current Load Step** window, shown in Fig. 4.34, opens.

1. Click the [A] **OK** button to initiate calculation. When the **Note** window, shown in Fig. 4.35, appears, then the calculation is finished.
2. Click the [B] **Close** button and to close the window. The **/STATUS Command** window, shown in Fig. 4.36, also opens but this window can be closed by clicking [C] the mark **X** at the upper right-hand corner of the window.

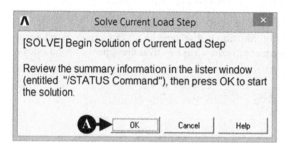

Fig. 4.34 Solve Current Load Step window.

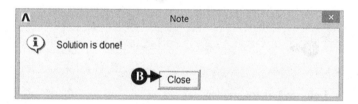

Fig. 4.35 Note window.

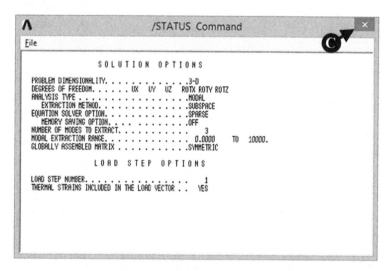

Fig. 4.36 /STATUS Command window.

4.2.5   Postprocessing

4.2.5.1   Read the calculated results of the first mode of vibration

From the ANSYS Main Menu, select **General Postproc** → **Read Results** → **First Set**.

4.2.5.2   Plot the calculated results

From the ANSYS Main Menu, select **General Postproc** → **Plot Results** →
Deformed Shape.

The window **Plot Deformed Shape**, shown in Fig. 4.37, opens.

1. Select [A] **Def + Undeformed** and click [B] **OK**.
2. Calculated result for the first mode of vibration is displayed (ANSYS Graphic Window) as
   shown in Fig. 4.38. The resonant frequency is shown as FRQE at the upper left-hand side on
   the window.

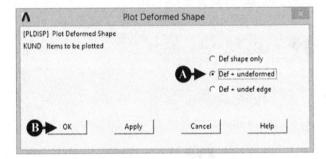

Fig. 4.37 Plot Deformed Shape window.

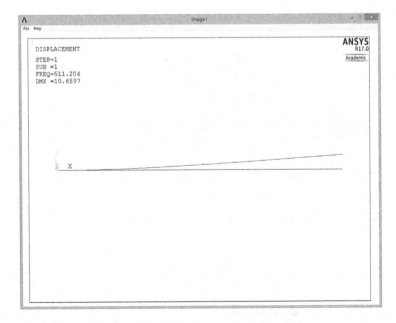

Fig. 4.38 Window for calculated result (the first mode of vibration).

4.2.5.3 Read the calculated results from the second and third modes of vibration

From the ANSYS Main Menu, select **General Postproc → Read Results → Next Set**.

Follow the same steps outlined in Section 4.2.5.2 and calculate the results for the second and third modes of vibration. The results are plotted in Figs 4.39 and 4.40.

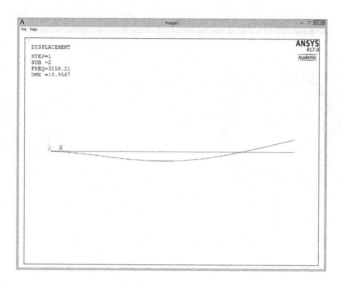

Fig. 4.39 Window for calculated result (the second mode of vibration).

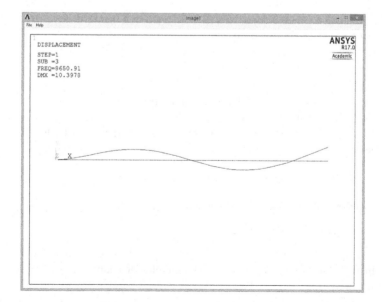

Fig. 4.40 Window for calculated result (the third mode of vibration).

4.3   Modal analysis of a suspension for hard disc drive

4.3.1   Problem description

A suspension of HDD has many resonant frequencies with various vibration modes
and it is said that the vibration mode with large radial displacement causes the tracking
error. So the suspension has to be operated with frequencies of less than this resonant
frequency.

Obtain the resonant frequencies and determine the vibration mode with large radial
displacement of the HDD suspension as shown in Fig. 4.41.

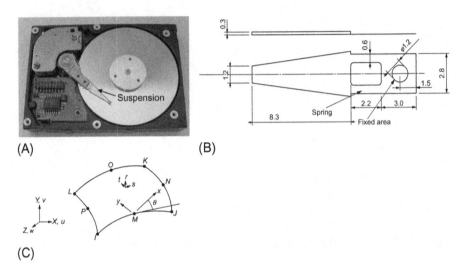

(C)

Fig. 4.41 Example 4.2 Mode analysis for a HDD suspension (A) hard disk drive, (B) drawing of
analysed suspension, and (C) shell element.
(From http://www-5.ibm.com/es/press/fotos/informaticapersonal/i/hdd.jpg.)

Material: Steel.
Young's modulus, $E = 206$ GPa, Poisson's ratio $\nu = 0.3$.
Density $\rho = 7.8 \times 10^3$ kg/m^3.
Boundary condition: All freedoms are constrained at the edge of a hole formed in
the suspension.

4.3.2   Create a model for analysis

4.3.2.1   Element type selection

In this example, the three-dimensional elastic shell is selected for calculations as
shown in Fig. 4.41. Shell element is very suitable for analysing the characteristics
of thin material.

From the ANSYS Main Menu, select **Preprocessor → Element Type → Add/Edit/Delete**.

Then the **Element Types** window, as shown in Fig. 4.42, is opened.

1. Click [A] **add**. Then the **Library of Element Types** window as shown in Fig. 4.43 opens.
2. Select [B] **Shell** in the **Library of Element Types** table, and then select [C] **3D 8node 281**.
3. Ensure the **Element type reference number** is set to **1** and click the [D] **OK** button. Then the window **Library of Element Types** is closed.
4. Click the [E] **Close** button in the window of Fig. 4.44.

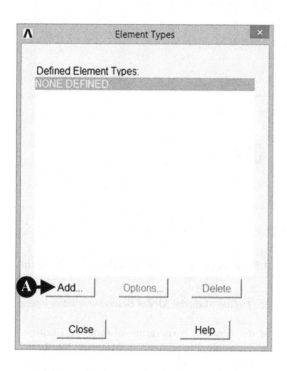

Fig. 4.42 **Element Types** window.

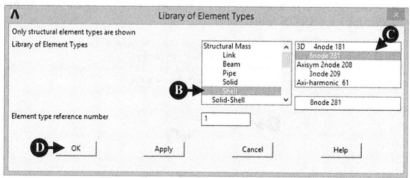

Fig. 4.43 **Library of Element Types** window.

Fig. 4.44 Element Types window.

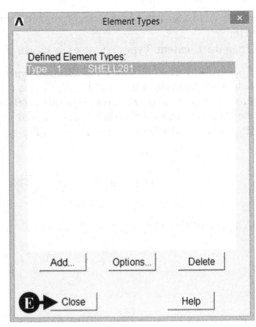

4.3.2.2 Material properties

This section describes the procedure of defining the material properties of shell element.

From the ANSYS Main Menu, select **Preprocessor → Material Props → Material Models**.

1. Click the above buttons in order and the **Define Material Model Behavior** window opens, as shown in Fig. 4.9.
2. Double-click the following terms in the window: **Structural → Linear → Elastic → Isotropic**. Then the window **Linear Isotropic Properties for Material Number 1**, Fig. 4.45, opens.
3. Input Young's Modulus of **206e9** to the [A] **EX** box and Poisson ratio of **0.3** to the [B] **PRXY** box. Then click the [C] **OK** button.

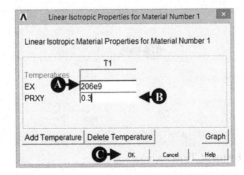

Fig. 4.45 Linear Isotropic Properties for Material Number 1 window.

Next, define the value of density of material.

1. Double-click the term **Density**, and the window **Density for Material Number 1**, Fig. 4.46, opens.
2. Input the value of Density, **7800**, to the [D] **DENS** box and click the [E] **OK** button. Finally close the window **Define Material Model Behavior** by clicking the **X** mark at the upper right end.

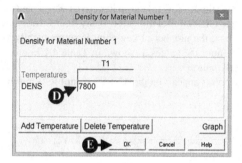

Fig. 4.46 Density for Material Number 1 window.

4.3.2.3 Sections for shell element

From the ANSYS Main Menu, select **Preprocessor → Sections → Shell → Lay-up → Add/Edit**.

1. The window **Create and Modify Shell Sections**, shown in Fig. 4.47, opens. Input the thickness of Shell section, **5.e-5**, to [A] **Thickness** and click the [B] **OK** button to close the window.

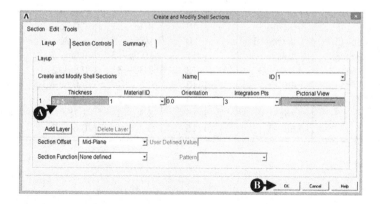

Fig. 4.47 Create and Modify Shell Sections window.

4.3.2.4   Create keypoints

To draw a suspension for analysis, the method using keypoints on the window are described in this section.

From the ANSYS Main Menu, select **Preprocessor → Modeling → Create → Keypoints → In Active CS**.

1. The window **Create Keypoints in Active Coordinate System**, shown in Fig. 4.48, opens.
2. Input [A] **0, -0.6e-3, 0** to the **X, Y, Z Location in active CS** box, and then click the [B] **Apply** button. Do not click the **OK** button at this stage. If any figure is not inputted into the [C] **NPT Key Point number** box, the number of key point is automatically assigned in order.
3. In the same window, input the values as shown in Table 4.1 in order. When all values are inputted, click the **OK** button.
4. Then all inputted keypoints appears on the **ANSYS Graphic Window** as shown in Fig. 4.49.

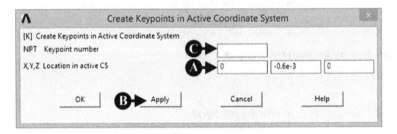

Fig. 4.48 Create Keypoint in Active Coordinate System window.

Table 4.1 X, Y, and Z Coordinates of Key Points for Suspension

KP No.	X	Y	Z	KP No.	X	Y	Z
1	0	$-0.6e-3$		9	$8.3e-3$	$-0.8e-3$	
2	$8.3e-3$	$-2.15e-3$		10	$10.5e-3$	$-0.8e-3$	
3	$8.3e-3$	$-1.4e-3$		11	$10.5e-3$	$0.8e-3$	
4	$13.5e-3$	$-1.4e-3$		12	$8.3e-3$	$0.8e-3$	
5	$13.5e-3$	$1.4e-3$		13	0	$-0.6e-3$	$0.3e-3$
6	$8.3e-3$	$1.4e-3$		14	$8.3e-3$	$-2.15e-3$	$0.3e-3$
7	$8.3e-3$	$2.15e-3$		15	$8.3e-3$	$2.15e-3$	$0.3e-3$
8	0	$0.6e-3$		16	0	$0.6e-3$	$0.3e-3$

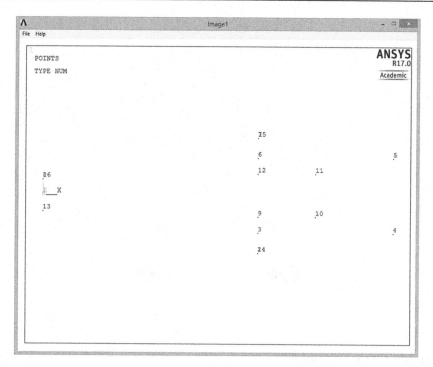

Fig. 4.49 ANSYS Graphic Window.

4.3.2.5 Create areas for suspension

Areas are created from keypoints by performing the following steps.

From the ANSYS Main Menu, select **Preprocessor** → **Modeling** → **Create** → **Area** → **Arbitrary** → **Through KPs**.

1. The **Create Area thru KPs** window, shown in Fig. 4.50, opens.
2. Pick the keypoints [A] **9, 10, 11,** and **12** in Fig. 4.51 in order and click the [B] **Apply** button in Fig. 4.50. An area is created on the window as shown in Fig. 4.51.
3. By performing the same steps, other areas are made on the window. Click the keypoints listed in Table 4.2 and make other areas. When you make areas No. 3 and 4, you have to rotate the drawing of suspension, using **PlotCtrls—Pan Zoom Rotate** in the Utility Menu.
4. When all areas are made on the window, click the [C] **OK** button in Fig. 4.50. Then the drawing of the suspension in Fig. 4.52 appears.
5. From the ANSYS Main Menu, select **Preprocessor** → **Modeling** → **Create** → **Area** → **Circle** → **Solid Circle**.

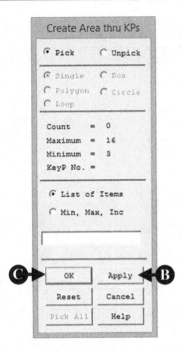

Fig. 4.50 **Create Area thru KPs** window.

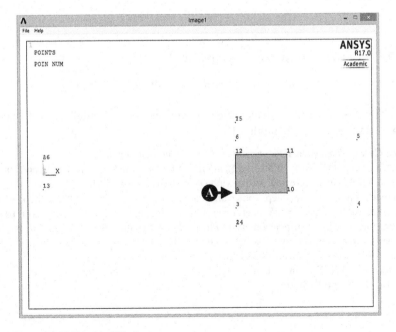

Fig. 4.51 ANSYS Graphic Window.

Table 4.2 **Keypoint Numbers for Making Areas of a Suspension**

Area No.	Keypoint Number
1	9, 10, 11, 12
2	1, 2, 3, 4, 5, 6, 7, 8
3	1, 2, 14, 13
4	8, 7, 15, 16

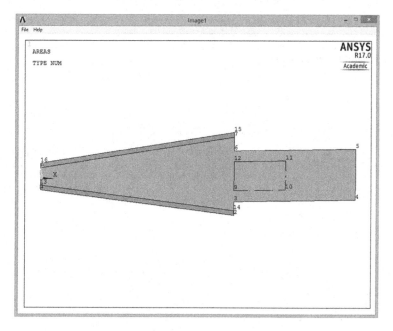

Fig. 4.52 ANSYS Graphic Window.

The **Solid Circle Area** window, shown in Fig. 4.53, opens. Input [D] the values of **12.0e-3**, **0**, and **0.6e-3** to the **X**, **Y**, and **Radius** boxes as shown in Fig. 4.53, respectively, and click the [E] **OK** button. Then the solid circle is made in the drawing of suspension as shown in Fig. 4.54.

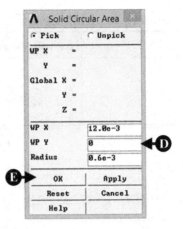

Fig. 4.53 Solid Circular Area window.

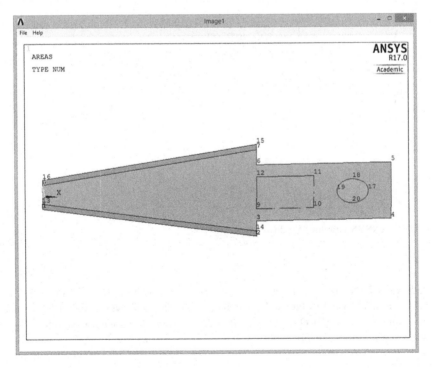

Fig. 4.54 ANSYS Graphic Window.

4.3.2.6 Boolean operation

In order to make a spring region and the fixed region of the suspension, the rectangular and circle areas in the suspension are subtracted by Boolean operation.

From the ANSYS Main Menu, select **Preprocessor** → **Modeling** → **Operate** → **Booleans** → **Subtract** → **Areas**.

1. The **Subtract Areas** window, shown in Fig. 4.55, opens.
2. Click [A] the area of the suspension in the **ANSYS Graphic Window** as shown in Fig. 4.56 and click the [B] **OK** button, as in Fig. 4.55. Then click the [C] rectangular and [D] circular areas as shown in Fig. 4.57 and click the [B] **OK** button, as in Fig. 4.55. The drawing of the suspension appears as shown in Fig. 4.58.
3. In analysis, all areas have to be glued.

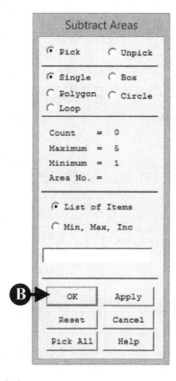

Fig. 4.55 Subtract Areas window.

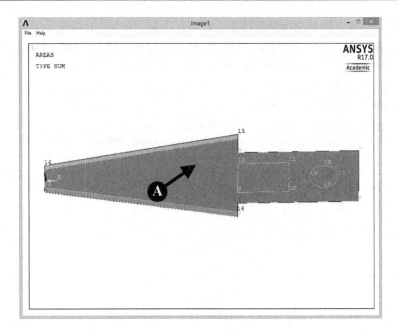

Fig. 4.56 ANSYS Graphic Window.

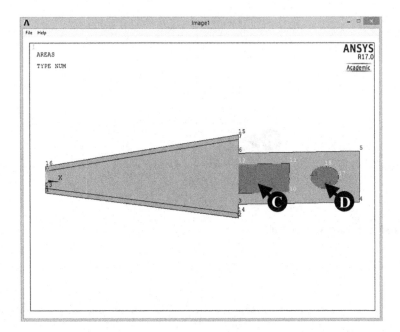

Fig. 4.57 ANSYS Graphic Window.

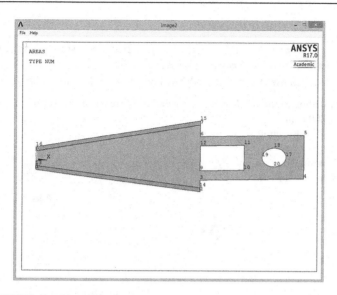

Fig. 4.58 ANSYS Graphic Window.

From the ANSYS Main Menu, select **Preprocessor → Modeling → Operate → Booleans → Glue → Areas**.

The **Glue Areas** window, shown in Fig. 4.59, opens. Click the [E] **Pick All** button.

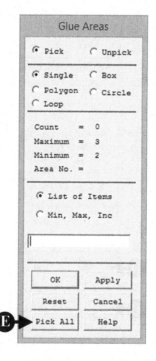

Fig. 4.59 Glue Areas window.

4.3.2.7   Create mesh in areas

From the ANSYS Main Menu, select **Preprocessor** → **Meshing** → **Size Cntrls** → **Manual Size** → **Areas** → **All Areas**.
The window **Element Sizes on All Selected Areas**, shown in Fig. 4.60, opens.

1. Input [A] **0.0002** to the **SIZE** box. This means that areas are divided by meshes of edge length of 0.0002 m.
2. Click the [B] **OK** button and close the window.

Fig. 4.60 Element Sizes on All Selected Areas window.

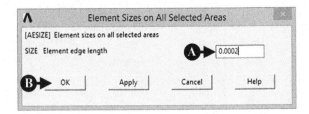

From the ANSYS Main Menu, select **Preprocessor** → **Meshing** → **Mesh** → **Areas** → **Free**.
The **Mesh Lines** window, shown in Fig. 4.61, opens.

1. Click [C] **Pick ALL** and then the [D] **OK** button to finish dividing the areas as shown in Fig. 4.62.

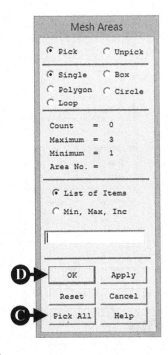

Fig. 4.61 Mesh Areas window.

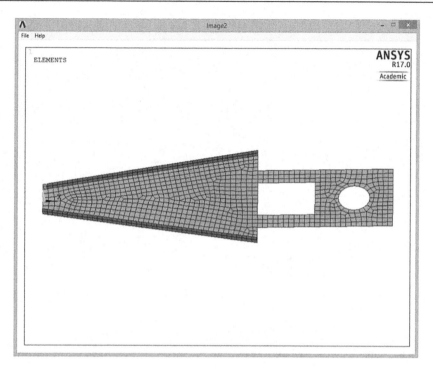

Fig. 4.62 ANSYS Graphic Window.

4.3.2.8 Boundary conditions

The suspension is fixed at the edge of the circle.

From the ANSYS Main Menu, select **Solution** → **Define Loads** → **Apply** → **Structural** → **Displacement** → **On Lines**.
   The window **Apply U, ROT on Nodes**, shown in Fig. 4.63, opens.

1. Pick [A] four lines included in the circle in Fig. 4.64 and click the [B] **OK** button in Fig. 4.63. Then the window **Apply U, ROT on Lines**, shown in Fig. 4.65, opens.
2. In order to set the boundary condition, select [C] **All DOF** in the **Lab2** box. Input [D] **0** to the **VALUE** box, and then click the [E] **OK** button. After these steps, the **ANSYS Graphic Window** is changed as seen in Fig. 4.66.

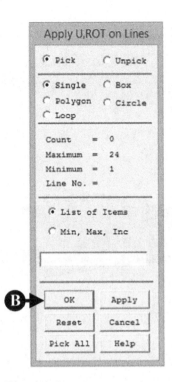

Fig. 4.63 Apply U, ROT on Lines window.

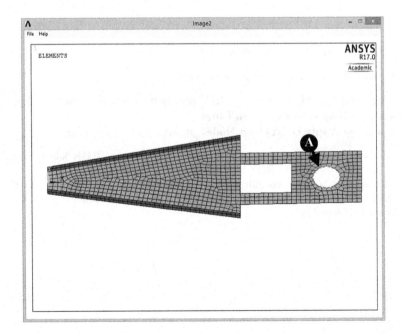

Fig. 4.64 ANSYS Graphic Window.

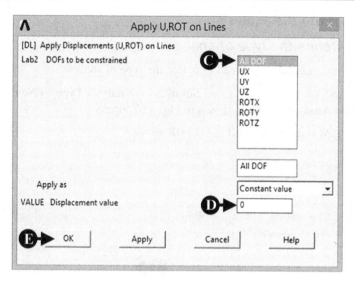

Fig. 4.65 Apply U, ROT on Lines window.

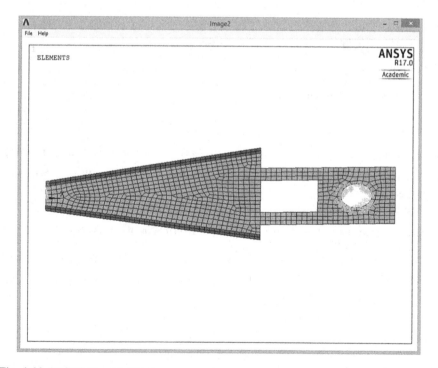

Fig. 4.66 ANSYS Graphic Window.

4.3.3 Execute the analysis

4.3.3.1 Define the type of analysis

The following steps are performed to define the type of analysis.

From the ANSYS Main Menu, select **Solution** → **Analysis Type** → **New Analysis**. The **New Analysis** window, shown in Fig. 4.67, opens.

1. Check [A] **Modal**, and then click the [B] **OK** button.

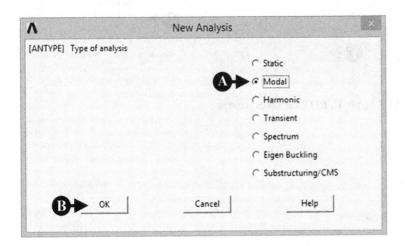

Fig. 4.67 **New Analysis** window.

In order to define the number of modes to extract, the following steps are performed.

From the ANSYS Main Menu, select **Solution** → **Analysis Type** → **Analysis Options**.
The **Modal Analysis** window, shown in Fig. 4.68, opens.

1. Check [C] **Subspace** of **MODOPT** and input [D] **10** in the **No. of modes to extract** box, then click the [E] **OK** button.
2. The window **Subspace Modal Analysis**, as shown in Fig. 4.69, then opens. Input [F] **20000** in the **FREQE** box and click the [G] **OK** button.

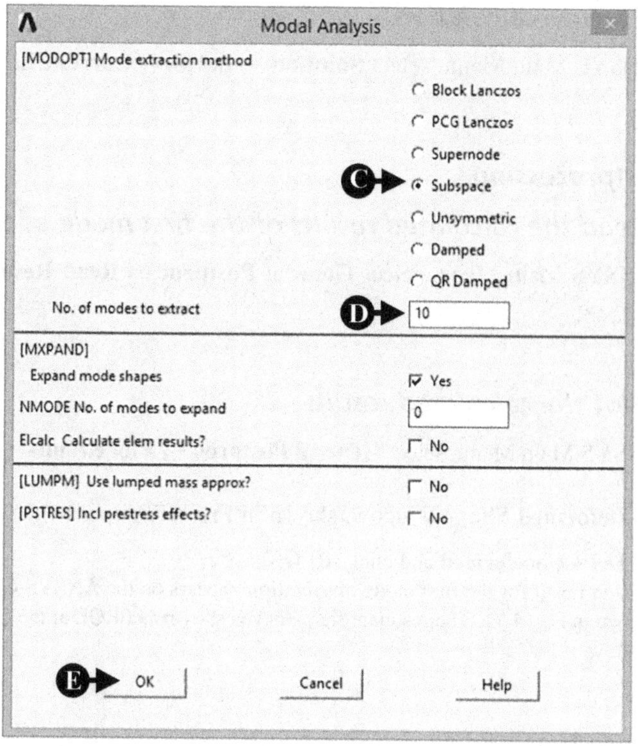

Fig. 4.68 Modal Analysis window.

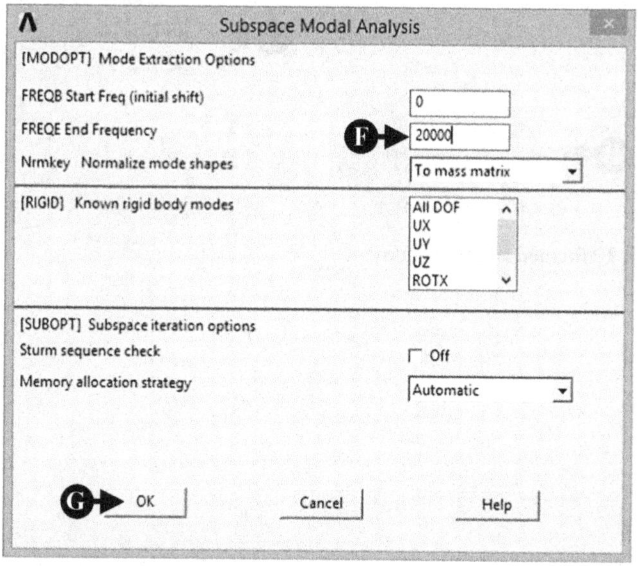

Fig. 4.69 Subspace Modal Analysis window.

4.3.3.2   Execute calculation

From the ANSYS Main Menu, select **Solution** → **Solve** → **Current LS**.

4.3.4   Postprocessing

4.3.4.1   Read the calculated results of the first mode of vibration

From the ANSYS Main Menu, select **General Postproc** → **Read Results** → **First Set**.

4.3.4.2   Plot the calculated results

From the ANSYS Main Menu, select **General Postproc** → **Plot Results** → **Deformed Shape**.

The **Plot Deformed Shape** window, shown in Fig. 4.70, opens.

1. Select [A] **Def + Undeformed** and click [B] **OK**.
2. The calculated result for the first mode of vibration appears on the **ANSYS Graphic Window** as shown in Fig. 4.71. The resonant frequency is shown as FRQE at the upper left side on the window.

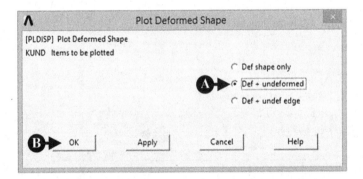

Fig. 4.70 Plot Deformed Shape window.

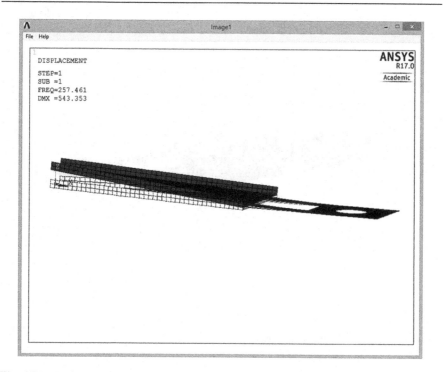

Fig. 4.71 The first mode of vibration.

4.3.4.3 Read the calculated results of higher modes of vibration

From the ANSYS Main Menu, select **General Postproc** → **Read Results** → **Next Set**.

Perform the same steps indicated in Section 4.2.4.2 and calculated results from the second mode to the sixth mode of vibration are displayed on the windows as shown in Figs 4.72–4.76.

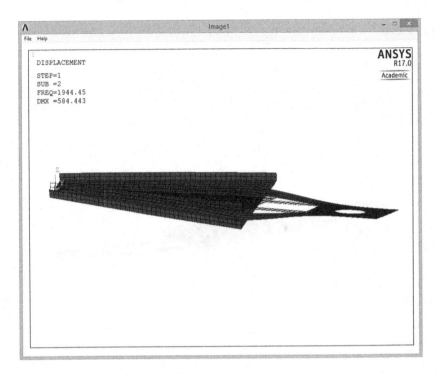

Fig. 4.72 The second mode of vibration.

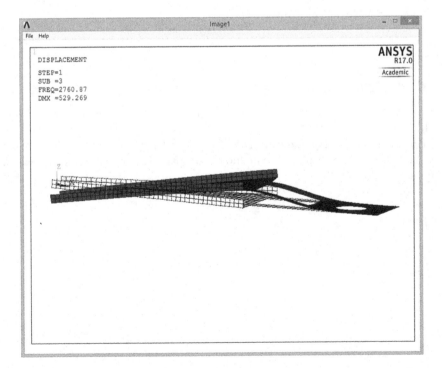

Fig. 4.73 The third mode of vibration.

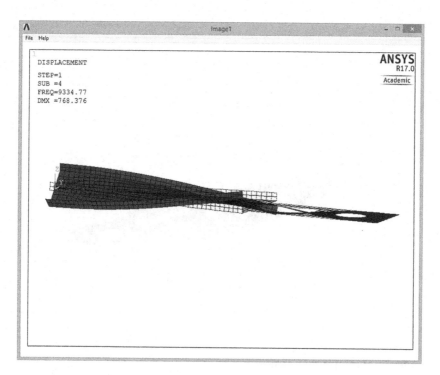

Fig. 4.74 The fourth mode of vibration.

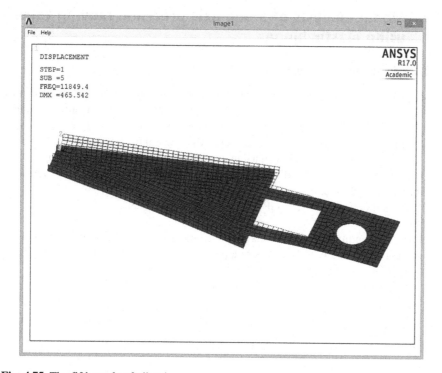

Fig. 4.75 The fifth mode of vibration.

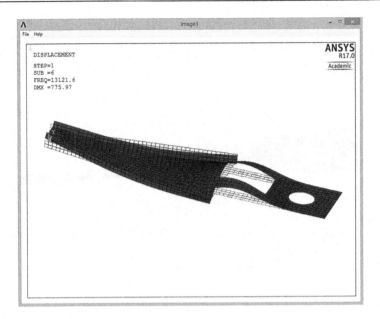

Fig. 4.76 The sixth mode of vibration.

4.4 Modal analysis of a one axis precision moving table using elastic hinges

4.4.1 Problem description

A one-axis table using elastic hinges has been often used in various precision equipment, and the position of a table is usually controlled at nanometer order accuracy using a piezoelectric actuator or a voice coil motor. Therefore, it is necessary to confirm the resonant frequency in order to determine the controllable frequency region.

Obtain the resonant frequency of a one-axis moving table using elastic hinges when the bottom of the table is fixed and a piezoelectric actuator is selected as an actuator.

Material: Steel, thickness of the table: 5 mm.

Young's modulus, $E = 206$ GPa, Poisson's ratio $\nu = 0.3$.

Density $\rho = 7.8 \times 10^3$ kg/m^3.

Boundary condition: All freedoms are constrained at the bottom of the table and the region A indicated in Fig. 4.77, where a piezoelectric actuator is glued.

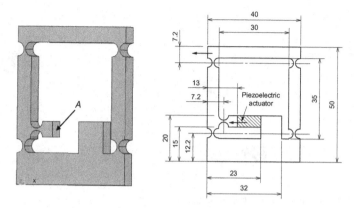

Fig. 4.77 Example 4.3: One axis moving table using elastic hinges.

4.4.2 Create a model for analysis

4.4.2.1 Select element type

In this example, the solid element is selected to analyse the resonant frequency of the moving table.

From the ANSYS Main Menu, select **Preprocessor** → **Element Type** → **Add/Edit/ Delete**.

Then the **Element Types** window, as shown in Fig. 4.78, opens.

1. Click the [A] **Add** button. Then **Library of Element Types** window as shown in Fig. 4.79 opens.

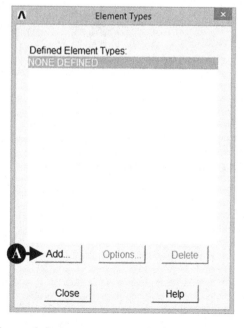

Fig. 4.78 Element Types window.

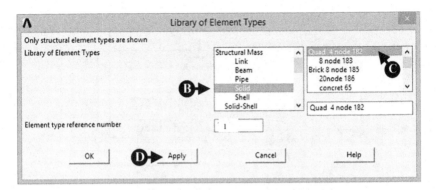

Fig. 4.79 Library of Element Types window.

2. Select [B] **Solid** in the **Library of Element Types** table, and then select [C] **Quad 4node 182**. Ensure the **Element type reference number** is set to **1** and click [D] **Apply**. Next, select [E] **Brick 8node 185** as shown in Fig. 4.80.
3. Ensure the **Element type reference number** is set to **2** and click the [F] **OK** button. Then the **Library of Element Types** window is closed.
4. Click the [G] **Close** button in the window of Fig. 4.81 where the names of selected elements are indicated.

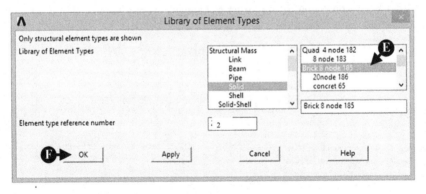

Fig. 4.80 Library of Element Types window.

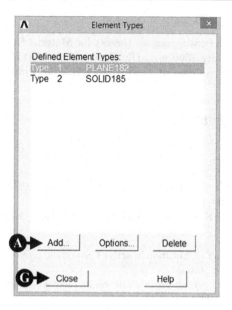

Fig. 4.81 Element Types window.

4.4.2.2 Material properties

This section describes the procedure of defining the material properties of solid element.

From the ANSYS Main Menu, select **Preprocessor** → **Material Props** → **Material Models**.

1. Click the above buttons in order and the window **Define Material Model Behavior** opens as shown in Fig. 4.82.
2. Double-click the following terms in the window: **Structural** → **Linear** → **Elastic** → **Isotropic**. Then the window **Linear Isotropic Properties for Material Number 1** opens.
3. Input Young's Modulus of **206e9** to the **EX** box and Poisson ratio of **0.3** to the **PRXY** box.

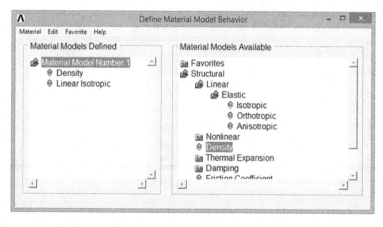

Fig. 4.82 Define Material Model Behavior window.

Next, define the value of density of material.

1. Double-click the term **Density**, and the **Density for Material Number 1** window opens.
2. Input the value of Density, **7800**, to the **DENS** box and click the **OK** button. Finally close the
   Define Material Model Behavior window by clicking the **X** mark at the upper right end.

4.4.2.3   Create keypoints

To draw the moving table for analysis, the method using keypoints on the window are
described in this section.

From   the   ANSYS   Main   Menu,   select   **Preprocessor** → **Modeling** →
Keypoints → **In Active CS**.

The window **Create Keypoints in Active Coordinate System**, shown in Fig. 4.83,
opens.

1. Input [A] **0, 0** to the **X, Y, Z Location in active CS** box, and then click the [B] **Apply** button.
   Do not click the **OK** button at this stage.
2. In the same window, input the values as shown in Table 4.3 in order. When all values are
   inputted, click the **OK** button.
3. Then all input keypoints appear on the **ANSYS Graphic Window**, as shown in Fig. 4.84.

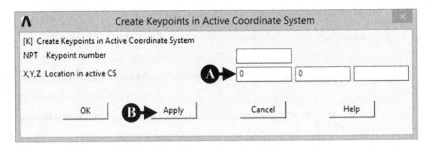

Fig. 4.83 Create Keypoints in Active Coordinate System window.

Table 4.3 Coordinates of KPs

KP No.	X	Y	KP No.	X	Y
1	0	0	9	0.005	0.015
2	0.04	0	10	0.013	0.015
3	0.04	0.05	11	0.013	0.02
4	0	0.05	12	0.005	0.02
5	0.005	0.01	13	0.023	0.01
6	0.035	0.01	14	0.032	0.01
7	0.035	0.045	15	0.032	0.02
8	0.005	0.045	16	0.023	0.02

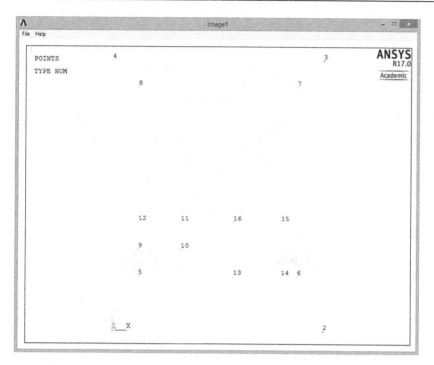

Fig. 4.84 ANSYS Graphic Window.

4.4.2.4 Create areas for the table

Areas are created from keypoints by carrying out the following steps.

From the ANSYS Main Menu, select **Preprocessor → Modeling → Create → Areas → Arbitrary → Through KPs**.
The window **Create Area thru KPs**, shown in Fig. 4.85, opens.

1. Pick the keypoints, **1**, **2**, **3**, and **4**, in order and click the [A] **Apply** button in Fig. 4.85. Then pick keypoints **5**, **6**, **7**, and **8**, and click the [B] **OK** button. Fig. 4.86 appears.

From the ANSYS Main Menu, select **Preprocessor → Modeling → Operate → Booleans → Subtract → Areas**.

2. Click the area of KP no. **1–4** and the **OK** button. Then click the area of KP no. **5–8** and the **OK** button. The drawing of the table appears as shown in Fig. 4.87.

From the ANSYS Main Menu, select **Preprocessor → Modeling → Create → Areas → Arbitrary → Through KPs**.

3. Pick the keypoints **9**, **10**, **11**, and **12** in order and click the [A] **Apply** button in Fig. 4.85. Then pick keypoints **13**, **14**, **15**, and **16** and click the [B] **OK** button. Fig. 4.88 appears.

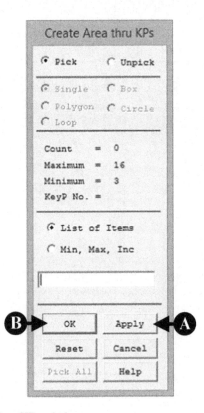

Fig. 4.85 Create Area thru KPs window.

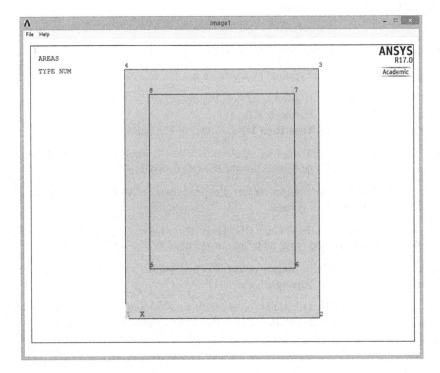

Fig. 4.86 ANSYS Graphic Window.

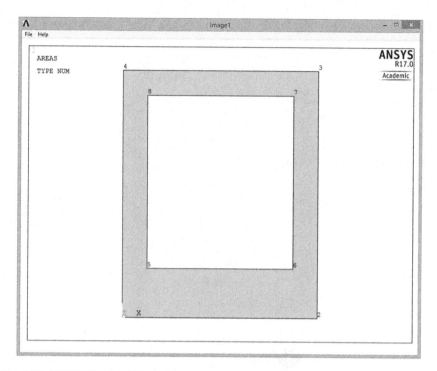

Fig. 4.87 ANSYS Graphic Window.

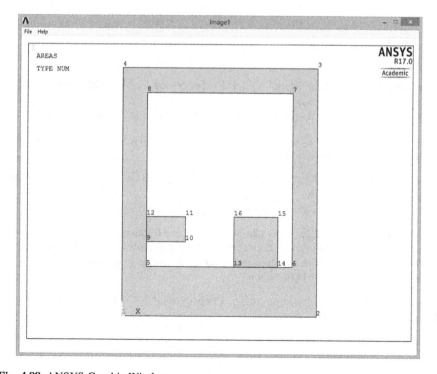

Fig. 4.88 ANSYS Graphic Window.

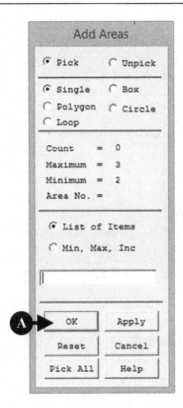

Fig. 4.89 Add Areas window.

From the ANSYS Main Menu, select **Preprocessor** → **Modeling** → **Operate** → **Booleans** → **Add** → **Areas**.

The window **Create Area thru KPs**, shown in Fig. 4.89, opens.

4. Pick three areas on the ANSYS graphic window and click the [A] **OK** button. Then three areas are added as shown in Fig. 4.90.

From the ANSYS Main Menu, select **Preprocessor** → **Modeling** → **Create** → **Area** → **Circle** → **Solid Circle**.

The **Solid Circle Area** window, shown in Fig. 4.91, opens.

5. Input the values of **0, 12.2e-3**, and **2.2e-3** to the **X, Y**, and **Radius** boxes as shown in Fig. 4.91 and click the [A] **Apply** button. Then continue to input the coordinates of the solid circle as shown in Table 4.4. When all values are inputted, the drawing of the table appears as shown in Fig. 4.92 Click the [B] **OK** button.

From the ANSYS Main Menu, select **Preprocessor** → **Modeling** → **Operate** → **Booleans** → **Subtract** → **Areas**.

1. Subtract all circular areas from the rectangular area by executing the above steps. Then Fig. 4.93 is displayed.

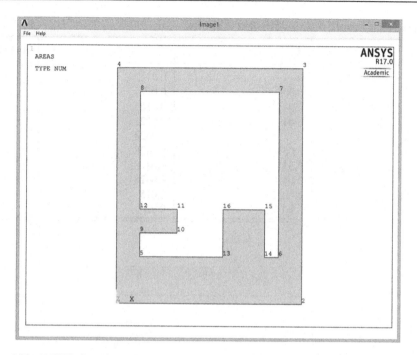

Fig. 4.90 ANSYS Graphic Window.

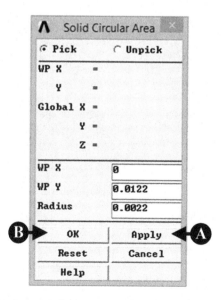

Fig. 4.91 Solid Circular Area window.

Table 4.4 **Coordinates of Solid Circles**

No.	X	Y	Radius
1	0	0.0122	0.0022
2	0.005	0.0122	
3	0.035	0.0122	
4	0.04	0.0122	
5	0	0.0428	
6	0.005	0.0428	
7	0.035	0.0428	
8	0.04	0.0428	
9	0.0072	0.015	
10	0.0072	0.02	

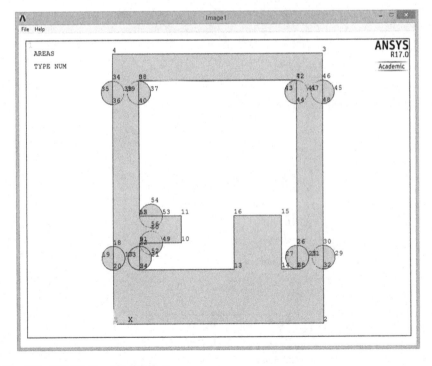

Fig. 4.92 ANSYS Graphic Window.

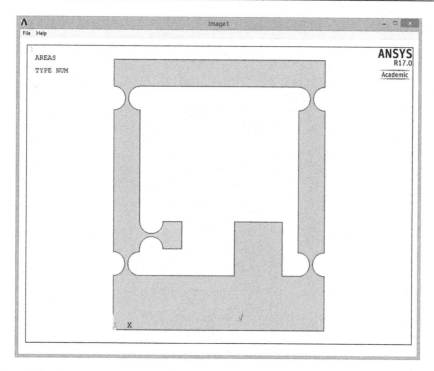

Fig. 4.93 ANSYS Graphic Window.

4.4.2.5 Create mesh in areas

From the ANSYS Main Menu, select **Preprocessor** → **Meshing** → **Mesh Tool**. The **Mesh Tool** window, shown in Fig. 4.94, opens.

1. Click [A] **Lines Set** and the window **Element Size on Picked Lines**, shown in Fig. 4.95, opens.
2. Click the [B] **Pick All** button and the window **Element Sizes on Picked Lines**, shown in Fig. 4.96, opens.
3. Input [C] **0.001** to the **SIZE** box and click the [D] **OK** button.
4. Click [E] **Mesh** of the **Mesh Tool** window Fig. 4.97, and then the **Mesh Areas** window, shown in Fig. 4.98, opens.
5. Pick the area of the table on the **ANSYS Graphic Window** and click the [F] **OK** button in Fig. 4.98. Then the meshed drawing of the table appears on the **ANSYS Graphic Window**, as shown in Fig. 4.99.

Next, by performing the following steps, the thickness of 5 mm and the mesh size are determined for the drawing of the table.

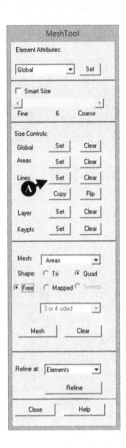

Fig. 4.94 **Mesh Tool** window.

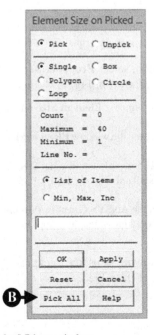

Fig. 4.95 **Element Size on Picked Lines** window.

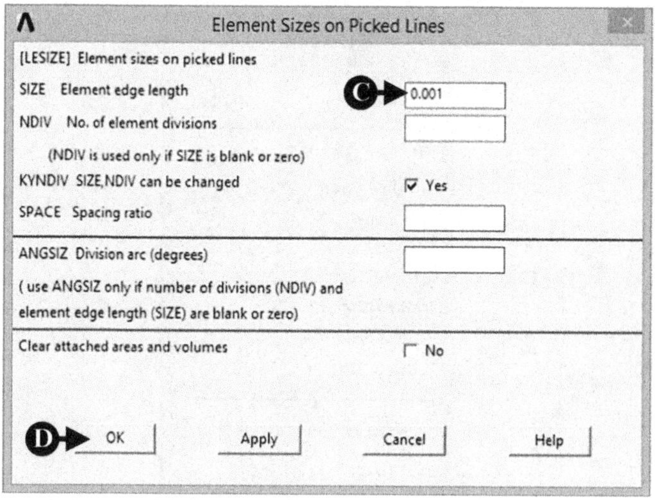

Fig. 4.96 Element Sizes on Picked Lines window.

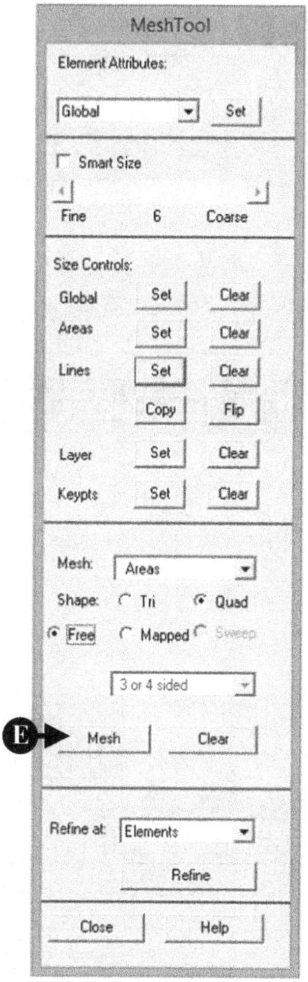

Fig. 4.97 Mesh Tool window.

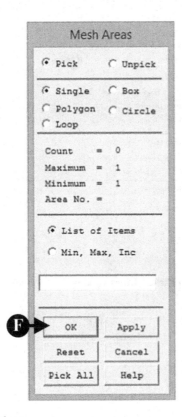

Fig. 4.98 **Mesh Areas** window.

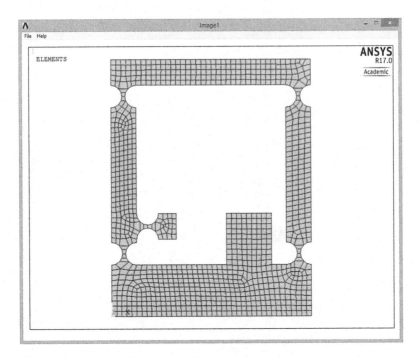

Fig. 4.99 ANSYS Graphic Window.

From the ANSYS Main Menu, select **Preprocessor** → **Modeling** → **Operate** → **Extrude** → **Elem Ext Opts**.

The window **Element Extrusion Options**, shown in Fig. 4.100, opens.

1. Input [A] **5** to the **VAL1** box. This means that the number of element division is 5 in the thickness direction. Then, click the [B] **OK** button.

From the ANSYS Main Menu, select **Preprocessor** → **Modeling** → **Operate** → **Extrude** → **Areas** → **By XYZ Offset**.

The window **Extrude Areas by Offset**, shown in Fig. 4.101, opens.

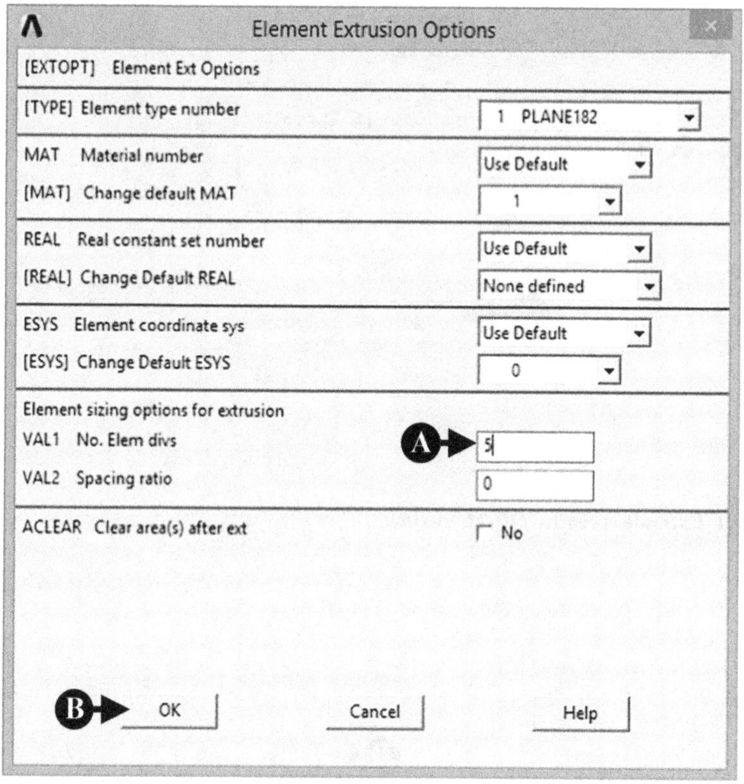

Fig. 4.100 Element Extrusion Options window.

1. Pick the area of the table on the **ANSYS Graphic Window** and click the [A] **OK** button.
2. In The window **Extrude Areas by XYZ Offset**, shown in Fig. 4.102, opens. Input [B] **0, 0, 0.005** to the **DX,DY,DZ** box and click the [C] **OK** button. Then, the drawing of the table meshed in the thickness direction appears as shown in Fig. 4.103.

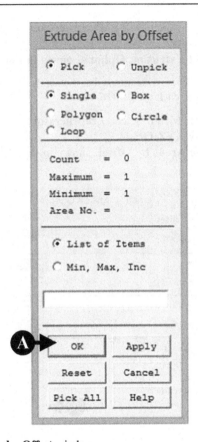

Fig. 4.101 Extrude Area by Offset window.

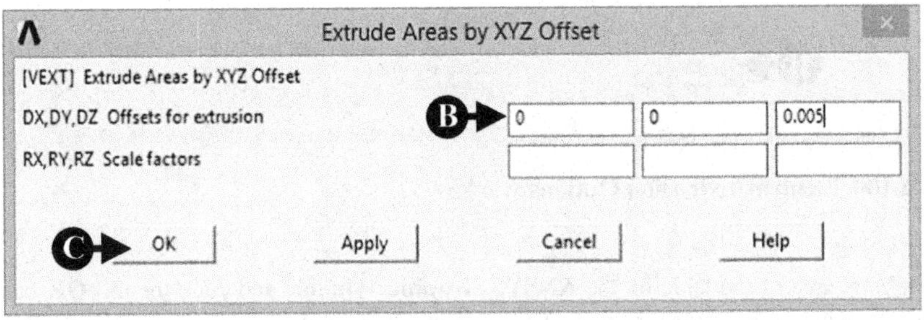

Fig. 4.102 Extrude Areas by XYZ Offset window.

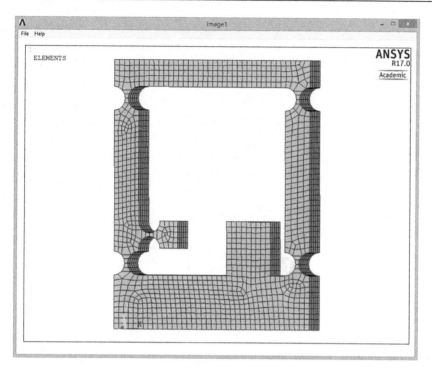

Fig. 4.103 ANSYS Graphic Window.

4.4.2.6 Boundary conditions

The table is fixed at both the bottom and the region A of the table.

From the ANSYS Main Menu, select **Solution → Define Loads → Apply → Structural → Displacement → On Areas**.
   The **Apply U, ROT on Areas** window, shown in Fig. 4.104, opens.

1. Pick [A] the side wall for a piezoelectric actuator in Fig. 4.105 and [B] the bottom of the table. Then click the [C] **OK** button and the window **Apply U, ROT on Areas**, shown in Fig. 4.106, opens.
2. Select [D] **All DOF** in the **Lab2** box and input [E] **0** to the **VALUE** box, then, click the [F] **OK** button. After these steps, the **ANSYS Graphic Window** is changed, as seen in Fig. 4.107.

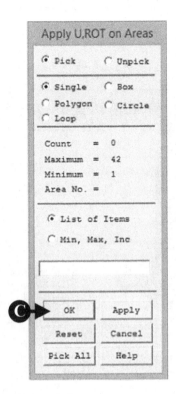

Fig. 4.104 Apply U, ROT on Areas window.

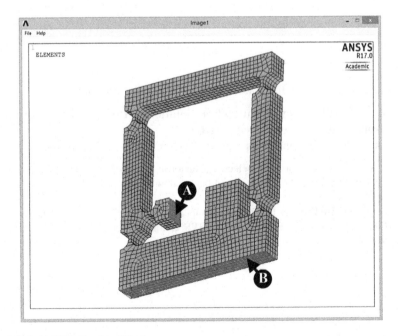

Fig. 4.105 ANSYS Graphic Window.

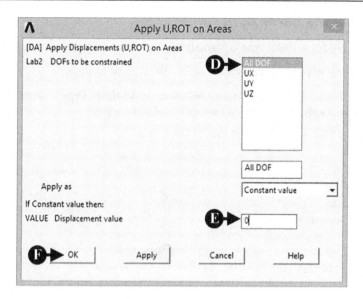

Fig. 4.106 Apply U, ROT on Areas window.

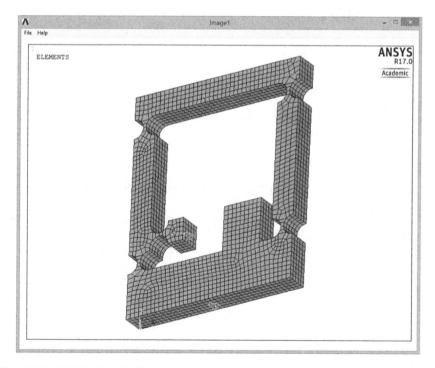

Fig. 4.107 ANSYS Graphic Window.

4.4.3   Execute the analysis

4.4.3.1   Define the type of analysis

The following steps are performed to define the type of analysis.

From the ANSYS Main Menu, select **Solution** → **Analysis Type** → **New Analysis**.
The **New Analysis** window, shown in Fig. 4.108, opens.

1. Check [A] **Modal**, and then click the [B] **OK** button.

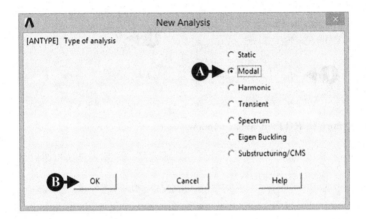

Fig. 4.108 New Analysis window.

In order to define the number of modes to extract, the following steps are performed.
From the ANSYS Main Menu, select **Solution** → **Analysis Type** → **Analysis Options**.
The **Modal Analysis** window, shown in Fig. 4.109, opens.

1. Check [A] **Subspace** of **MODOPT** and input [B] **3** in the **No. of modes to extract** box and click the [C] **OK** button.
2. Then, the **Subspace Modal Analysis** window as shown in Fig. 4.110 opens. Input [D] **5000** in the **FREQE** box and click the [E] **OK** button.

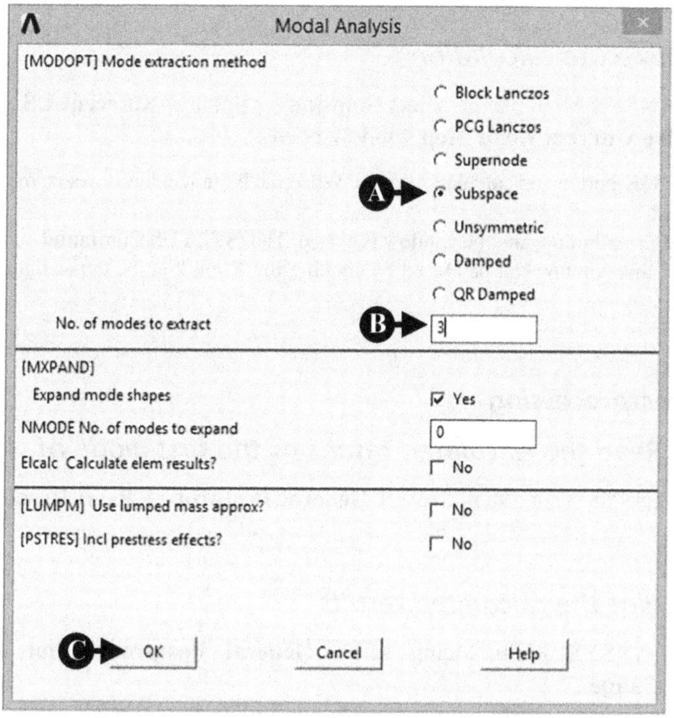

Fig. 4.109 Modal Analysis window.

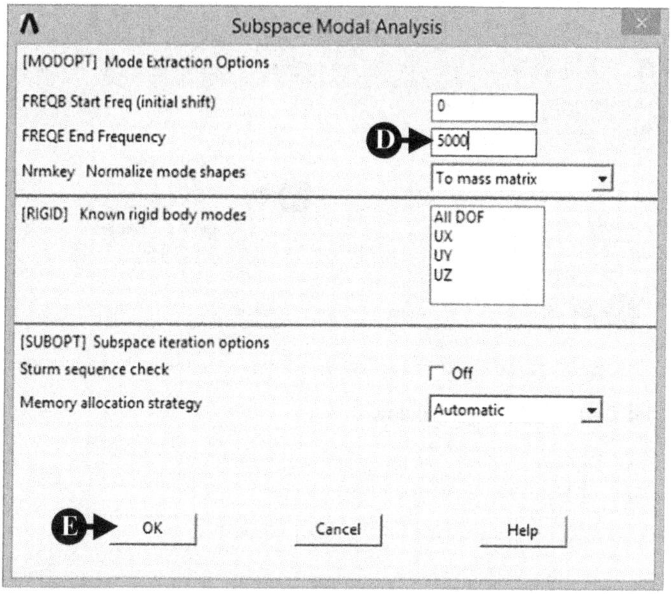

Fig. 4.110 Subspace Modal Analysis window.

4.4.3.2 Execute calculation

From the ANSYS Main Menu, select **Solution** → **Solve** → **Current LS**.
The **Solve Current Load Step** window opens.

1. Click the **OK** button and calculation starts. When the **Note** window appears, the calculation is finished.
2. Click the **Close** button, and the window is closed. The **/STATUS Command** window is also open, but this window can be closed by clicking the **X** mark at the upper right side of the window.

4.4.4 Postprocessing

4.4.4.1 Read the calculated results of the first mode of vibration

From the ANSYS Main Menu, select **General Postproc** → **Read Results** → **First Set**.

4.4.4.2 Plot the calculated results

From the ANSYS Main Menu, select **General Postproc** → **Plot Results** → **Deformed Shape**.
The **Plot Deformed Shape** window, shown in Fig. 4.111, opens.

1. Select [A] **Def + Undeformed** and click [B] **OK**.
2. The calculated result for the first mode of vibration appears on the **ANSYS Graphic Window**, as shown in Fig. 4.112.

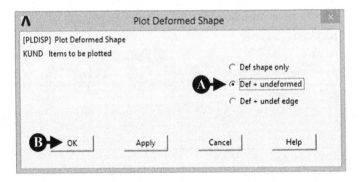

Fig. 4.111 Plot Deformed Shape window.

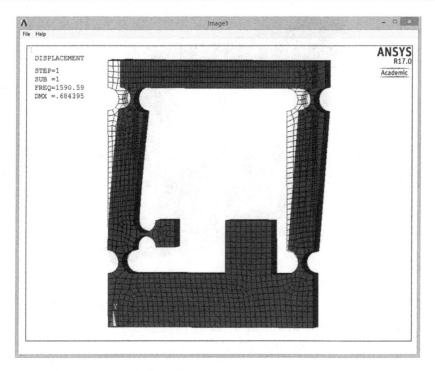

Fig. 4.112 ANSYS Graphic Window for the first mode.

4.4.4.3 Read the calculated results of the second and third modes of vibration

From the ANSYS Main Menu, select **General Postproc → Read Results → Next Set**.

Perform the same steps indicated in Section 4.4.4.2, and the calculated results for the higher modes of vibration are displayed on the windows as shown in Figs 4.113 and 4.114.

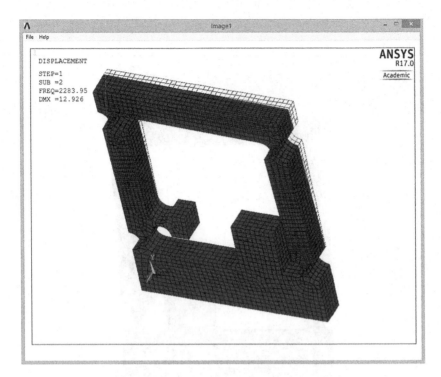

Fig. 4.113 ANSYS Graphic Window for the second mode.

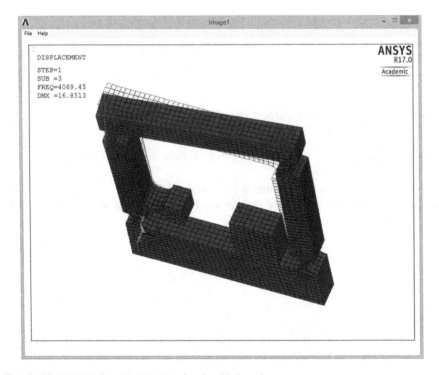

Fig. 4.114 ANSYS Graphic Window for the third mode.

4.4.4.4 Animate the vibration mode shape

In order to easily judge the vibration mode shape, the animation of mode shape can be used.

From the Utility Menu, select **PlotCtrls → Animate → Mode Shape**.
The **Animate Mode Shape** window, shown in Fig. 4.115, opens.

1. Input [A] **0.1** to the **Time delay** box and click the [B] **OK** button. Then the animation of the mode shape is indicated on the **ANSYS Graphic Window**.

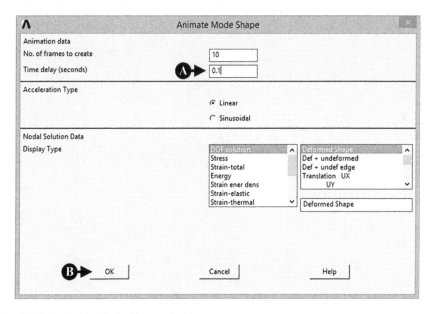

Fig. 4.115 Animate Mode Shape window.

Analysis for fluid dynamics

5

5.1 Introduction to FLUENT

Various fluids such as air and liquid are used as an operating fluid in a blower, a compressor and a pump. The shape of flow channel often determines the efficiency of these machines. In this chapter, the flow structures in a diffuser and the channel with a butterfly valve are examined by using FLUENT, which is a computational fluid dynamics (CFD) programme of Workbench. A diffuser is usually used for increasing the static pressure by reducing the fluid velocity and the diffuser can be easily found in a centrifugal pump, as shown in Fig. 5.1.

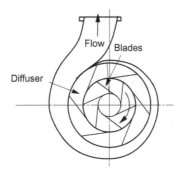

Fig. 5.1 Typical machines for fluid.

5.2 Analysis of flow structure in a diffuser

5.2.1 Problem description

Analyse the flow structure of an axisymmetric conical diffuser with diffuser angle $2\theta = 6$ degrees and expansion ratio $= 4$, as shown in Fig. 5.2.

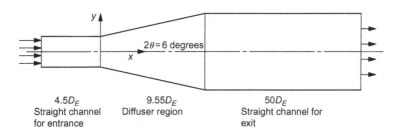

Fig. 5.2 Example 5.1, Axisymmetrical conical diffuser.

Engineering Analysis with ANSYS Software. https://doi.org/10.1016/B978-0-08-102164-4.00005-4

Shape of flow channel:

1. Diffuser shape is axisymmetric and conical, diffuser angle $2\theta = 6$ degrees, expansion ratio = 4;
2. Diameter of entrance of the diffuser $D_E = 0.2$ m;
3. Length of straight channel for entrance: $4.5\ D_E = 0.9$ m;
4. Length of diffuser region: $9.55\ D_E = 1.91$ m; and
5. Length of straight channel for exit: $50.0\ D_E = 10.0$ m.
   Operating fluid: Air (300 K)
   Flow field: Turbulence
   Velocity at the entrance: 20 m/s
   Reynolds Number: 2.54×10^5 (assumed to set the diameter of diffuser entrance to a representative length)

Boundary conditions:

1. Velocities in all directions are zero on all walls.
2. Pressure is equal to zero at the exit.
3. Velocity in the y direction is zero on the x axis.

5.2.2 Create a model for analysis using Workbench

5.2.2.1 Select kind of analysis

Double-click **Shortcut Workbench 17.0**.
   The window **Unsaved Project—Workbench**, Fig. 5.3, opens.

Fig. 5.3 Unsaved Project—Workbench window.

1. Drag the [A] **Fluid Flow (Fluent)** button into the [B] **Project schematic** window as shown in Fig. 5.4.

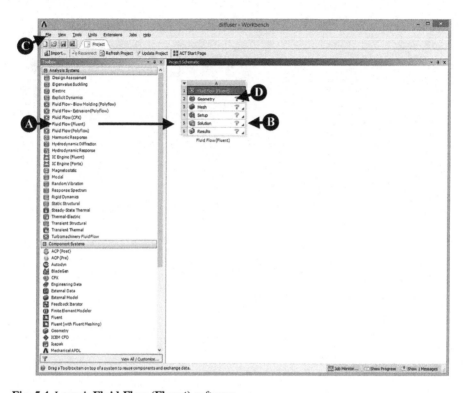

Fig. 5.4 Launch **Fluid Flow (Fluent)** software.

2. Save this project as **diffuser** by clicking [C] **File**.

5.2.2.2 Preparation for DesignModeler to create a geometry for analysis

1. Double-click [D] **Geometry** in the **Project Schematic** window. Then the **Graphics** window of DesignModeler opens, as shown in Fig. 5.5.
2. To change from a 3D screen to a 2D screen, place a cursor on the arrow of the z direction and then the arrow is highlighted. Click [A] the arrow in the z direction and then the screen is changed from 3D to 2D, as shown in Fig. 5.6.

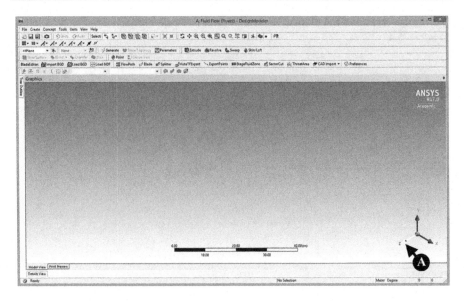

Fig. 5.5 DesignModeler window.

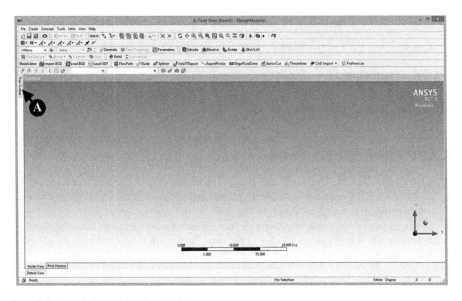

Fig. 5.6 2D window of DesignModeler.

3. Click the [A] **Tree Outline** button on the left-hand side of the **Graphics** window and select **Schetching** → **Settings**. Fig. 5.7 then opens.

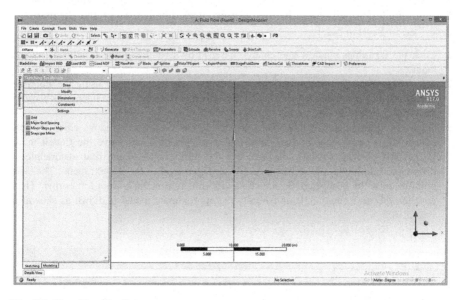

Fig. 5.7 Sketching Toolboxes.

4. Check the **Show in 2D** box in **Grid** and input **5 m** to the **Major Grid Spacing** box. Input **10** to the **Minor Steps per Major** box. The **Graphics** window is then changed, as shown in Fig. 5.8.

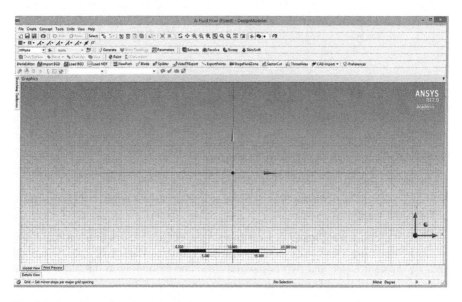

Fig. 5.8 Graphics window with grids.

5.2.2.3   Create a wire frame for diffuser

To draw a diffuser for analysis, the method using the drawing software, DesignModeler, is described in this section. The dimensions of the diffuser are described in Section 5.2.1.

1. Click the **Graphics** window and enlarge the grid by scrolling up the scrollwheel of the mouse.
2. Select **Sketching Toolboxes → Draw → Polyline** (see Fig. 5.9). By using the **Polyline** command, the upper half of the diffuser is described with approximate dimensions because the diffuser treated in this section is axisymmetric.
3. Click **Polyline** and the **Pencil** mark appears on the **Graphics** window. Click the left mouse button at the approximate position of the window and move the **Pencil** mark to the next position to draw the shape of the diffuser. When the first quadrangle is drawn, click the right mouse button and click **Open End** in the context menu. The start point to draw the quadrangle in the clockwise direction is the bottom left corner. Then describe the next quadrangle and finally the approximate model is drawn as shown in Fig. 5.10.

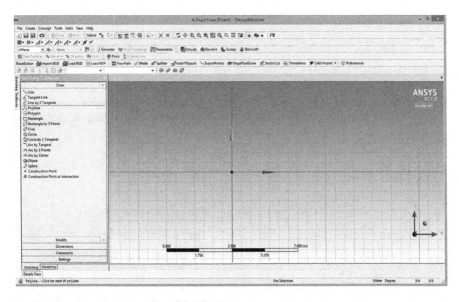

Fig. 5.9 Draw commands in **Sketching Toolboxes.**

4. Make constraints on the drawn lines by using the commands of **Vertical** and **Horizontal** in **Sketching Toolboxes**.

   Click **Sketching Toolboxes → Constraints → Vertical**. Click four vertical lines.
   Click **Sketching Toolboxes → Constraints → Horizontal**. Click three horizontal lines on the centre line and two horizontal lines corresponding to the diffuser wall. Then the shape of the diffuser is changed as seen in Fig. 5.11.

5. Use the commands of **dimensions** to input the dimensions of the diffuser.

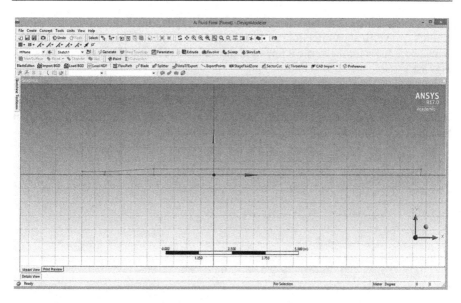

Fig. 5.10 Wire frame drawing for a diffuser described in the window.

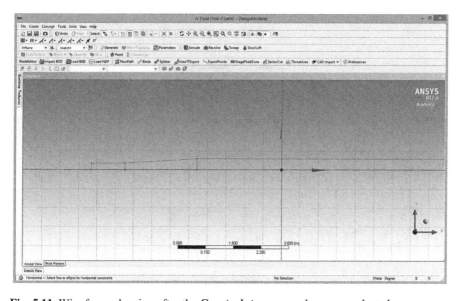

Fig. 5.11 Wire frame drawing after the **Constraints** commands were conducted.

Click **Sketching Toolboxes → Dimensions → Vertical**.

Pick two horizontal parallel lines at the inlet of the diffuser and then move the mouse to the left. Click the left mouse button at the proper position and the variable **V1** indicating the dimension appears as shown in Fig. 5.12. Pick two horizontal parallel lines at the outlet and the variable **V2** appears.

Click **Sketching Toolboxes** → **Dimensions** → **Horizontal**.

Pick two vertical parallel lines at the inlet and the variable **H3** appears. Conduct the same procedures for the vertical lines and the variables **H4** and **H5** appear as shown in Fig. 5.13.

Click the cursor on the **Details View** button and the table as shown in Fig. 5.14 appears. Change the values of **H3**, **H4**, **H5**, **V1**, and **V2** according to the shape of flow channel. Click [A] the row **H3** and input **0.9** to the left column and click the **Return** button. Then input the following values to the columns in turn.

H3 → 0.9, **H4** → 1.91, **H5** → 10, **V1** → 0.1, **V2** → 0.2.

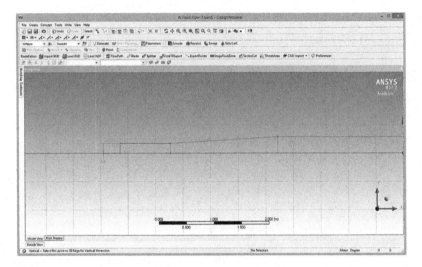

Fig. 5.12 Setting of the dimensions for the drawing.

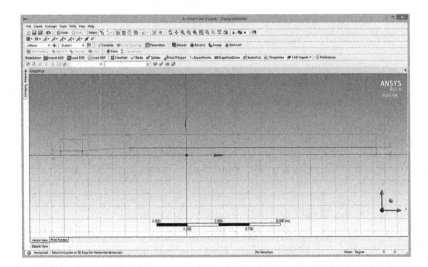

Fig. 5.13 Wire frame drawing of the diffuser after the **Dimensions** commands were completed.

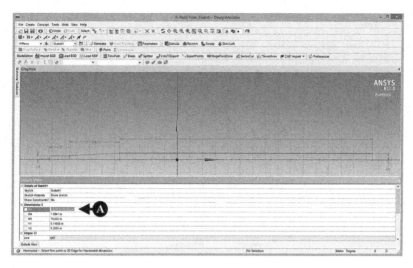

Fig. 5.14 Input **Details View** table.

5.2.2.4 Create surfaces for diffuser

Surfaces are created from **Sketches** by performing the following steps.

From the DesignModeler Main Menu, select **Concept → Surfaces from Sketches → Details View**.

1. The **Details View** table, Fig. 5.15, opens.
2. Click the [A] **Base Object** button and click [B] **Sketch1**. Then click the [C] **Apply** button.
3. Click [D] **Operation** and change from **Add Material** to **Add Frozen**.
4. Click the **Generate** button and the **Graphics** window is changed as seen in Fig. 5.16.

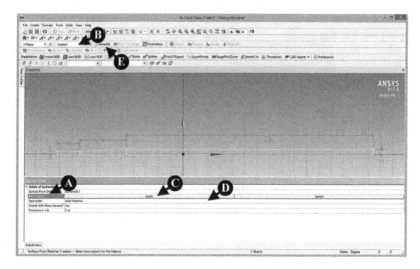

Fig. 5.15 Procedure of making surfaces from **Sketch1**.

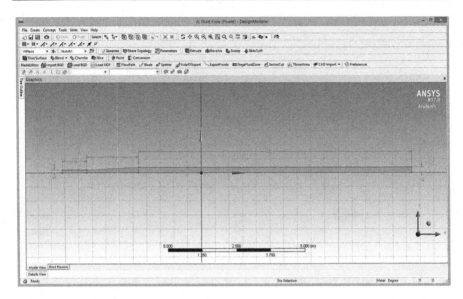

Fig. 5.16 Surfaces of the diffuser.

5. Select **Tree Outline** → + button of **3 Parts, 3 Bodies**. Fig. 5.17 opens.
6. Press the **Ctrl** button on the keyboard and click [A] three **Surface Body** commands. Click the right mouse button and click **Form New Parts** in the drop-down list. Then Fig. 5.18 opens, in which **3 Parts, 3 Bodies** is changed to [B] **1 Part, 3 Bodies**.

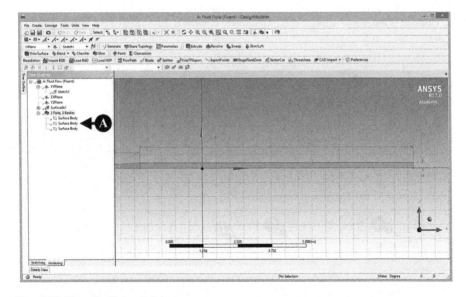

Fig. 5.17 Tree Outline window.

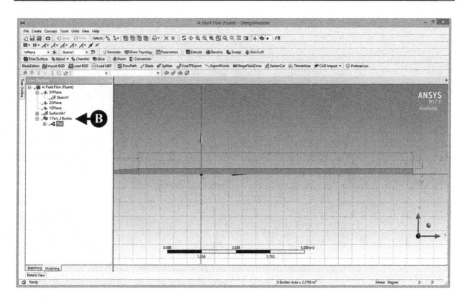

Fig. 5.18 Tree Outline window.

5.2.3   Create mesh in surface for diffuser

1. Double-click [A] **Mesh** in the **Project Schematic** window of Fig. 5.19. Then the **Meshing** window as shown in Fig. 5.20 opens, after clicking the arrow in the z direction.
2. Click [A] **Mesh** in the **Outline** window and the **Details of 'Mesh'** table opens, as seen in Fig. 5.21.

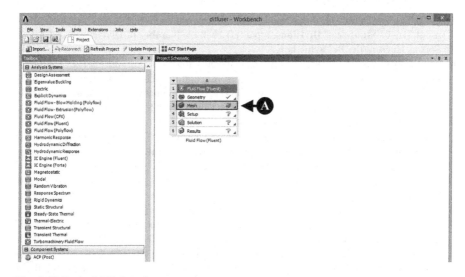

Fig. 5.19 Launch **Mesh** software.

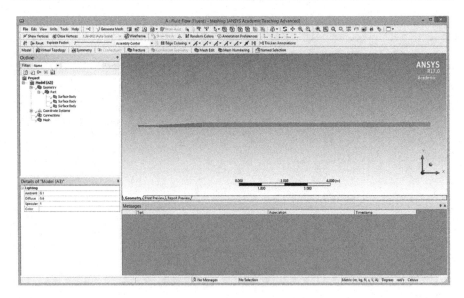

Fig. 5.20 2D drawing for meshing.

3. Select **Sizing** → **Size Function**. **Curvature** is changed to [B] **Adaptive** (see Fig. 5.22).
4. Click [C] **Edge** →, [D] **Mesh Control**, then click **Sizing** in the pull-down list. Then the **Details of 'Sizing'** table appears, as seen in Fig. 5.23.

Press the **Ctrl** key and select [E] two horizontal lines of the diffuser at the inlet. Click the **No Selection** button of **Geometry** and then the **Apply** button. The **Details of 'Edge Sizing'** table as seen in Fig. 5.24 appears.

5. Click **Type** in **Details of 'Edge Sizing'** of the frame shown in Fig. 5.25. Change [G] **Element Size** to **Number of Divisions**. Input [H] **10** in the **Number of Divisions** box. Change **Soft** to **Hard** in the **Behavior** box.
6. Click **Mesh Control** and then click **Sizing** in the pull-down list. Press the **Ctrl** key and select two lines of the diffuser area. Click the **No Selection** button of **Geometry** and then the **Apply** button. Click **Type** in **Details of 'Edge Sizing'**. Change **Element Size** to **Number of Divisions**. Input **20** in the **Number of Divisions** box. Change **Soft** to **Hard** in the **Behavior** box.
7. In the same way of step 6, click **Mesh Control** and then select **Sizing** in the pull-down list. Press the **Ctrl** key and select two horizontal lines at the outlet. Click the **No Selection** button of **Geometry** and then the **Apply** button. Click **Type** in **Details of 'Edge Sizing'**. Change **Element Size** to **Number of Divisions**, and input **80** in the **Number of Divisions** box. Change **Soft** to **Hard** in the **Behavior** box. Then the drawing in the window is changed as seen in Fig. 5.26.
8. Next, the meshes of four vertical lines are controlled. In the vertical lines, the mesh distance is not the same interval and set to be shorter, as it is closer to the outside wall of the diffuser. Click **Mesh Control** → **Sizing** in the pull-down list. Press the **Ctrl** key and select three lines: the second, third, and fourth lines of the diffuser area from the left end. Click the **No Selection** button of **Geometry**, and then the **Apply** button. Click **Type** in **Details of 'Edge Sizing'**. Change **Element Size** to **Number of Divisions**. Input **50** in the **Number**

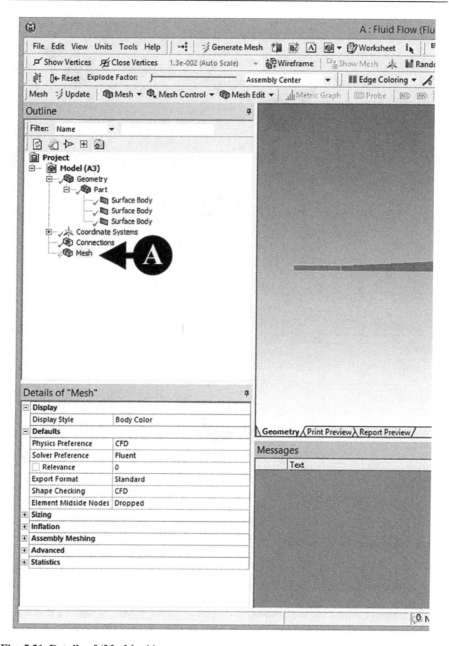

Fig. 5.21 Details of 'Mesh' table.

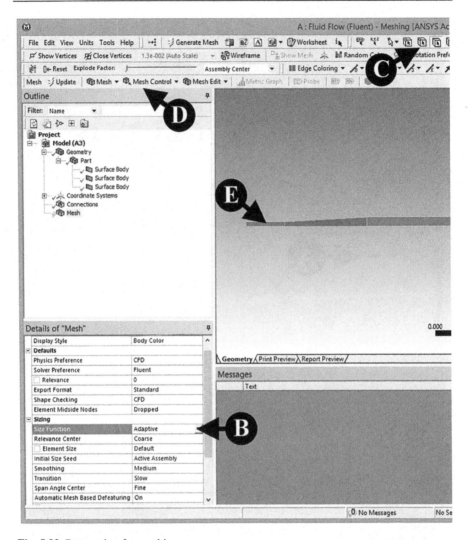

Fig. 5.22 Preparation for meshing.

of Divisions box. Change **Soft** to **Hard** in the **Behavior** box. Click **Bias Type** and select [- - —— ————]. Input **5** in the **Bias Factor** box.

9. Click **Mesh Control** and select **Sizing** in the pull-down list. Select one line, the first line of the diffuser area from the left end. Click the **No Selection** button of **Geometry**, then click the **Apply** button. Click **Type** in **Details of 'Edge Sizing'**. Change **Element Size** to **Number of Divisions**, and input **50** in the **Number of Divisions** box. Change **Soft** to **Hard** in the **Behavior** box. Click **Bias Type** and select [- - —— ————]. Input **5** in the **Bias Factor** box and then click the **Reverse Bias** button. Click the first line in the window and click the **Apply** button in the **Reverse Bias** box. Then the **Outline** window is changed as seen in Fig. 5.27.

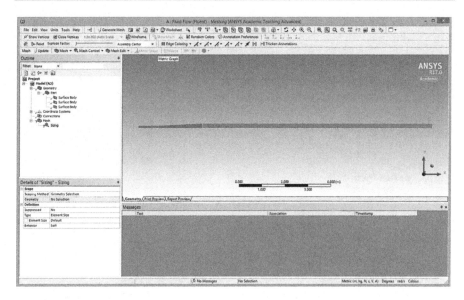

Fig. 5.23 Details of 'Sizing' table.

10. Click **Mesh Control** → **Face Meshing** in the pull-down list. Press the **Ctrl** key and select three areas of the diffuser. Click the **Apply** button of **Geometry** in the **Details of 'Face Meshing'** window. The window as seen in Fig. 5.28 appears.
11. Click [A] **Generate Mesh** and then [B] **Show Mesh**. Then the window as seen in Fig. 5.29 appears by enlarging the drawing.

5.2.4 Boundary conditions

In this section, boundary conditions listed in problem description are set.

From the **Outline** window, click **Model (A3)** with the right mouse button. Select **Insert** → **Named Selection**.

1. From the **Outline** window, click **Model (A3)** with the right mouse button. Click **Insert** and then **Named Selection**. Click the vertical line of the diffuser at the inlet, then click the **Apply** button in the **Geometry** box.
2. Click the right mouse button on the word **Selection** in the **Outline** window. Click the **Rename** button. Change the name from **Insert** to **Inlet**.
3. From the **Outline** window, click **Model (A3)** with the right mouse button. Click **Insert** and then **Named Selection**. Press the **Ctrl** key and click three lines of the diffuser outside the wall. Click the **Apply** button in the **Geometry** box.
4. Click the right mouse button on the word **Selection** in the **Outline** window. Click the **Rename** button. Change the name from **Selection** to **Wall**.
5. From the **Outline** window, click **Model (A3)** with the right mouse button. Click **Insert** and then **Named Selection**. Click the vertical line at the outlet of the diffuser. Click the **Apply** button in the **Geometry** box.
6. Click the right mouse button on the word **Selection** in the **Outline** window. Click the **Rename** button and change the name from **Selection** to **Outlet**.

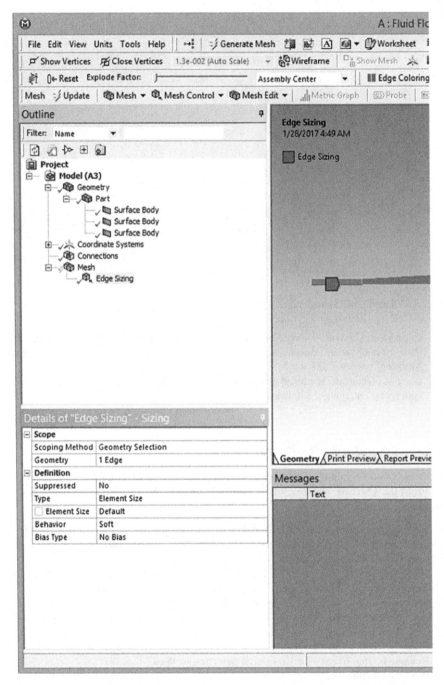

Fig. 5.24 Details of 'Edge Sizing' table.

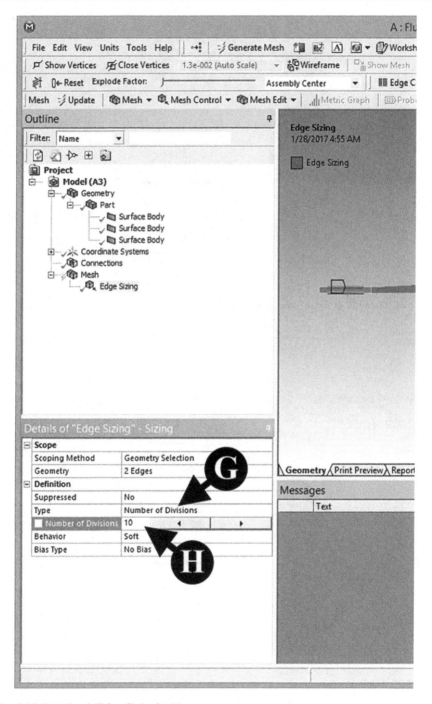

Fig. 5.25 Details of 'Edge Sizing' table.

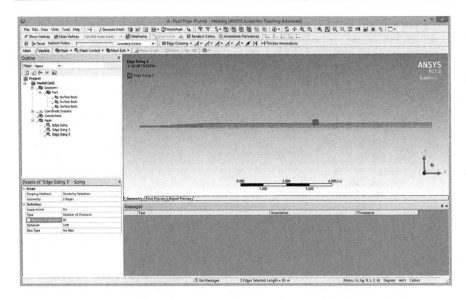

Fig. 5.26 Drawing of the diffuser in the window.

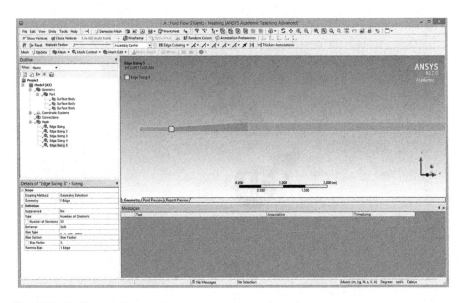

Fig. 5.27 Outline window.

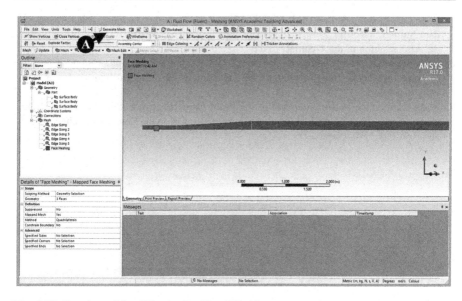

Fig. 5.28 Drawing of the diffuser after **Face Meshing** command.

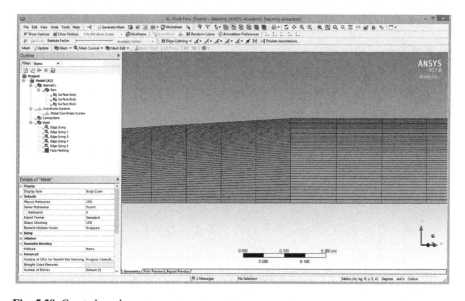

Fig. 5.29 Created mesh.

7. From the **Outline** window, click **Model (A3)** with the right mouse button. Click **Insert** and then **Named Selection**. Press the **Ctrl** key and click three center lines of the diffuser. Click the **Apply** button in the **Geometry** box.

8. Click the right mouse button on the word **Selection** in the **Outline** window. Click the **Rename** button and change the name from **Selection** to **Axis**. The **Outline** window is changed as seen in Fig. 5.30.

9. Right-click on the **Setup** command of the **Fluid Flow (Fluent)** box in the **Project Schematic** window. Click **Edit** in the pull-down list. Click **OK** in the **Fluent Launcher** window and then Fig. 5.31 appears.

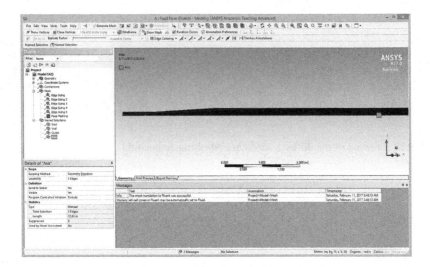

Fig. 5.30 Outline window.

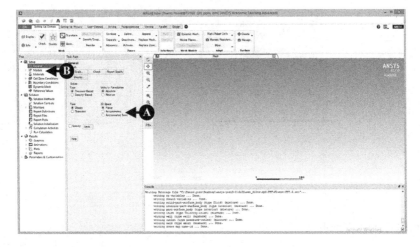

Fig. 5.31 Setup window.

5.2.5   Setup for solution and excursion of the analysis

1. Check [A] **Axisymmetric** in the **Task Page** window.
2. Click [B] **Models** in the **Tree** window, and the frame shown in Fig. 5.32 appears. Double-click [C] **Energy-off**, check **Energy Equation** and click **OK**.
3. Click **Viscous—Laminar**, check **k-epsilon (2 eqn)** and click **OK**.
4. Click [D] **Materials** and confirm that **Fluid** is **air**.

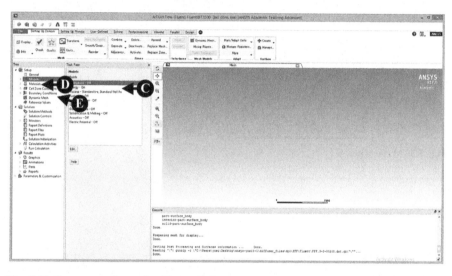

Fig. 5.32 Setup procedure using **Tree** window.

5. Click [E] **Boundary Conditions**, and Fig. 5.33 appears. Click **Inlet** in **Zone** and then **Velocity-Inlet** in the pull-down list, and Fig. 5.34 appears. Input [A] **20** in the **Velocity Magnitude (m/s)** box and click **OK**.
6. Click **Boundary Conditions** then **Outlet** in **Zone**. Click **Pressure-Outlet** in the pull-down list and Fig. 5.35 appears. Confirm [A] that **Gauge Pressure (pascal)** is zero. Click [B] **Thermal** and confirm [C] that **Backflow Total Temperature (k)** is **300** in Fig. 5.36. Click the **OK** button.
7. In a similar way to step 6, click **Boundary Conditions** then **Wall** in **Zone**. Click **Wall** in the pull-down list and Fig. 5.37 appears. Confirm that **Stationary Wall** and **No Slip** are checked. Click **Thermal** and confirm that **Heat Flus (w/m2)** is **0**. Click the **OK** button.
8. Click **Solution Initialization** in the **Tree** window. Click the **Initialize** button in the **Tree Page** window.
9. Click **Calculation Activities**.
10. Click **Run Calculation**, and Fig. 5.38 appears. Input [A] **200** in the **Number of Iterations** box and click the **Calculate** button. Fig. 5.39 appears when the calculation is done. Click **OK**.

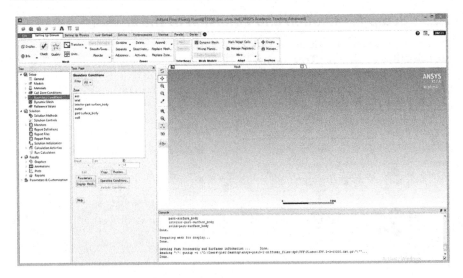

Fig. 5.33 Tree window.

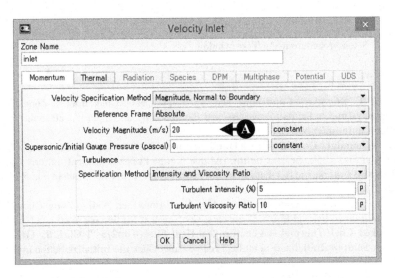

Fig. 5.34 Velocity inlet window.

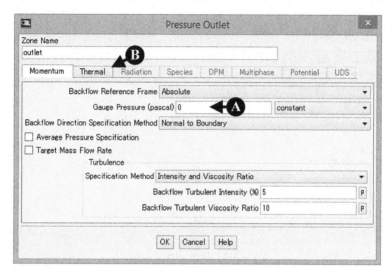

Fig. 5.35 Pressure Outlet window.

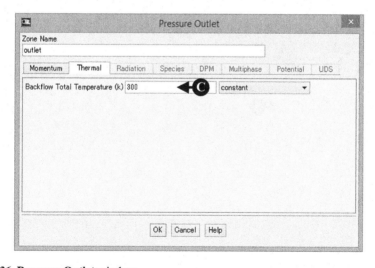

Fig. 5.36 Pressure Outlet window.

![Wall window dialog box]
Wall

Zone Name
wall

Adjacent Cell Zone
solid-part-surface_body

| Momentum | Thermal | Radiation | Species | DPM | Multiphase | UDS | Wall Film | Potential |

Wall Motion   Motion
  ○ Stationary Wall   ☑ Relative to Adjacent Cell Zone
  ○ Moving Wall

Shear Condition
  ● No Slip
  ○ Specified Shear
  ○ Specularity Coefficient
  ○ Marangoni Stress

Wall Roughness
  Roughness Height (m) 0   constant ▼
  Roughness Constant   0.5   constant ▼

OK   Cancel   Help

Fig. 5.37 Wall window.

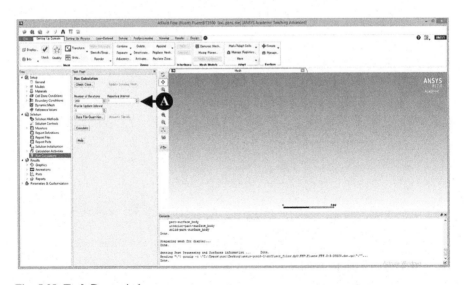

Fig. 5.38 Task Page window.

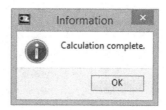

Fig. 5.39 Information window.

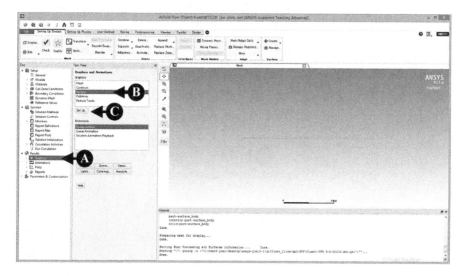

Fig. 5.40 Tree window.

5.2.6 Postprocessing

1. Click [A] **Graphics** in the **Tree** window, and Fig. 5.40 appears. Click [B] **Vectors** and then the [C] **Set Up** button. Fig. 5.41 appears. Click the [D] **Display** button.

Then velocity vectors of air flow in the diffuser appear in the window, as seen in Fig. 5.42.

5.2.6.1 Create XY plot of air flow at a cross section of the diffuser

By using the [A] **Zoom to Area** button, enlarge the graphics of air flow in the diffuser as seen in Fig. 5.43.

1. Click [A] **Plots** in the **Tree** window. Click [B] **XY Plot** in **Task Page** and then the **Set Up** button. Then Fig. 5.44 appears.
2. Input [A] **0** to the **X** box and [B] **1** to the **Y** box. Click [C] **New Surface** → **Line/Rake**. Fig. 5.45 appears.

Fig. 5.41 Vectors window.

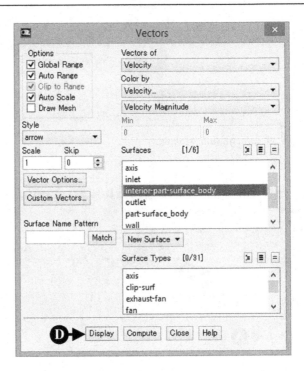

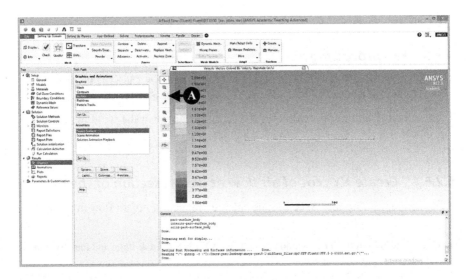

Fig. 5.42 Calculated result window with velocity vectors.

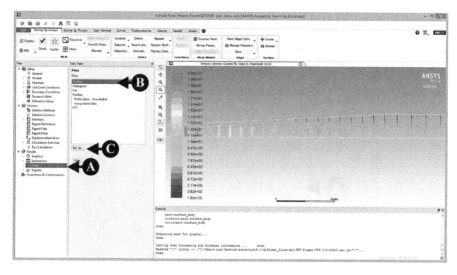

Fig. 5.43 Enlarged view of calculated result with velocity vectors.

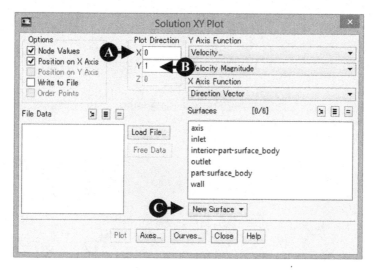

Fig. 5.44 Solution XY Plot window.

3. Click [A] **Select Points with Mouse**. According to the comment, select two points [B], [C] by clicking the right mouse button, as seen in Fig. 5.46. The **Line/Rake Surface** window appears (see Fig. 5.47).

4. Input [A] **-3.0** to the **x0** and **x1** boxes, [B] **0** to the **y0** box, and [C] **0.2** to the **y1** box. Click the [D] **Create** button.

5. Click [A] **line-6** of **Surfaces** in Fig. 5.48. Click the [B] **Plot** button, and Fig. 5.49 appears.

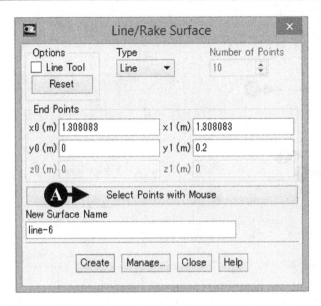

Fig. 5.45 Line/Rake Surface window.

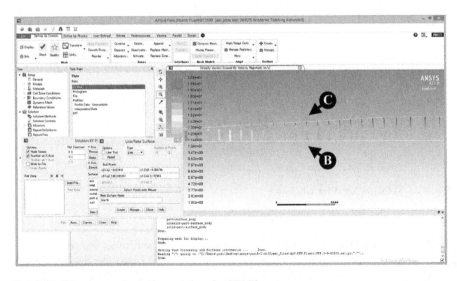

Fig. 5.46 Selection method of two points for XY Plot.

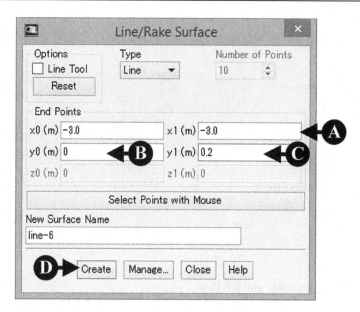

Fig. 5.47 Line/Rake Surface window.

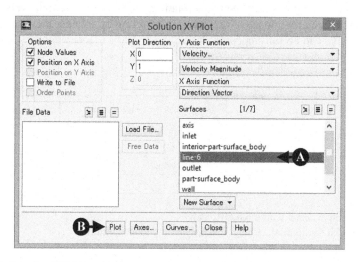

Fig. 5.48 Solution XY Plot window.

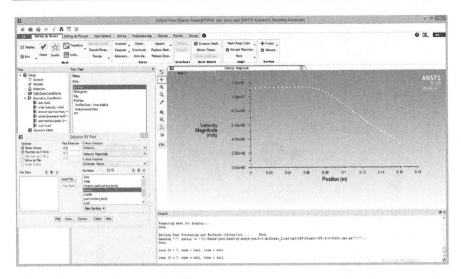

Fig. 5.49 Calculation result with **XY Plot**.

5.3 Analysis of flow structure in a channel with a butterfly valve

5.3.1 Problem description

Analyse the flow structure around a butterfly valve as shown in Fig. 5.50. The butterfly valve has a metal disc mounted on a rotational rod. In operation, the valve can be opened incrementally to control flow rate. In addition, by rotating the disc a quarter turn, the valve can be fully open or closed.

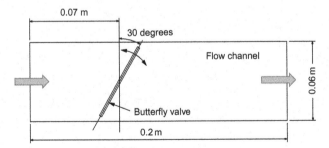

Fig. 5.50 Flow channel with a butterfly valve.
(From www.kitz.co.jp/product/bidg-jutaku/kyutouyou/index.html.)

Shape of flow channel:

1. Diameter of flow channel $D_E = 0.06$ m.
2. Diameter of a butterfly valve $D_E = 0.06$ m.
3. Tilt angle of a butterfly valve $\alpha = 30$ degrees.

Operating fluid: water
Flow field: turbulent
Velocity at the entrance: 0.01 m/s

Boundary conditions:

1. Velocities in all directions are zero on all walls.
2. Pressure is equal to zero at the exit.
3. Velocity in the y direction is zero on the x axis.

5.3.2 Create a model for analysis using Workbench

5.3.2.1 Select kind of analysis (Fluent)

Double-click **Shortcut Workbench 17.0**
The window **Unsaved Project—Workbench** appears

1. Drag the **Fluid Flow (Fluent)** button into the **Project schematic** window.
2. Save this project as **valve** by clicking **File**.

5.3.2.2 Preparation for DesignModeler to create a geometry for analysis

1. Double-click **Geometry** in the **Project Schematic** window. Then the **Graphics** window of DesignModeler opens.
2. To change from a 3D screen to a 2D screen, place a cursor on the arrow of the z direction and then the arrow is highlighted. Click the arrow in the z direction; the screen is changed from 3D to 2D.
3. Click the **Tree Outline** button on the left-hand side of the **Graphics** window, then select **Schetching** → **Settings**.
4. Check the **Show in 2D** box and input **0.5 m** to the **Major Grid Spacing** box. Input **10** to the **Minor Steps per Major** box.

5.3.2.3 Create a wire frame for a butterfly valve

To draw a butterfly valve for analysis, DesignModeler is described in this section. The dimensions of the valve are described in Section 5.3.1.

1. Click the **Graphics** window and enlarge the grid by scrolling up the wheel button of the mouse (see Fig. 5.51).
2. Select **Schetching Toolboxes** → **Draw** → **Rectangle**. By using the **Rectangle** command, the channel of the valve is described with approximate.
3. Select **Schetching Toolboxes** → **Draw** → **Rectangle by 3 Points**. Click **Rectangle by 3 Points** and the **Pencil** mark appears on the **Graphics** window. Click the left mouse button at the approximate position in the channel and move the **Pencil** mark to the next two positions to draw the shape of the valve. Then the lines as seen in Fig. 5.52 are created.
4. Use the **Dimensions** command to input the dimensions of the valve.

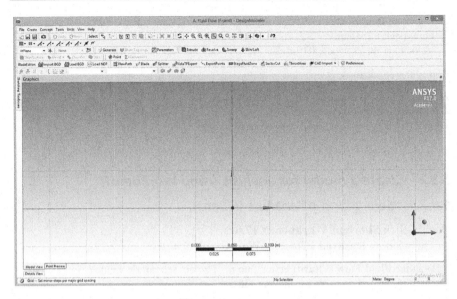

Fig. 5.51 Graphics window with grids.

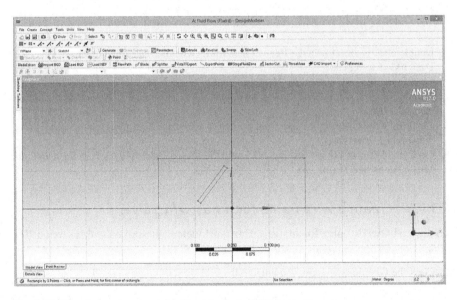

Fig. 5.52 Wire frame drawing of a flow channel with a butterfly valve.

Click **Sketching Toolboxes** → **Dimensions** → **Vertical**.

Pick two horizontal parallel lines at the inlet of the channel and then move the mouse to the left. Click the left mouse button at the proper position and the variable **V1**. Pick the bottom horizontal line and the lowest point of the edge of the valve and the variable **V2** appears.

Click **Sketching Toolboxes** → **Dimensions** → **Horizontal**.

Pick two vertical parallel lines at both end of the channel and the variable **H3** appears. Pick the vertical line at the outlet of the channel and the lowest point of the edge of the valve and the variable **H4** appears.

Click **Sketching Toolboxes** → **Dimensions** → **Length/Distance**.

To determine the length of the valve, click two parallel end lines of the valve and the variable **L5** appears. In the same way, click two parallel lines of the valve to determine the thickness of the valve and the variable **L6** appears.

Click **Sketching Toolboxes** → **Dimensions** → **Angle**.

To determine the angle of the valve, click the bottom line first and then the line of the valve towards the outlet of the channel. The variable **A7** appears.

Click **Details View** and input the following values of the **Dimensions:7** table, as seen in Fig. 5.53.

A7 → 60, **H3** → 0.2, **H4** → 0.1441, **L5** → 0.06, **L6** → 0.002, **V1** → 0.06, **V2** → 0.0035.

5.3.2.4   Create a surface for a channel with a butterfly valve

Select **Sketching Toolboxes** → **Modeling** → **XYPlane** → **Sketch1**. Highlight the geometry of **Sketch1** in the **Graphics** window.

Select **Concept** → **Surfaces from sketches** → **Generate**. Then a surface is made in Sketch1 as seen in Fig. 5.54.

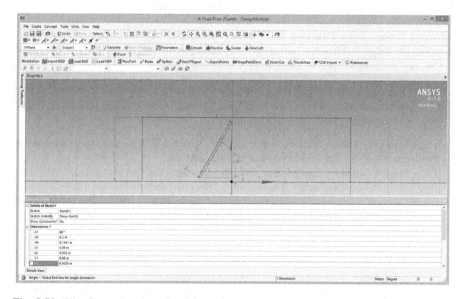

Fig. 5.53 Wire frame drawing after **Dimensions** commands and **Details View** table.

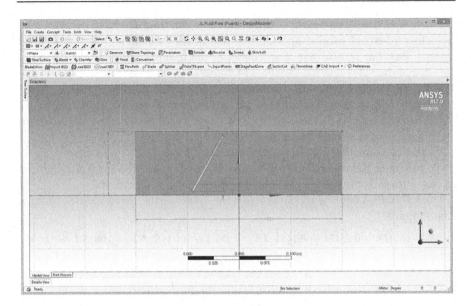

Fig. 5.54 Wire frame drawing after making a surface.

5.3.2.5   Create meshes for a channel with a butterfly valve

1. Double-click [A] **Mesh** in the **Project Schematic** window of Fig. 5.55. Then the **Meshing** window opens, as shown in Fig. 5.56. Click the arrow [A] in the z direction.
2. Click [B] **Mesh** in the **Outline** window and the **Details of 'Mesh'** table opens, as seen in Fig. 5.57.
3. Select **Sizing** → **Size Function**. **Curvature** is changed to [A] **Adaptive** (see Fig. 5.58).
4. Click [B] **Edge**, [C] **Mesh Control** and then **Sizing** in the pull-down list. The **Details of 'Sizing'** table then appears as seen in Fig. 5.59.

Press the Ctrl key and select [D] two horizontal lines of the wall of the channel. Click the No Selection button of Geometry and then the Apply button. The Details of 'Edge Sizing' table appears, as seen in Fig. 5.60.

5. Click **Type** in **Details of 'Edge Sizing'** of Fig. 5.60. Change [E] **Element Size** to **Number of Divisions**. Input [F] **200** in the **Number of Divisions** box.
6. Click **Mesh Control** → **Sizing** in the pull-down list. Press the **Ctrl** key and select two vertical lines at the inlet and outlet of the channel. Select two lines for the valve surfaces. Click the **No Selection** button of **Geometry**, then the **Apply** button. Click **Type** in **Details of 'Edge Sizing'** and change **Element Size** to **Number of Divisions**. Input **40** in the **Number of Divisions** box and click **Bias Type** → - - ——— - - → **Bias Option** → **Bias Factor**. Input **4** in the Bias Factor box.
7. Click **Mesh Control** → **Sizing** in the pull-down list. Press the **Ctrl** key and select two lines at the valve edges. Click the **No Selection** button of **Geometry** and then the **Apply** button. Click **Type** in **Details of 'Edge Sizing'**. Change **Element Size** to **Number of Divisions**. Input **2** in the **Number of Divisions** box.

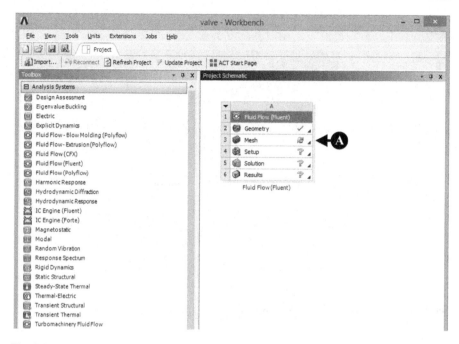

Fig. 5.55 Launch **Mesh** software.

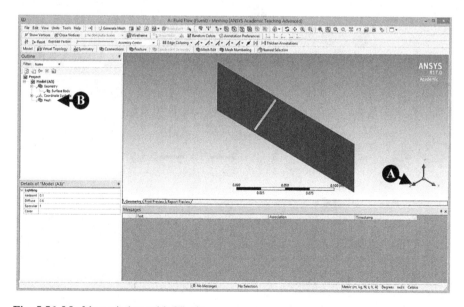

Fig. 5.56 Meshing window with 3D view.

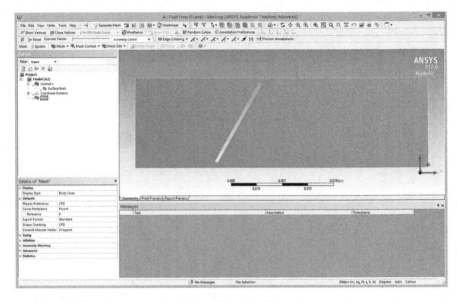

Fig. 5.57 2D drawing and **Details of 'Mesh'** table.

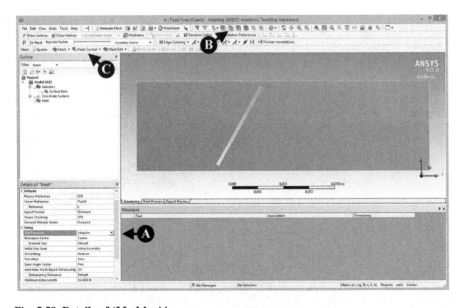

Fig. 5.58 **Details of 'Mesh'** table.

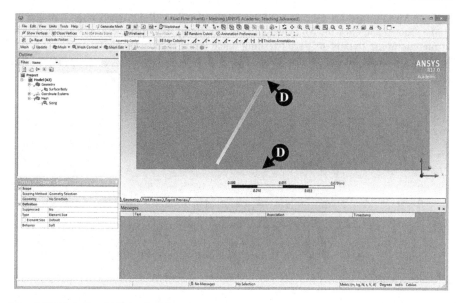

Fig. 5.59 Details of 'Sizing' table.

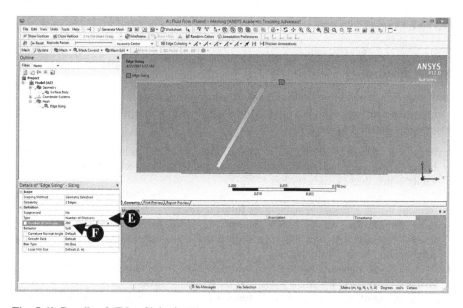

Fig. 5.60 Details of 'Edge Sizing' table.

8. Click **Mesh Control** and then select **Face Meshing** in the pull-down list. Select a surface of the channel and click the **No Selection** button of **Geometry**, then the **Apply** button. Click **method** in the **Details of 'Face Meshing'** table and change to **Quadrilaterals**.
9. Click **Generate Mesh → Show Mesh**. Then Fig. 5.61 appears.

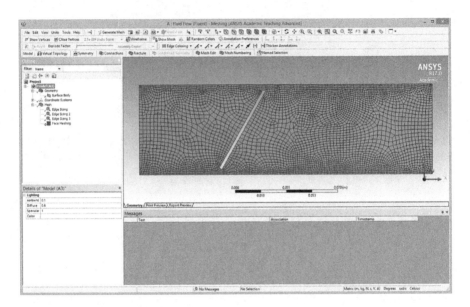

Fig. 5.61 Drawing with created mesh.

5.3.3 Boundary conditions

In this section, boundary conditions listed in the problem description are set.

From the **Outline** window, click **Model (A3)** with the right mouse button. Click **Insert** then **Named Selection**.

1. From the **Outline** window, click **Model (A3)** with the right mouse button, then **Insert → Named Selection**. Click the vertical line of the channel at the inlet. Click the **Apply** button in the **Geometry** box.
2. Click the right mouse button on the word **Selection** in the **Outline** window. Click the **Rename** button. Change the name from **Insert** to **Inlet**.
3. From the **Outline** window, click **Model (A3)** with the right mouse button, then **Insert → Named Selection**. Click the vertical line at the outlet of the diffuser. Click the **Apply** button in the **Geometry** box.
4. Click the right mouse button on the word **Selection** in the **Outline** window. Click the **Rename** button and change the name from **Selection** to **Outlet**.
5. From the **Outline** window, click **Model (A3)** with the right mouse button, then **Insert → Named Selection**. Press the **Ctrl** key and select two horizontal lines of the channel and four lines of the butterfly valve. Click the **Apply** button in the **Geometry** box.

6. Click the right mouse button on the word **Selection** in the **Outline** window. Click the **Rename** button. Change the name from **Selection** to **Wall**. The **Outline** window is changed, as seen in Fig. 5.62.

7. Right-click on the **Setup** command of the **Fluid Flow (Fluent)** box in the **Project Schematic** window. Double-click **Setup** and then Fig. 5.63 appears. Click **OK** and Fig. 5.64 opens.

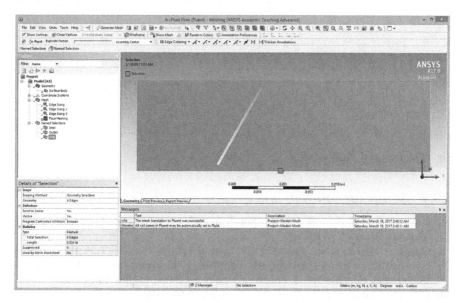

Fig. 5.62 Changed **Outline** window.

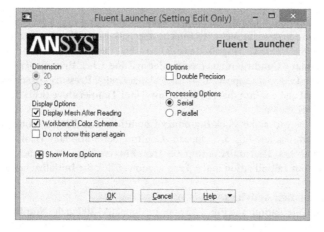

Fig. 5.63 Fluent Launcher window.

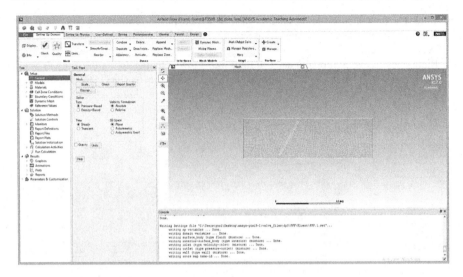

Fig. 5.64 Tree and **Mesh** windows.

5.3.4 Setup for solution and excursion of the analysis

1. Confirm the following conditions of **Solver** in the **Task Page** window: **Pressure-Based**, **Absolute**, **Steady**, and **Planer**.
2. Click [A] **Models** in the **Tree** window, as seen in Fig. 5.65. The **Task Page** window appears. Double-click [B] **Energy-off**, check **Energy Equation** and click the **OK** button.
3. Double-click [C] **Viscous—Laminar** and check **k-epsilon (2 eqn)**. Click **OK**.
4. Click [D] **Materials** and double-click **Fluid**. Fig. 5.66 opens.

Click [A] **Fluent Database**, and Fig. 5.67 opens. Select [B] **water-liquid** in the **Fluent Fluid Materials** table and click [C] **Copy** then [D] **Close**. Click [E] **Change/Create** in the **Create/Edit Materials** window, then click **Close**.

5. Click [A] **Boundary Conditions** in the **Tree** window, and Fig. 5.68 appears. Click [B] **inlet** in **Zone** and select **Velocity-Inlet** in the pull-down list. Fig. 5.69 appears. Input [C] **0.01** in the **Velocity Magnitude (m/s)** box, and then **OK**.
6. Click **Boundary Conditions** and then **Outlet** in **Zone**. Click **Pressure-Outlet** in the pull-down list, and Fig. 5.70 appears. Confirm [A] that **Gauge Pressure (pascal)** is zero. Click [B] **Thermal** and confirm that [C] **Backflow Total Temperature (k)** is **300** in Fig. 5.71. Click the **OK** button.
7. In a similar way to step 6, click **Boundary Conditions** then **Wall** in **Zone**. Click **Wall** in the pull-down list, and Fig. 5.72 appears. Confirm that **Stationary Wall** and **No Slip** are checked, and click **Thermal**. Confirm that **Heat Flus (w/m2)** is **0** and click the **OK** button.
8. Click **Solution Initialization** in the **Tree** window. Click the **Initialize** button in the **Tree Page** window.
9. Click **Calculation Activities**.
10. Click **Run Calculation**, and Fig. 5.73 appears. Input [A] **200** in the **Number of Iterations** box and click the [B] **Calculate** button. Fig. 5.74 appears when the calculation is done. Click **OK**.

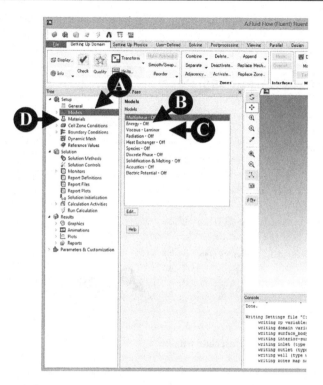

Fig. 5.65 Tree window.

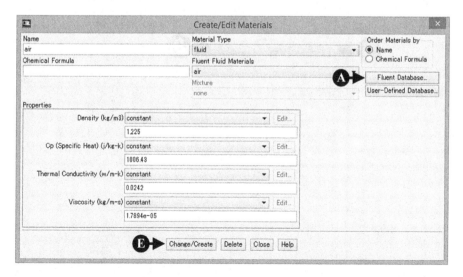

Fig. 5.66 Create/Edit Materials window.

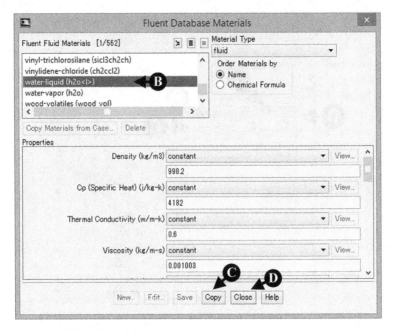

Fig. 5.67 **Fluent Database Materials** window.

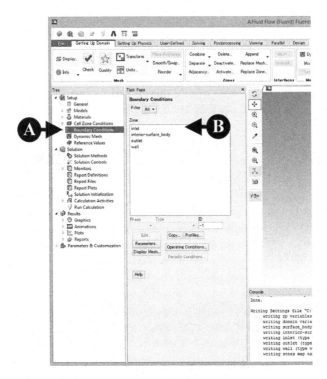

Fig. 5.68 **Task Page** of **Boundary Conditions** window.

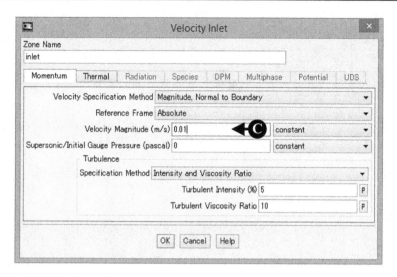

Fig. 5.69 Velocity Inlet window.

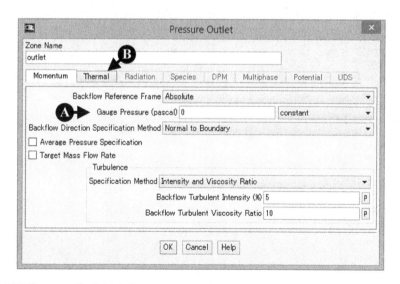

Fig. 5.70 Pressure Outlet window.

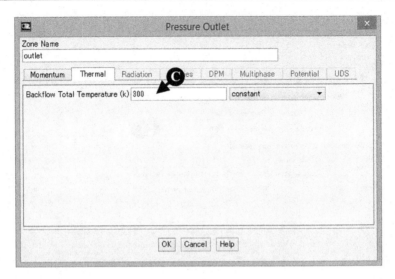

Fig. 5.71 Pressure Outlet window.

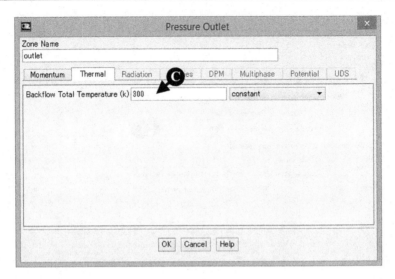

Fig. 5.72 Wall window.

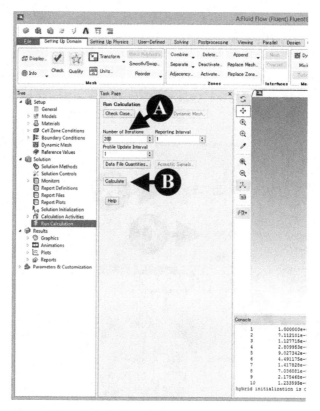

Fig. 5.73 **Task Page** of **Run Calculation** window.

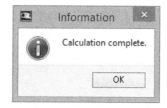

Fig. 5.74 **Information** of **Calculation complete** window.

5.3.5 Postprocessing

1. Click [A] **Graphics** in the **Tree** window, and Fig. 5.75 appears. Click [B] **Vectors** and [C] **Set Up**. Fig. 5.76 appears. Input [D] **4** to the **Scale** box and click the [E] **Display** button.

Then velocity vectors of air flow in the butterfly valve appear in the window as seen in Fig. 5.77.

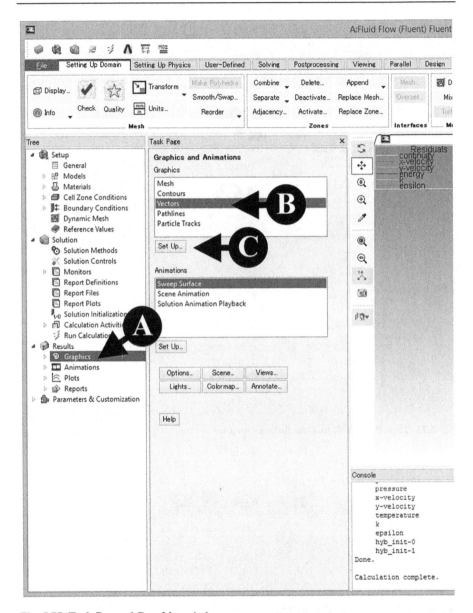

Fig. 5.75 Task Page of **Graphics** window.

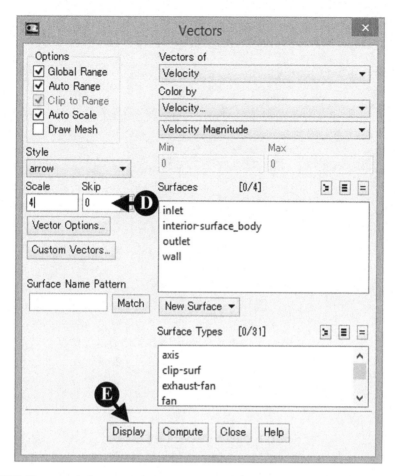

Fig. 5.76 **Vectors** window.

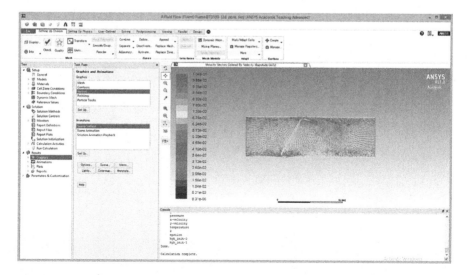

Fig. 5.77 Calculation results window.

Application of ANSYS to thermo-mechanics

6.1 General characteristic of heat transfer problems

The transfer of heat is normally from a high temperature object to a lower temperature object. Heat transfer changes the internal energy of both systems involved according to the first law of thermodynamics.

Heat may be defined as energy in transit. An object does not possess 'heat'; the appropriate term for the microscopic energy in an object is internal energy. This internal energy may be increased by transferring energy to the object from a higher temperature (hotter) object—this is properly called heating.

A convenient definition of temperature is that it is a measure of the average translation kinetic energy associated with the disordered microscopic motion of atoms and molecules. The flow of heat is from a high temperature region towards a lower temperature region. The details of the relationship to molecular motion are dealt with by the kinetic theory. The temperature defined from kinetic theory is called the kinetic temperature. Temperature is not directly proportional to internal energy since temperature measures only the kinetic energy part of the internal energy, so two objects with the same temperature do not, in general, have the same internal energy.

Internal energy is defined as the energy associated with the random, disordered motion of molecules. It is separated in scale from the macroscopic ordered energy associated with moving objects. It also refers to the invisible microscopic energy on the atomic and molecular scale. For an ideal monoatomic gas, this is just the translational kinetic energy of the linear motion of the 'hard sphere' type atoms, and the behaviour of the system is well described by the kinetic theory. However, for polyatomic gases there is rotational and vibrational kinetic energy as well. Then in liquids and solids there is potential energy associated with the intermolecular attractive forces.

Heat transfer by means of molecular agitation within a material without any motion of the material as a whole is called conduction. If one end of a metal rod is at a higher temperature, then energy will be transferred down the rod towards the colder end, because the higher speed particles will collide with the slower ones with a net transfer of energy to the slower ones. For heat transfer between two plane surfaces, such as heat loss through the wall of a house, the rate of conduction could be estimated from.

Engineering Analysis with ANSYS Software. https://doi.org/10.1016/B978-0-08-102164-4.00006-6

$$\frac{Q}{t} = \frac{\kappa A (T_{\text{hot}} - T_{\text{cold}})}{d}$$

where the left-hand side concerns rate of conduction heat transfer; κ is thermal conductivity of the barrier; A is area through which heat transfer takes place; T is temperature; and d is the thickness of barrier.

Another mechanism for heat transfer is convection. Heat transfer by mass motion of a fluid such as air or water when the heated fluid is caused to move away from the source of heat, carrying energy with it, is called convection. Convection above a hot surface occurs because hot air expands, becomes less dense and thus rises. Convection can also lead to circulation in a liquid, as in the heating of a pot of water over a flame. Heated water expands and becomes more buoyant. Cooler, denser water near the surface descends, and patterns of circulation can be formed.

Radiation is heat transfer by the emission of electromagnetic waves, which carry energy away from the emitting object. For ordinary temperatures (less than red-hot), the radiation is in the infrared region of the electromagnetic spectrum. The relationship governing radiation from hot objects is called the Stefan–Boltzmann law:

$$P = e\sigma A \left(T^4 - T^4_{\,c} \right)$$

where P is net radiated power; A is radiating area; σ is Stefan's constant; e is emissivity coefficient; T is temperature of radiator; and T_c is temperature of surroundings.

6.2 Heat transfer through two adjacent walls

6.2.1 Problem description

A furnace with dimensions of its cross-section specified in Fig. 6.1 is constructed from two materials. The inner wall is made of concrete with a thermal conductivity, $k_c = 0.01$ W/m K. The outer wall is constructed from bricks with a thermal conductivity, $k_b = 0.0057$ W/m K. The temperature within the furnace is 673 K and the convection heat transfer coefficient $k_1 = 0.208$ W/m^2 K. The outside wall of the furnace is exposed to the surrounding air, which is at 253 K and the corresponding convection heat transfer coefficient, $k_2 = 0.068$ W/m^2 K.

Determine the temperature distribution within the concrete and brick walls under steady-state conditions. Also determine the heat fluxes through each wall.

This is a 2D problem and will be modelled using GUI facilities.

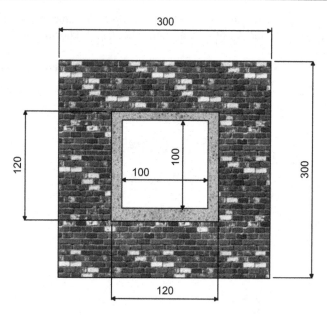

Fig. 6.1 Cross-section of the furnace.

6.2.2 Construction of the model

From the ANSYS Main Menu, select **Preferences**. This frame is shown in Fig. 6.2.

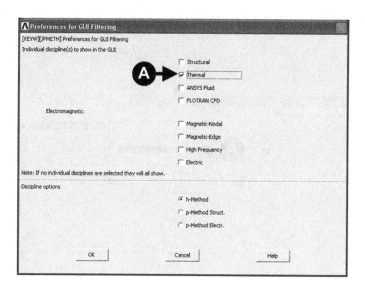

Fig. 6.2 Preferences: thermal.

Depending on the nature of analysis to be attempted, an appropriate analysis type should be selected. In the problem considered here, [A] **Thermal** was selected, as shown in Fig. 6.2.

From the ANSYS Main Menu, select **Preprocessor → Element Type → Add/ Edit/Delete**. The frame shown in Fig. 6.3 appears.

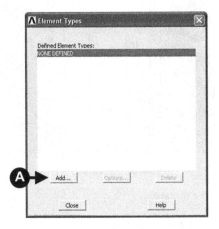

Fig. 6.3 Element types selection.

Clicking the [A] **Add** button activates a new set of options, which are shown in Fig. 6.4.

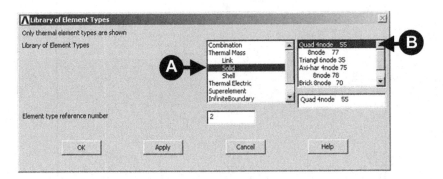

Fig. 6.4 Library of element types.

Fig. 6.4 shows that for the problem considered, the following were selected: [A] **Thermal Mass → Solid** and [B] **4node 55**. This element is referred to as Type 1 PLANE55.

From the ANSYS Main Menu, select **Preprocessor → Material Props → Material Models**. Fig. 6.5 shows the resulting frame.

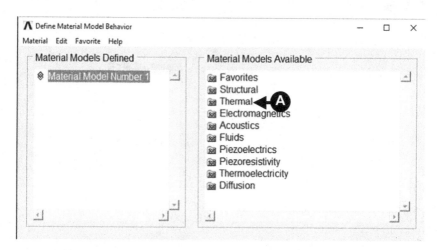

Fig. 6.5 Define material model behaviour.

From the right-hand column, select [A] **Thermal → Conductivity → Isotropic**. As a result, the frame shown in Fig. 6.6 appears. Thermal conductivity [A] **KXX** = 0.01 W/m K was selected, as shown in Fig. 6.6.

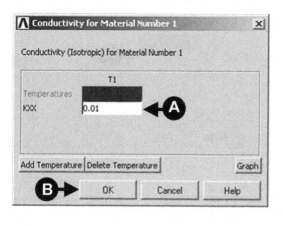

Fig. 6.6 Conductivity for Material Number 1.

Clicking the [B] **OK** button ends input into Material Number 1. In the frame shown in Fig. 6.7, select from the top menu [A] **Material** → **New Model**. A database for Material Number 2 is created.

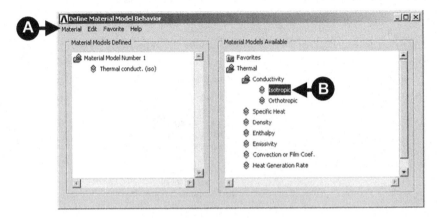

Fig. 6.7 Define material model behaviour.

As in the case of Material Number 1, select [B] **Thermal** → **Conductivity** → **Isotropic**. The frame shown in Fig. 6.8 appears. Enter [A] **KXX** = 0.0057 W/m K and click the [B] **OK** button as shown in Fig. 6.8.

Fig. 6.8 Conductivity for Material Number 2.

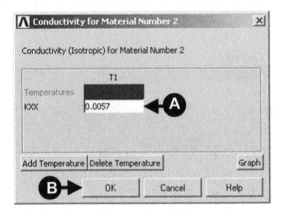

In order to create numbered primitives from the ANSYS Utility Menu, select **PlotCtrls** → **Numbering** and check the box area numbers.

From the ANSYS Main Menu, select **Preprocessor** → **Modelling** → **Create** → **Areas** → **Rectangle** → **By Dimensions**. Fig. 6.9 shows the resulting frame.

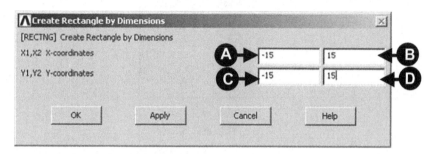

Fig. 6.9 Create rectangle by dimensions.

Input [A] X1 = − 15; [B] X2 = 15; [C] Y1 = − 15 and [D] Y2 = 15 to create an outer wall perimeter area as shown in Fig. 6.9. Next, the perimeter of the inner wall is created in the same way. Fig. 6.10 shows a frame with appropriate entries.

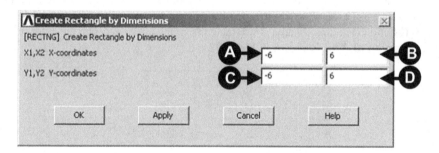

Fig. 6.10 Create rectangle by dimensions.

In order to generate the brick wall area of the chimney, subtract the two areas that have been created. From the ANSYS Main Menu, select **Preprocessor** → **Modelling** → **Operate** → **Booleans** → **Subtract** → **Areas**. Fig. 6.11 shows the resulting frame.

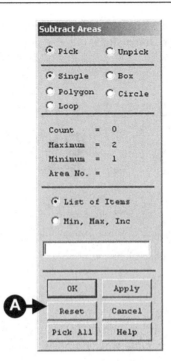

Fig. 6.11 Subtract areas.

First, select the larger area (outer brick wall) and click the [A] **OK** button in the frame of Fig. 6.11. Next, select the smaller area (inner concrete wall) and click the [A] **OK** button. The smaller area is subtracted from the larger one, and the outer brick wall is produced. This is shown in Fig. 6.12.

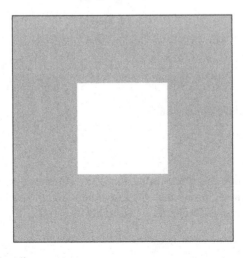

Fig. 6.12 Brick wall outline.

Using a similar approach, the inner concrete wall is constructed. From the ANSYS Main Menu, select **Preprocessor** → **Modelling** → **Create** → **Areas** → **Rectangle** → **By Dimensions**. Fig. 6.13 shows the resulting frame with inputs: [A] X1 = − 6; [B] X2 = 6; [C] Y1 = − 6 and [D] Y2 = 6. Pressing the [E] **OK** button creates rectangular area A1.

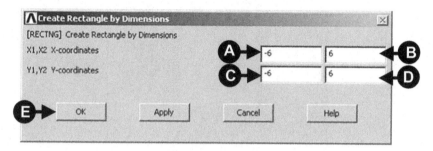

Fig. 6.13 Create rectangle by dimensions.

Again, from the ANSYS Main Menu, select **Preprocessor** → **Modelling** → **Create** → **Areas** → **Rectangle** → **By Dimensions**. A frame with inputs: [A] X1 = − 5; [B] X2 = 5; [C] Y1 = − 5 and [D] Y2 = 5 is shown in Fig. 6.14.

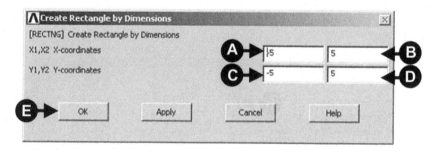

Fig. 6.14 Create rectangle by dimensions.

Clicking the [E] **OK** button creates rectangular area A2. As before, to create the concrete area of the furnace, subtract area A2 from area A1. From the ANSYS Main Menu, select **Preprocessor** → **Modelling** → **Operate** → **Booleans** → **Subtract** → **Areas**. The frame shown in Fig. 6.11 appears. Select area A1 first and click the [A] **OK** button.

Next, select area A2 and click the [A] **OK** button. As a result, the inner concrete wall is created. This is shown in Fig. 6.15.

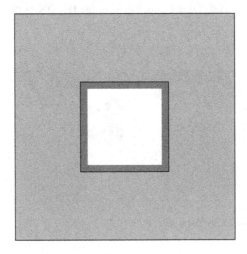

Fig. 6.15 Outline of brick and concrete walls.

From the ANSYS Main Menu, select **Preprocessor** → **Meshing** → **Size Cntrls** → **ManualSize** → **Global** → **Size**. As a result of this selection, the frame shown in Fig. 6.16 appears.

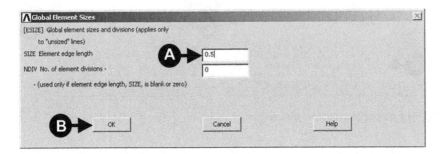

Fig. 6.16 Global element sizes.

Set the input for the element edge length as [A] **SIZE** = 0.5 and click the [B] **OK** button.

Because the outer brick wall and inner concrete wall were created as separate entities, it is necessary to 'glue' them together so they share lines along their common boundaries. From the ANSYS Main Menu, select **Preprocessor → Modelling → Operate → Boolean → Glue → Areas**. The frame shown in Fig. 6.17 appears.

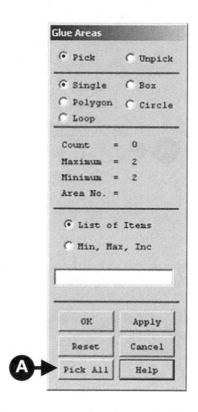

Fig. 6.17 Glue areas.

Select the [A] **Pick All** option in the frame of Fig. 6.17 to glue the outer and inner wall areas. Before meshing occurs, it is necessary to specify material numbers for the concrete and the brick walls.

From the ANSYS Main Menu, select **Preprocessor → Meshing → Mesh Attributes → Picked Areas**. The frame shown in Fig. 6.18 is created.

Area Attributes
Pick
Unpick
Single
Box
Polygon
Circle
Loop
Count = 0
Maximum = 2
Minimum = 1
Area No. =
List of Items
Min, Max, Inc
A
OK
Apply
Reset
Cancel
Pick All
Help

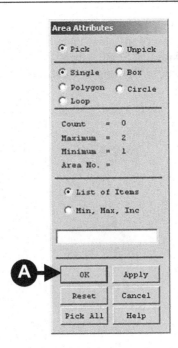

Fig. 6.18 Area attributes.

Select the first concrete wall area and click the [A] **OK** button in the frame of Fig. 6.18. A new frame is produced, as shown in Fig. 6.19.

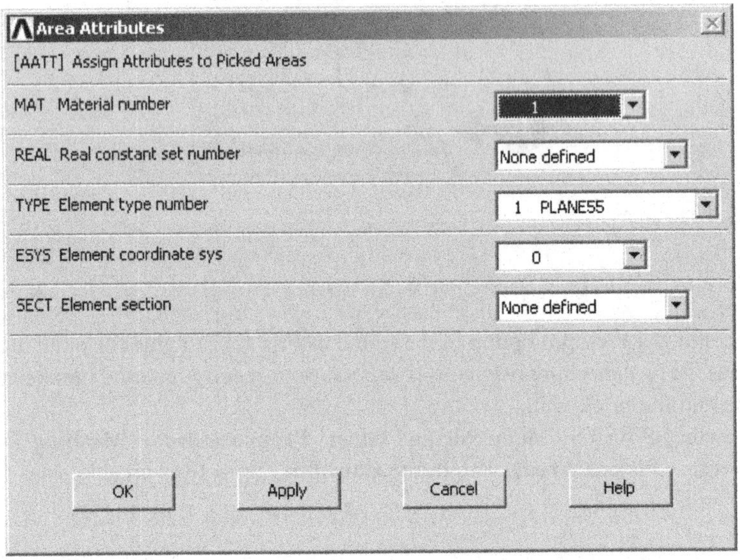

Area Attributes
[AATT] Assign Attributes to Picked Areas
MAT Material number
1
REAL Real constant set number
None defined
TYPE Element type number
1 PLANE55
ESYS Element coordinate sys
0
SECT Element section
None defined
OK
Apply
Cancel
Help

Fig. 6.19 Area attributes (concrete wall).

Material Number 1 is assigned to the concrete inner wall, as shown in Fig. 6.19.

Next, assign Material Number 2 to the brick outer wall following the procedure outlined above and select the brick outer wall. Fig. 6.20 shows a frame with an appropriate entry.

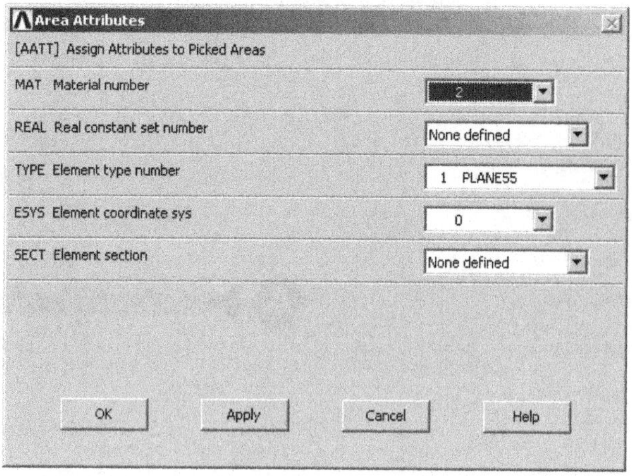

Area Attributes
[AATT] Assign Attributes to Picked Areas
MAT Material number
2
REAL Real constant set number
None defined
TYPE Element type number
1 PLANE55
ESYS Element coordinate sys
0
SECT Element section
None defined
OK
Apply
Cancel
Help

Fig. 6.20 Area attributes.

Now meshing of both areas can be carried out. From the ANSYS Main Menu, select **Preprocessor** → **Meshing** → **Mesh** → **Areas** → **Free**. The frame shown in Fig. 6.21 appears.

Mesh Areas
Pick
Unpick
Single
Box
Polygon
Circle
Loop
Count = 0
Maximum = 2
Minimum = 1
Area No. =
List of Items
Min, Max, Inc
OK
Apply
Reset
Cancel
A
Pick All
Help

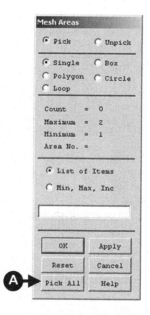

Fig. 6.21 Mesh areas.

Select [A] **Pick All** option shown in Fig. 6.21 to mesh both areas.

In order to see both areas meshed, from the Utility Menu, select **Plot-Ctrls → Numbering**. In the resulting frame, shown in Fig. 6.22, select [A] **Material Numbers** and click the [B] **OK** button.

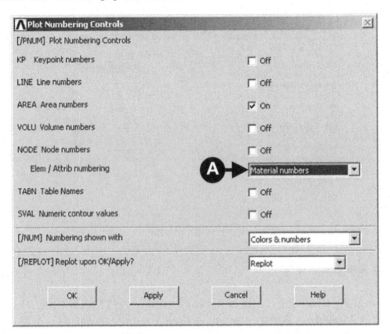

Fig. 6.22 Plot numbering controls.

As a result, both walls with mesh are displayed (see Fig. 6.23).

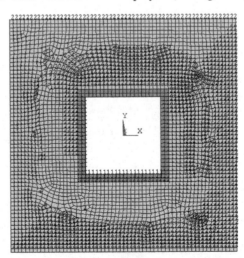

Fig. 6.23 Outer and inner wall of the furnace meshed.

6.2.3 Solution

Before a solution can be run, boundary conditions have to be applied. From the ANSYS Main Menu, select **Solution → Define Loads → Apply → Thermal → Convection → On Lines**. This selection produces a frame shown in Fig. 6.24.

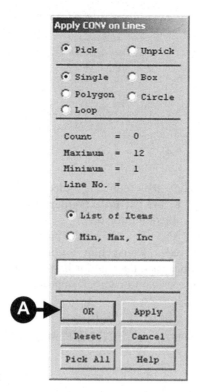

Fig. 6.24 Apply CONV on lines.

First, pick the convective lines (facing inside the furnace) of the concrete wall and press the [A] **OK** button. The frame shown in Fig. 6.25 is created.

Fig. 6.25 Apply CONV on lines
(the inner wall).

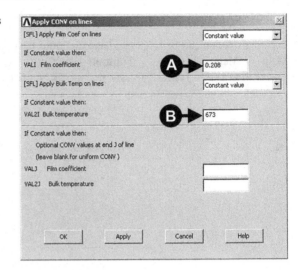

As seen in Fig. 6.25, the following selections were made: [A] **Film coefficient** = 0.208 W/m² K and [B] **Bulk temperature** = 673 K, as specified for the concrete wall in the problem formulation.

Again from the ANSYS Main Menu, select **Solution** → **Define Loads** → **Apply** → **Thermal** → **Convection** → **On Lines**. The frame shown in Fig. 6.24 appears. This time, pick the exterior lines of the brick wall and press the [A] **OK** button. The frame shown in Fig. 6.26 appears.

Fig. 6.26 Apply COVN on lines
(the outer wall).

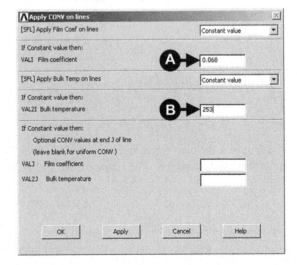

For the outer brick wall, the following selections were made (see the frame in Fig. 6.26): [A] **Film coefficient** $= 0.068$ W/m^2 K and [B] **Bulk temperature** $= 253$ K as specified for the brick wall in the problem formulation.

Finally, to see the applied convective boundary conditions from the Utility Menu, select **PlotCtrls → Symbols**. The frame shown in Fig. 6.27 appears.

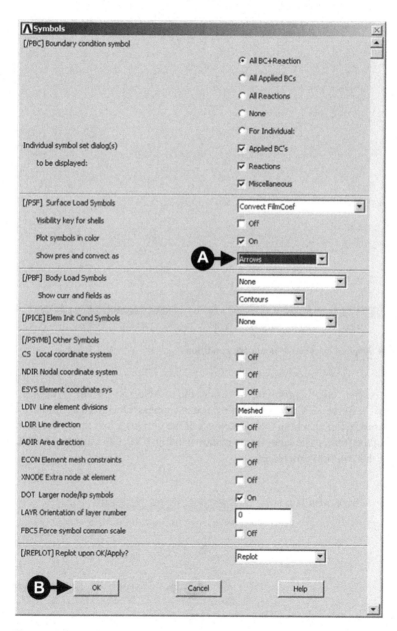

Fig. 6.27 Symbols.

In the frame shown in Fig. 6.27, select [A] **Show pres and convect as** = **Arrows** and click the [B] **OK** button.

From the Utility Menu, select **Plot → Lines** to produce an image shown in Fig. 6.28.

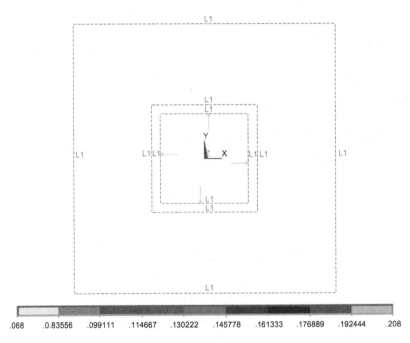

.068   .0.83556   .099111   .114667   .130222   .145778   .161333   .176889   .192444   .208

Fig. 6.28 Applied convective boundary conditions.

To solve the problem, from the ANSYS Main Menu, select **Solution → Solve → Current LS**. Two frames appear. One gives a summary of solution options. After checking the correctness of the options, close this using the menu at the top of the frame. The other frame is shown in Fig. 6.29. Clicking the [A] **OK** button initiates the solution process.

Fig. 6.29 Solve current load step.

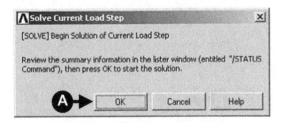

6.2.4 Postprocessing

The end of the successful solution process is denoted by the message 'Solution is done'. The postprocessing phase can be started. It is first necessary to obtain information about temperatures and heat fluxes across the furnace's walls.

From the ANSYS Main Menu, select **General Postproc → Plot Results → Contour Plot → Nodal Solu**. The frame shown in Fig. 6.30 appears.

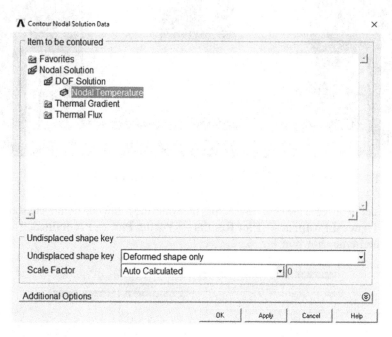

Fig. 6.30 Contour nodal solution data.

Selections made are shown in Fig. 6.30. Clicking the [A] **OK** button results in the graph shown in Fig. 6.31.

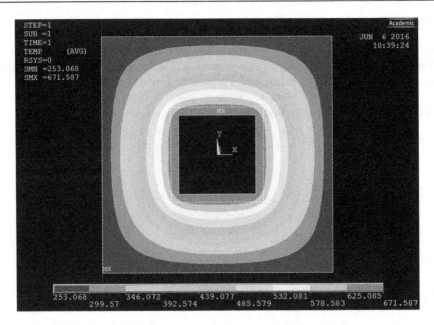

Fig. 6.31 Temperature distribution in the furnace as a contour plot.

In order to observe the heat flow across the walls the following command should be issued: **General Postproc → Plot Results → Vector Plot → Predefined**. This produces the frame shown in Fig. 6.32.

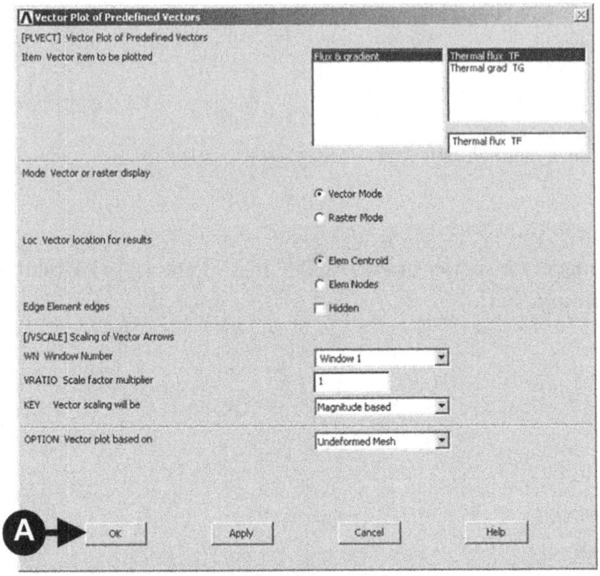

Fig. 6.32 Vector plot of predefined vectors.

Clicking the [A] **OK** button produces a graph as shown in Fig. 6.33.

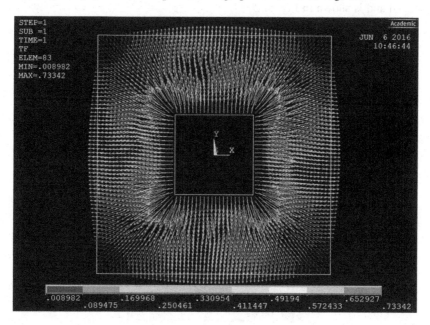

Fig. 6.33 Heat flow across the wall plotted as vectors.

In order to observe temperature variations across the walls, it is necessary to define the path along which the variations are going to be determined. From the Utility Menu, select **Plot: Areas**. Next, from the ANSYS Main Menu, select **General Post-proc → Path Operations → Define Path → On Working Plane**. The resulting frame is shown in Fig. 6.34.

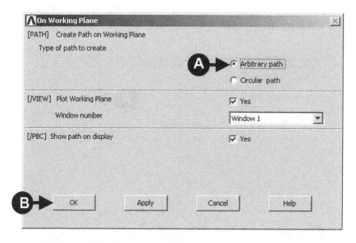

Fig. 6.34 On working plane (definition of the path).

By activating the [A] **Arbitrary path** button and clicking [B] **OK**, another frame is produced and is shown in Fig. 6.35.

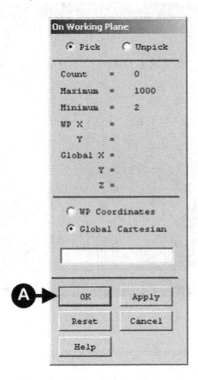

Fig. 6.35 On working plane (selection of two points defining the path).

Two points should be selected by clicking on the inner line of the concrete wall and moving in Y-direction at the right angle, clicking on the outer liner of the brick wall. As a result of clicking the [A] **OK** button, the frame shown in Fig. 6.36 appears.

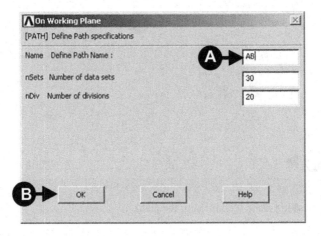

Fig. 6.36 On working plane (path name: AB).

In the box [A] **Define Path Name**, write **AB** and click the [B] **OK** button.

From the ANSYS Main Menu, select **General Postproc → Path Operations → Map onto Path**. The frame shown in Fig. 6.37 appears.

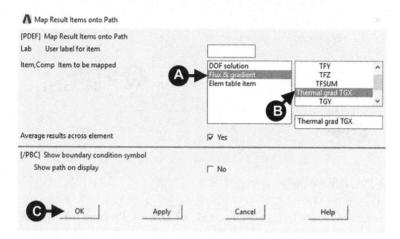

Fig. 6.37 Map results items onto path (AB path).

In Fig. 6.37, the following selections are made: [A] **Flux & gradient**, [B] **Thermal grad TGX**; then click the [C] **OK** button. Repeating the steps described above, recall the frame shown in Fig. 6.37. This time, select the following: [A] **Flux & gradient**, [B] **Thermal grad TGY**, and click the [C] **OK** button. Finally, recall the frame shown in Fig. 6.37 and select [A] **Flux & gradient** and [B] **Thermal grad TGSUM** as shown in Fig. 6.38, then click the [C] **OK** button.

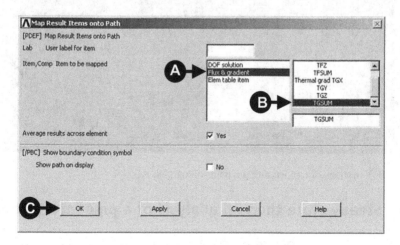

Fig. 6.38 Map results items onto path (AB path).

From the ANSYS Main Menu, select **General Postproc** → **Path Operations** → **Plot Path Item** → **On Graph**. The frame shown in Fig. 6.39 appears.

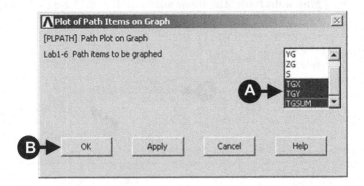

Fig. 6.39 Plot of path items on graph.

The selections made [A] are highlighted in Fig. 6.39. Pressing the [B] **OK** button results in a graph as shown in Fig. 6.40.

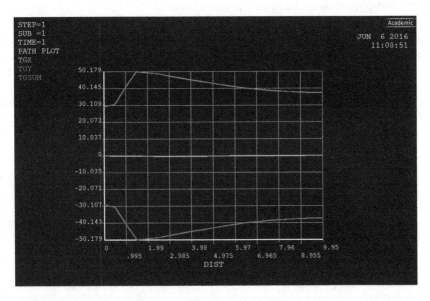

Fig. 6.40 Variations of temperature gradients along path AB.

6.3 Steady-state thermal analysis of a pipe intersection

6.3.1 Description of the problem

A cylindrical tank is penetrated radially by a small pipe at a point on its axis remote from the ends of the tank, as shown in Fig. 6.41.

Fig. 6.41 Pipe intersection.

The inside of the tank is exposed to fluid with a temperature of 232°C. The pipe experiences a steady flow of fluid with a temperature of 38°C, and the two flow regimes are isolated from each other by means of a thin tube. The convection (film) coefficient in the pipe varies with the metal temperature and is thus expressed as a material property. The objective is to determine the temperature distribution at the pipe-tank junction.

The following data describing the problem are given:

- inside diameter of the pipe = 8 mm;
- outside diameter of the pipe = 10 mm;
- inside diameter of the tank = 26 mm;
- outside diameter of the tank = 30 mm;
- inside bulk fluid temperature, tank = 232°C;
- inside convection coefficient, tank = 4.92 W/m² °C;
- inside bulk fluid temperature, pipe = 38°C; and
- inside convection (film) coefficient, pipe = −2.

Table 6.1 provides information about variation of the thermal parameters with temperature.

Table 6.1 Variation of the thermal parameters with temperature

	Temperature (°C)				
	21	**93**	**149**	**204**	**260**
Convection coefficient (W/m² °C)	41.918	39.852	34.637	27.06	21.746
Density (kg/m³)	7889	7889	7889	7889	7889
Conductivity (J/s m °C)	0.2505	0.267	0.2805	0.294	0.3069
Specific heat (J/kg °C)	6.898	7.143	7.265	7.448	7.631

The assumption is made that the quarter symmetry is applicable and that, at the terminus of the model (longitudinal and circumferential cuts in the tank), there is sufficient attenuation of the pipe effects such that these edges can be held at 232°C.

The solid model is constructed by intersecting the tank with the pipe and then removing the internal part of the pipe using a Boolean operation.

Boundary temperatures along with the convection coefficients and bulk fluid temperatures are dealt with in the solution phase, after which a static solution is executed.

Temperature contours and thermal flux displays are obtained in postprocessing.

Details of steps taken to create the model of the pipe intersecting with the tank are outlined below.

6.3.2 Preparation for model building

From the ANSYS Main Menu, select **Preferences**. This frame is shown in Fig. 6.42.

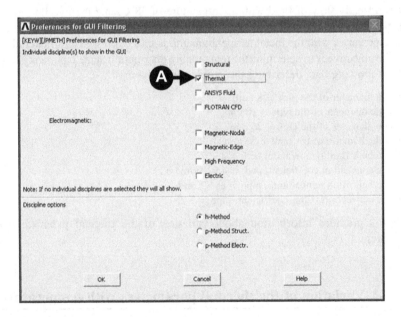

Fig. 6.42 Preferences: thermal.

Depending on the nature of analysis to be attempted, an appropriate analysis type should be selected. In the problem considered here, [A] **Thermal** was selected, as shown in Fig. 6.42.

From the ANSYS Main Menu, select **Preprocessor** and then **Element Type** and **Add/Edit/Delete**. The frame shown in Fig. 6.43 appears.

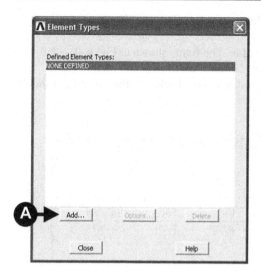

Fig. 6.43 Element types selection.

Clicking the [A] **Add** button activates a new set of options, which are shown in Fig. 6.44.

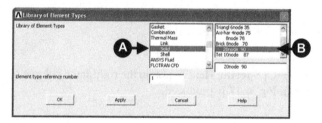

Fig. 6.44 Library of element types.

Fig. 6.44 indicates that for the problem considered here, the following was selected: [A] **Thermal Mass** → **Solid** and [B] **20node 90**.

From the ANSYS Main Menu, select **Material Props** and then **Material Models**. Fig. 6.45 shows the resulting frame.

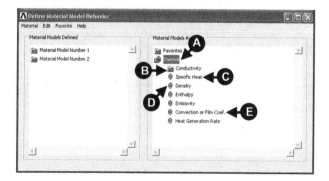

Fig. 6.45 Define material model behaviour.

From the options listed on the right-hand side, select [A] **Thermal** as shown in Fig. 6.45.

Next select [B] **Conductivity, Isotropic**. The frame shown in Fig. 6.46 is built up by pressing the **Add Temperature** button. When all five temperatures and corresponding KXX are entered in accordance with Table 6.1, the **OK** button should be pressed.

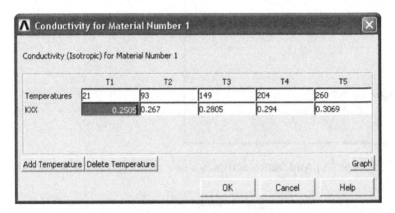

Fig. 6.46 Conductivity for Material Number 1.

By selecting the [C] **Specific Heat** option in the right-hand column (see Fig. 6.45), the frame shown in Fig. 6.47 is produced.

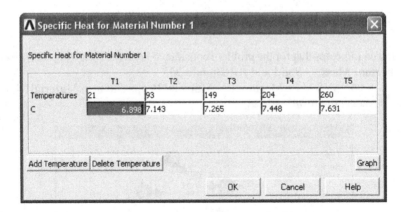

Fig. 6.47 Specific heat for Material Number 1.

Using a similar approach to that described above for conductivity, appropriate values of specific heat versus temperature, taken from Table 6.1, are typed as shown in Fig. 6.47.

The next material property to be defined is density. According to Table 6.1, density is constant for all temperatures used. Therefore, selecting [D] **Density** from the right-hand column (see Fig. 6.45) results in the frame shown in Fig. 6.48.

A density of 7888.8 kg/m^3 is typed in the boxes for various temperatures shown in Fig. 6.48.

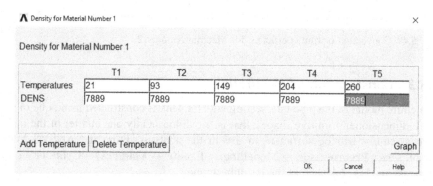

Fig. 6.48 Density for Material Number 1.

All the above properties were used to characterise Material Number 1. The convection of the film coefficient is another important parameter characterising the system being analysed. However, this is a property belonging not to Material Number 1 (material of the tank and pipe) but to a thin film formed by the liquid on solid surfaces. It is a different entity, and is therefore called Material Number 2. Therefore, from the top menu **Material** (shown in Fig. 6.45), select **New Model** number 2. Next, [E] **Convection or Film Coef.** (see Fig. 6.45) should be selected and the frame shown in Fig. 6.49 created. Appropriate values of the film coefficient for various temperatures, taken from Table 6.1, are introduced as shown in Fig. 6.49. The consecutive temperatures T1, T2, T3, T4 and T5 and corresponding specific heat values are obtained by pressing the **Add Temperature** button.

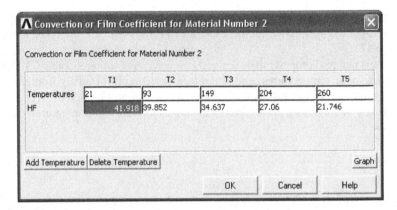

Fig. 6.49 Convection or film coefficient for Material Number 2.

6.3.3 Construction of the model

The entire model of the pipe intersecting with the tank is constructed using one of the three-dimensional primitive shapes, that is, a cylinder. Only one quarter of the tank-pipe assembly will be sufficient to use in the analysis. From the ANSYS Main Menu, select **Preprocessor → Modelling → Create → Volumes → Cylinder → By Dimensions**. Fig. 6.50 shows the resulting frame.

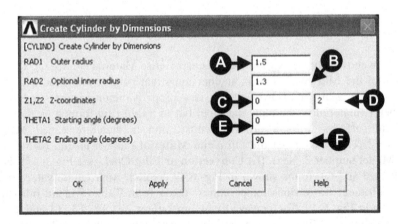

Fig. 6.50 Create cylinder by dimensions.

In Fig. 6.50, as shown, the following inputs are made: [A] **RAD1** = 1.5 cm; [B] **RAD2** = 1.3 cm; [C] **Z1** = 0; [D] **Z2** = 2 cm; [E] **THETA1** = 0; [F] **THETA2** = 90.

As the pipe axis is at right angles to the cylinder axis, it is necessary to rotate the working plane to the pipe axis by 90 degrees. This is achieved by selecting from the

Utility Menu **WorkPlane** → **Offset WP by Increments**. The resulting frame is shown in Fig. 6.51.

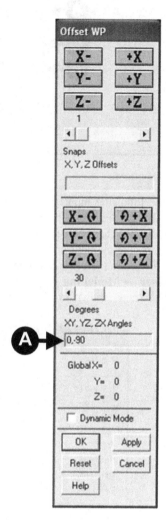

Fig. 6.51 Offset WP by increments.

In Fig. 6.51, the input is shown as [A] **XY** = 0; **YZ** = −90 and the ZX is left unchanged from default value. Next, from the ANSYS Main Menu, select **Preprocessor** → **Modelling** → **Create** → **Volumes** → **Cylinder** → **By   Dimensions**. Fig. 6.52 shows the resulting frame.

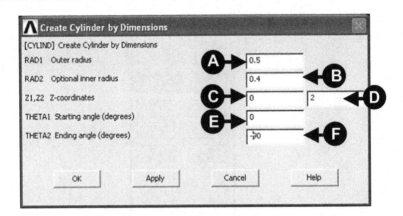

Fig. 6.52 Create cylinder by dimensions.

In Fig. 6.52, as shown, the following inputs are made: [A] **RAD1** = 0.5 cm; [B] **RAD2** = 0.4 cm; [C] **Z1** = 0; [D] **Z2** = 2 cm; [E] **THETA1** = 0; [F] **THETA2** = − 90.

After that the working plane should be set to the default setting by inputting in Fig. 6.51 **YZ** = 90 this time. As the cylinder and the pipe are separate entities, it is necessary to overlap them in order to make one component. From the ANSYS MainMenu, select **Preprocessor** → **Modelling** → **Create** → **Operate** → **Booleans** → **Overlap** → **Volumes**. The frame shown in Fig. 6.53 is created.

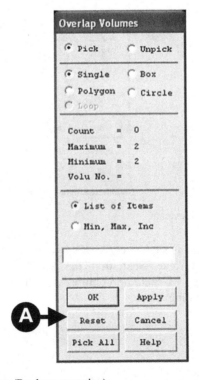

Fig. 6.53 Overlap volumes (Boolean operation).

Pick both elements (cylinder and pipe) and press the [A] **OK** button to execute the selection. Next, activate volume numbering, which will be of use when carrying out further operations on volumes. This is done by selecting from the Utility Menu **Plot-Ctrls** → **Numbering** and checking the **VOLU** option in the resulting frame.

Finally, a three-dimensional view of the model should be set by selecting the following from the Utility Menu: **PlotCtrls** → **View Settings** → **Viewing Direction**. The resulting frame is shown in Fig. 6.54.

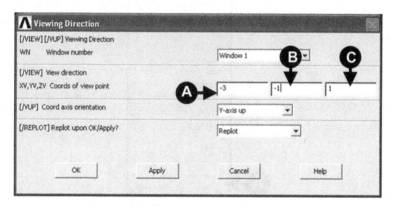

Fig. 6.54 View settings.

The following inputs should be made (see Fig. 6.54): [A] **XV** = − 3; [B] **YV** = − 1; [C] **ZV** = 1 to plot the model, as shown in Fig. 6.55. However, this is not the only possible view of the model, and any other preference may be chosen.

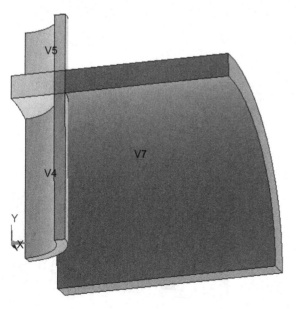

Fig. 6.55 Quarter symmetry model of the tank-pipe intersection.

Certain volumes of the models, shown in Fig. 6.55, are redundant and should be deleted. From the ANSYS Main Menu, select **Preprocessor** → **Modelling** → **Delete** → **Volumes and Below**. Fig. 6.56 shows the resulting frame.

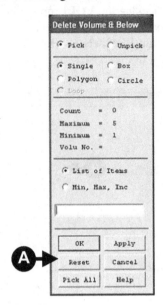

Fig. 6.56 Delete volumes and below.

Volumes V4 and V3 (a corner of the cylinder) should be picked and the [A] **OK** button pressed to implement the selection. After the delete operation, the model should look like that shown in Fig. 6.57.

Fig. 6.57 Quarter symmetry
model of the tank-pipe intersection
after VDELE command.

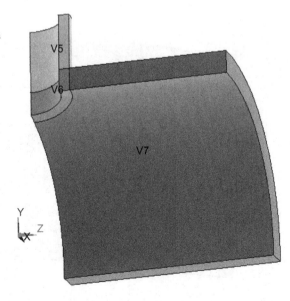

Finally, volumes V5, V6 and V7 should be added to create a single volume required for further analysis. From the ANSYS Main Menu, select **Preprocessor: Modelling: Operate: Booleans: Add: Volumes**. The resulting frame asks for picking volumes to be added. Pick all three volumes—V5, V6, and V7—and click the **OK** button to implement the operation. Fig. 6.58 shows the model of the pipe intersecting the cylinder as one volume V1.

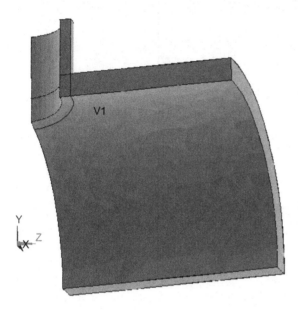

Fig. 6.58 Quarter symmetry model of the tank-pipe intersection represented by a single volume V1.

Meshing of the model usually begins with setting the size of elements to be used. From the ANSYS Main Menu, select **Meshing → Size Cntrls → SmartSize → Basic**. A frame, shown in Fig. 6.59, appears.

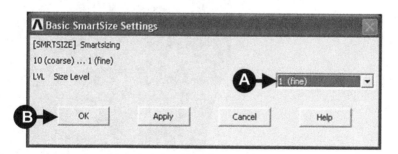

Fig. 6.59 Basic smartsize settings.

For the case considered, [A] **Size Level**—1(fine) was selected, as shown in Fig. 6.59. Clicking the [B] **OK** button implements the selection. Next, from the

ANSYS Main Menu, select **Mesh** → **Volumes** → **Free**. The frame shown in Fig. 6.60 appears. Select the volume to be meshed and click the [A] **OK** button.

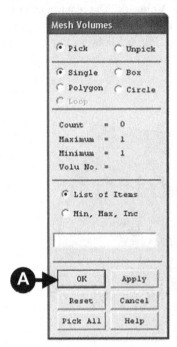

Fig. 6.60 Mesh volumes frame.

The resulting network of elements is shown in Fig. 6.61.

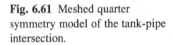

Fig. 6.61 Meshed quarter symmetry model of the tank-pipe intersection.

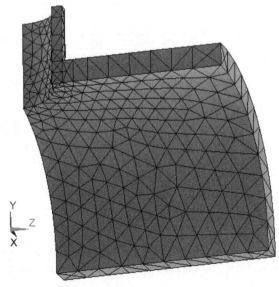

6.3.4 Solution

The meshing operation ends the model construction and the Preprocessor stage. The solution stage can now be started. From the ANSYS Main Menu, select **Solution → Analysis Type → New Analysis**. Fig. 6.62 shows the resulting frame.

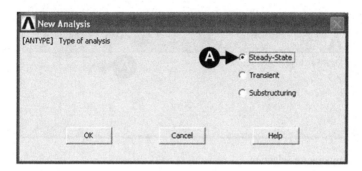

Fig. 6.62 New analysis window.

Activate the [A] **Steady-State** button. Next, select **Solution → Analysis Type → Analysis Options**. In the resulting frame, shown in Fig. 6.63, select the [A] **Program chosen** option.

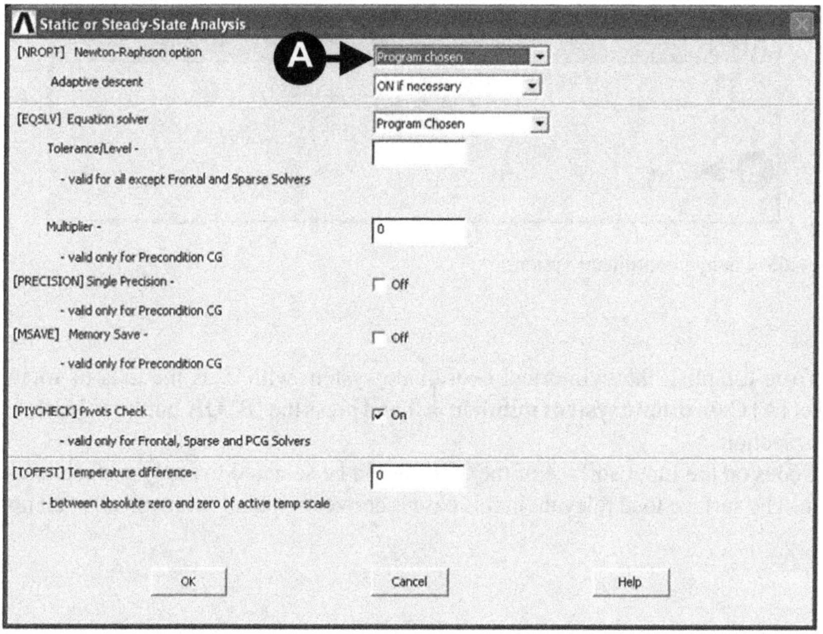

Fig. 6.63 Analysis options.

To set the starting temperature of 232°C at all nodes, select **Solution** → **Define Loads** → **Apply** → **Thermal** → **Temperature** → **Uniform Temp**. Fig. 6.64 shows the resulting frame. Input [A] **Uniform temperature** = 232°C, as shown in Fig. 6.64.

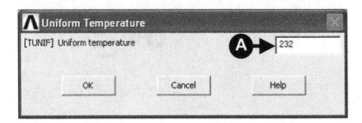

Fig. 6.64 Temperature selection.

From the Utility Menu, select **WorkPlane: Change Active CS to** → **Specified Coord Sys**. As a result, the frame shown in Fig. 6.65 appears.

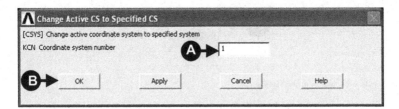

Fig. 6.65 Change coordinate system.

To re-establish the cylindrical coordinate system with Z as the axis of rotation, select [A] **Coordinate system number** = 1 and press the [B] **OK** button to implement the selection.

Nodes on the inner surface of the tank should be selected to apply surface loads to them. The surface load relevant in this case is convection load acting on all at all nodes located on the inner surface of the tank. From the Utility Menu, choose **Select → Entities**. The frame shown in Fig. 6.66 appears.

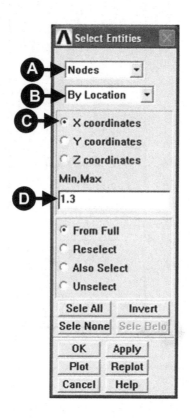

Fig. 6.66 Select entities.

From the first pull-down menu, select [A] **Nodes**; from the second pull-down menu, select [B] **By location**. Also, activate [C] **X coordinates** button and enter [D] **Min,Max** = 1.3 (inside radius of the tank). All four required steps are shown in Fig. 6.66. When the subset of nodes on the inner surface of the tank is selected, then the convection load at all nodes has to be applied. From the ANSYS Main Menu, select **Solution → Define Load → Apply → Thermal → Convection → On nodes**. The resulting frame is shown in Fig. 6.67.

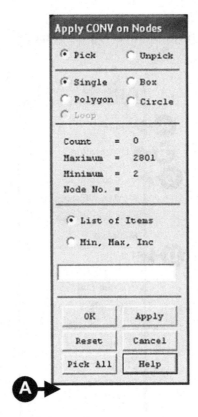

Fig. 6.67 Apply thermal convection on nodes.

Press [A] **Pick All** to call up another frame as shown in Fig. 6.68.

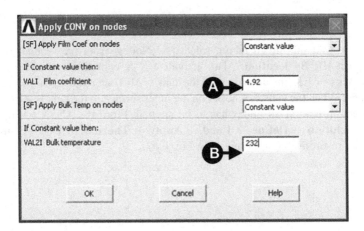

Fig. 6.68 Select all nodes.

Inputs into the frame of Fig. 6.68 are shown as: [A] **Film coefficient** = 4.92 and [B] **Bulk temperature** = 232. Both quantities are taken from the problem description.

From the Utility Menu, choose **Select → Entities** in order to select a subset of nodes located at the far edge of the tank. The frame shown in Fig. 6.69 appears.

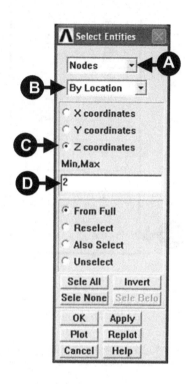

Fig. 6.69 Select entities.

From the first pull-down menu, select [A] **Nodes**; from the second pull-down menu, select [B] **By location**. Also, activate [C] **Z coordinates** button and [D] enter **Min,Max** = 2 (the length of the tank in Z-direction). All four required steps are shown in Fig. 6.69. Next, constraints at nodes located at the far edge of the tank (additional subset of nodes just selected) have to be applied. From the ANSYS Main Menu, select **Solution → Define Loads → Apply → Thermal → Temperature → On Nodes**. The frame shown in Fig. 6.70 appears.

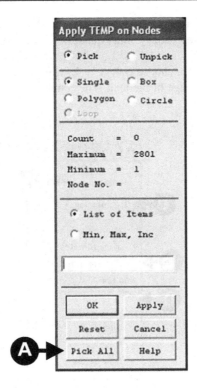

Fig. 6.70 Select all nodes.

Click the [A] **Pick All** button as shown in Fig. 6.70. This action brings up another frame, as shown in Fig. 6.71.

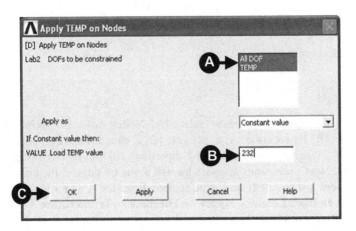

Fig. 6.71 Apply temperature to all nodes.

Activate both [A] **All DOF** and **TEMP** and input [B] **TEMP value** $= 232°C$, as shown in Fig. 6.71. Finally, click [C] **OK** to apply temperature constraints on nodes at the far edge of the tank. The steps outlined above should be followed to apply constraints at nodes located at the bottom of the tank. From the Utility Menu, choose **Select: Entities**. The frame shown in Fig. 6.72 appears.

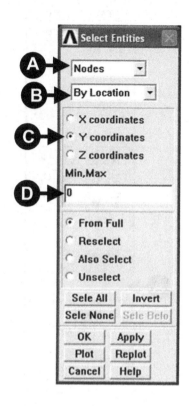

Fig. 6.72 Select entities.

From the first pull-down menu, select [A] **Nodes**; from the second pull-down menu, select [B] **By location**. Also, activate the [C] **Y coordinates** button and [D] enter **Min,Max** $= 0$ (location of the bottom of the tank in Y-direction). All four required steps are shown in Fig. 6.72. Next, constraints at the nodes located at the bottom of the tank (additional subset of nodes selected above) must be applied. From the ANSYS Main Menu, select **Solution → Define Loads → Apply → Thermal → Temperature → On Nodes**. The frame shown in Fig. 6.70 appears. As shown in Fig. 6.70, click [A] **Pick All** to bring up the frame shown in Fig. 6.71. As before, activate both [A] **All DOF** and **TEMP** and input [B] **TEMP value** $= 232°C$. Clicking [C] **OK** applies the temperature constraints on the nodes at the bottom of the tank.

Now, it is necessary to rotate the working plane (WP) to the pipe axis. From the Utility Menu, select **WorkPlane → Offset WP by Increments**. Fig. 6.73 shows the resulting frame.

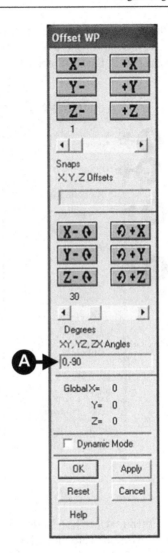

Fig. 6.73 Offset WP by increments.

In the degrees box, input [A] **XY** $= 0$ and **YZ** $= -90$ as shown. Having WP rotated to the pipe axis, a local cylindrical coordinate system must be defined at the origin of the working plane. From the Utility Menu, select **WorkPlane** $\rightarrow$ **Local Coordinate Systems** $\rightarrow$ **Create local CS** $\rightarrow$ **At WP Origin**. The resulting frame is shown in Fig. 6.74.

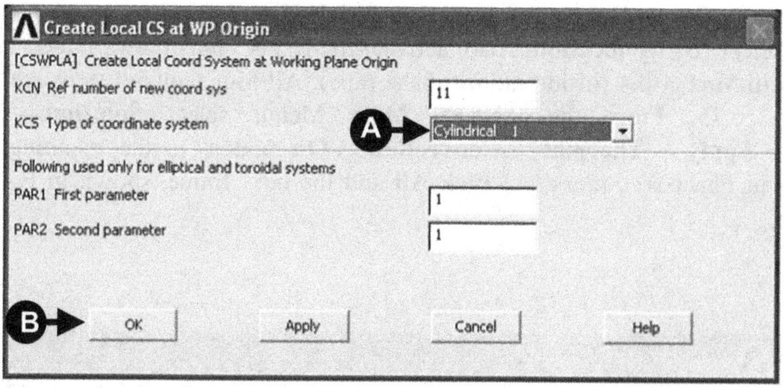

Fig. 6.74 Create local CS.

From the pull-down menu, select [A] **Cylindrical 1** and click the [B] **OK** button to implement the selection. The analysis involves nodes located on the inner surface of the pipe. To include this subset of nodes, from the Utility Menu, choose **Select → Entities**. Fig. 6.75 shows the resulting frame.

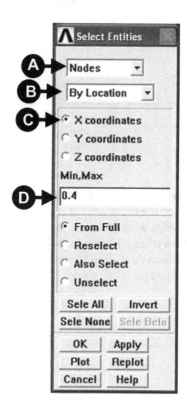

Fig. 6.75 Select entities.

From the first pull-down menu, select [A] **Nodes**; from the second pull-down menu, select [B] **By location**. Also, activate the [C] **X coordinates** button and [D] enter **Min,Max** = 0.4 (inside radius of the pipe). All four required steps are shown in Fig. 6.75. From the ANSYS Main Menu, select **Solution** → **Define Load** → **Apply** → **Thermal** → **Convection** → **On nodes**. In the resulting frame (shown in Fig. 6.67), press [A] **Pick All** and the next frame, shown in Fig. 6.76, appears.

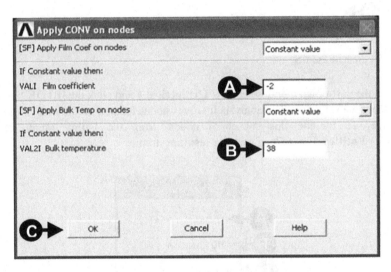

Fig. 6.76 Apply convection on nodes.

Input [A] **Film coefficient** = −2 and [B] **Bulk temperature** = 38, as shown in Fig. 6.76. Pressing the [C] **OK** button implements the selections. The final action is to select all entities involved with a single command. Therefore, from the Utility Menu, choose **Select** → **Everything**. For the loads to be applied to tank and pipe surfaces in the form of arrows from the Utility Menu, select **PlotCtrls** → **Symbols**. The frame in Fig. 6.77 shows the required selection: [A] **Arrows**.

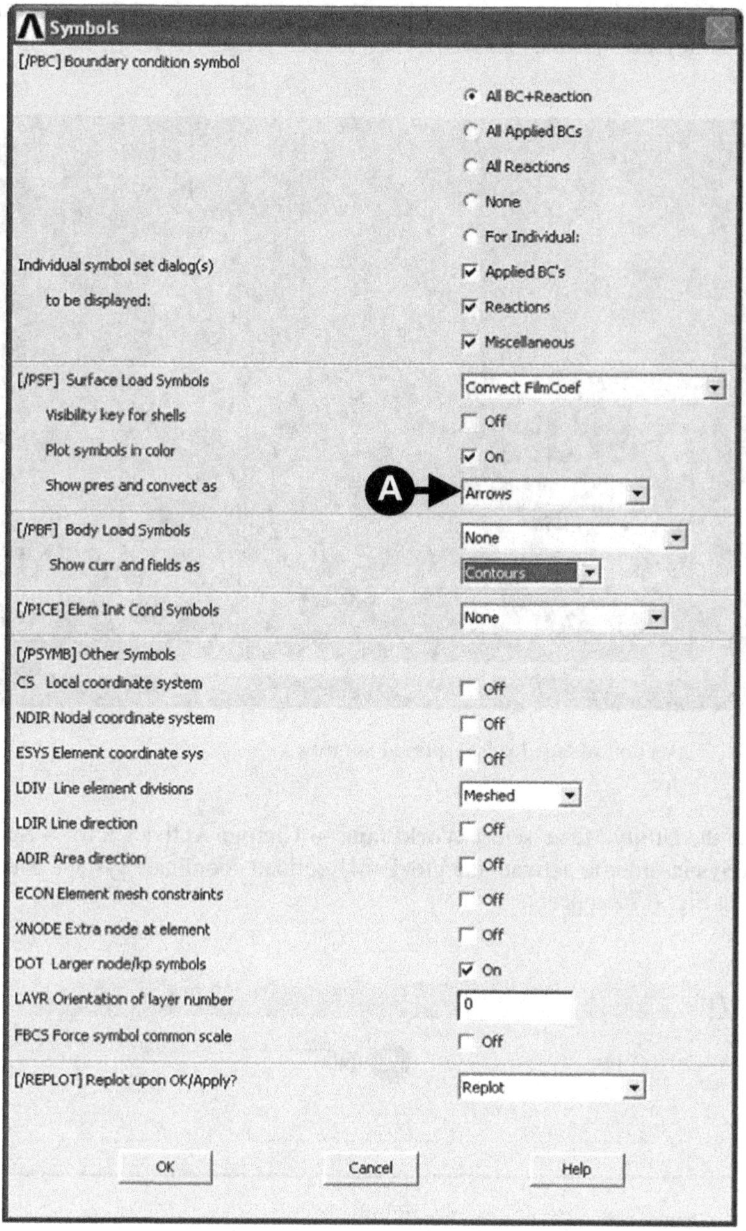

Fig. 6.77 Select symbols.

From the Utility Menu, selecting **Plot: Nodes** results in Fig. 6.78, where surface loads at nodes are shown as arrows.

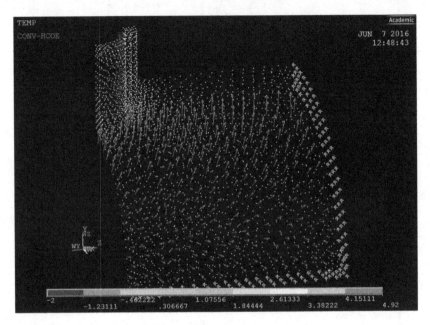

Fig. 6.78 Convection surface loads displayed as arrows.

From the Utility Menu, select **WorkPlane → Change Active CS to → Specified Coord Sys** in order to activate the previously defined coordinate system. The frame shown in Fig. 6.79 appears.

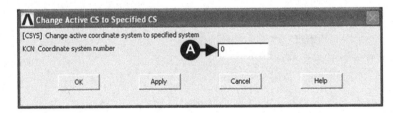

Fig. 6.79 Change active CS to specified CS.

Input [A] **KCN** (coordinate system number) = 0 to return to Cartesian system. Additionally, from the ANSYS Main Menu, select **Solution → Analysis Type → Sol'n Controls**. As a result, the frame shown in Fig. 6.80 appears.

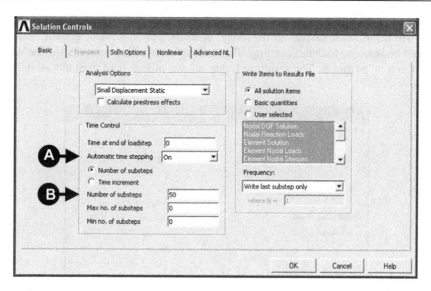

Fig. 6.80 Solution controls.

Input the following: [A] **Automation time stepping** = On and [B] **Number of substeps** = 50, as shown in Fig. 6.80. Finally, from the ANSYS Main Menu, select **Solve → Current LS** and in the dialog box that appears, click the **OK** button to start the solution process.

6.3.5 Postprocessing stage

Once the solution is complete, the next stage is to display results in a form required to answer questions posed by the formulation of the problem.

From the Utility Menu, select **PlotCtrls → Style → Edge Options**. Fig. 6.81 shows the resulting frame.

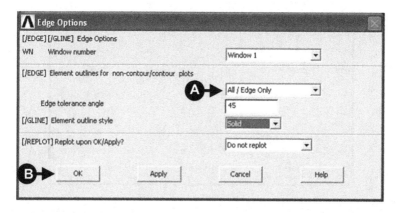

Fig. 6.81 Edge option.

Select [A] **All/Edge only** and press the [B] **OK** button to implement the selection; this will result in the display of the 'edge' of the object only. Next, graphic controls should be returned to default setting. This is done by selecting from the Utility Menu **PlotCtrls → Symbols**. The resulting frame, shown in Fig. 6.82, contains all default settings.

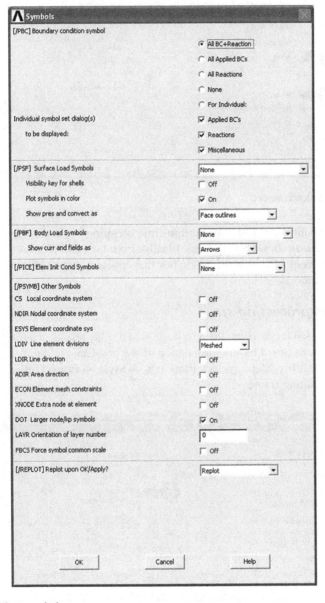

Fig. 6.82 Select symbols.

The first plot is to show temperature distribution as continuous contours. From the ANSYS Main Menu, select **General Postproc → Plot Results → Contour Plot → Nodal Solu**. The resulting frame is shown in Fig. 6.83.

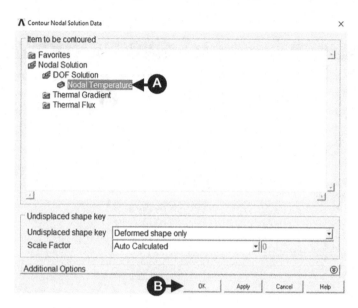

Fig. 6.83 Nodal solution.

Select [A] **Temperature** and press the [B] **OK** button as shown in Fig. 6.83. The resulting temperature map is shown in Fig. 6.84.

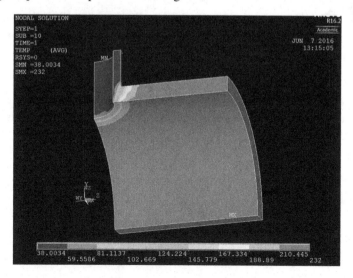

Fig. 6.84 Temperature map on inner surfaces of the tank and the pipe.

The next display of results concerns thermal flux at the intersection between the tank and the pipe. From the ANSYS Main Menu, select **General Postproc** → **Plot Results** → **Vector Plot** → **Predefined**. The resulting frame is shown in Fig. 6.85.

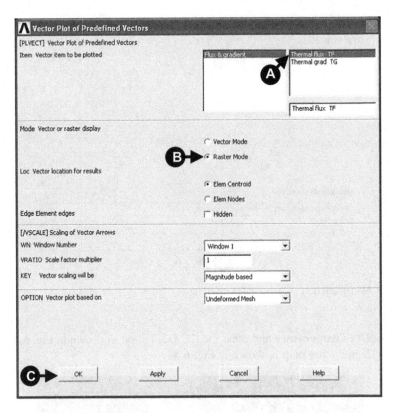

Fig. 6.85 Vector plot selection.

In Fig. 6.85, select [A] **Thermal flux TF** and [B] **Raster Mode**. Pressing the [C] **OK** button implements selections and produces thermal flux as vectors. This is shown in Fig. 6.86.

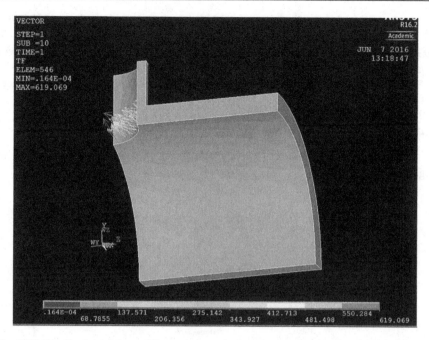

Fig. 6.86 Distribution of thermal flux vectors at the intersection between the tank and the pipe.

6.4 Heat dissipation through a developed surface

6.4.1 Problem description

Ribbed or developed surfaces, also called fins, are frequently used to dissipate heat. There are many examples of their use in practical engineering applications, such as computers, electronic systems, and radiators, to mention a few.

Fig. 6.87 shows a typical configuration and geometry of a fin made of aluminium with thermal conductivity coefficient $k = 170$ W/m K.

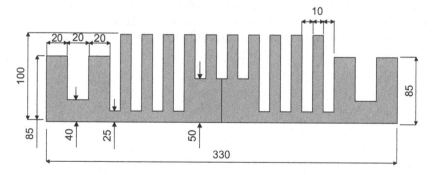

Fig. 6.87 Cross-section of the fin.

The bottom surface of the fin is exposed to a constant heat flux of $q = 1000$ W/m. Air flows over the developed surface, keeping the surrounding temperature at 293 K. The heat transfer coefficient between the fin and the surrounding atmosphere is $h = 40$ W/m^2 K.

The task is to determine the temperature distribution within the developed surface.

6.4.2 Construction of the model

From the ANSYS Main Menu, select **Preferences** to call up a frame shown in Fig. 6.88.

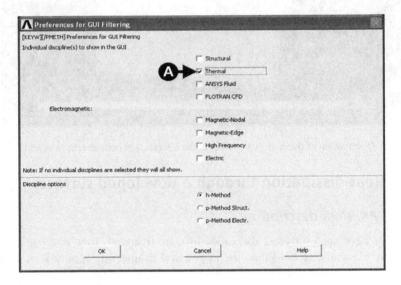

Fig. 6.88 Preferences: thermal.

Because the problem to be solved is asking for temperature distribution, [A] **Thermal** is selected, as indicated in the figure. Next, from the ANSYS Main Menu, select **Preprocessor → Element Type → Add/Edit/Delete**. The frame shown in Fig. 6.89 appears.

Fig. 6.89 Define element type.

Click the [A] **Add** button to call up another frame as shown in Fig. 6.90.

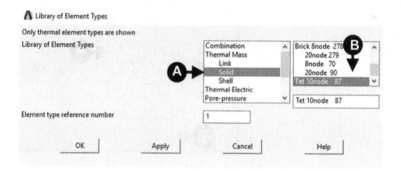

Fig. 6.90 Library of element types.

In Fig. 6.90, the following selections are made: [A] **Thermal Mass** → **Solid** and [B] **Tet 10 node 87**. From the ANSYS Main Menu, select **Preprocessor** → **Material Props** → **Material Models**. Fig. 6.91 shows the resulting frame.

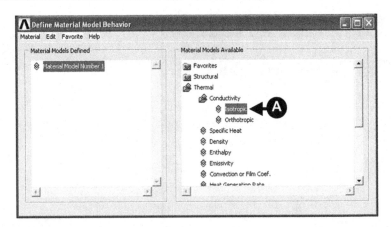

Fig. 6.91 Define material model behaviour.

From the right-hand column, select [A] **Thermal** → **Conductivity** → **Isotropic**. In response to this selection, another frame, shown in Fig. 6.92, appears.

Fig. 6.92 Conductivity coefficient.

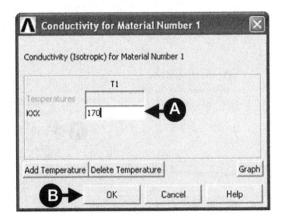

Thermal conductivity [A] **KXX** = 170 W/m K is entered and the [B] **OK** button is clicked to implement the entry, as shown in the figure.

The model of the developed area will be constructed using primitives, and it is useful to have them numbered. Thus, from the ANSYS Utility Menu, select **Plot-Ctrls** → **Numbering** and check the [A] box area numbers, as shown in Fig. 6.93.

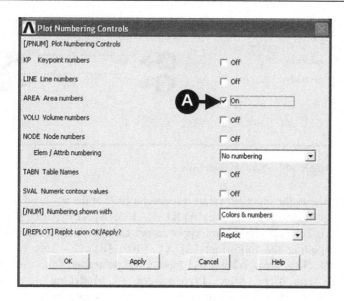

Fig. 6.93 Numbering controls.

From the ANSYS Main Menu, select **Preprocessor** → **Modelling** → **Create** → **Areas** → **Rectangle** → **By Dimensions**. Fig. 6.94 shows the resulting frame.

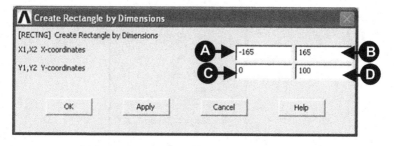

Fig. 6.94 Create rectangle by dimensions.

Input [A] X1 = −165; [B] X2 = 165; [C] Y1 = 0; [D] Y2 = 100 to create a rectangular area (A1) within which the fin will be comprised. Next, create two rectangles at the left and right upper corners, to be cut off from the main rectangle. From the ANSYS Main Menu, select **Preprocessor** → **Modelling** → **Create** → **Areas** → **Rectangle** → **By Dimensions**. Fig. 6.95 shows the resulting frame.

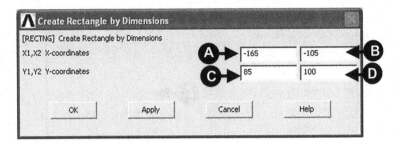

Fig. 6.95 Rectangle with specified dimensions.

Fig. 6.95 shows the inputs to create a rectangle (A2) at the left-hand upper corner of the main rectangle (A1). They are: [A] **X1** $= -165$; [B] **X2** $= -105$; [C] **Y1** $= 85$; [D] **Y2** $= 100$. To create right-hand upper corner rectangles (A3), repeat the above procedure, inputting the following: [A] X1 $= 105$; [B] X2 $= 165$; [C] Y1 $= 85$; [D] Y2 $= 100$. Now, areas A2 and A3 have to be subtracted from area A1. From the ANSYS Main Menu, select **Preprocessor** $\rightarrow$ **Modelling** $\rightarrow$ **Operate** $\rightarrow$ **Booleans** $\rightarrow$ **Subtract** $\rightarrow$ **Areas**. Fig. 6.96 shows the resulting frame.

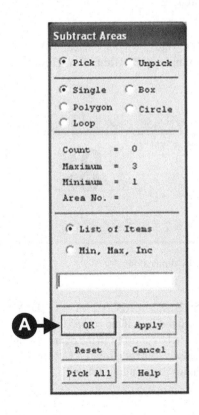

Fig. 6.96 Subtract areas.

First, select area A1 (large rectangle) to be subtracted from and click the [A] **OK** button. Next, select two smaller rectangles A2 and A3, and click the [A] **OK** button. A new area A4 is created with two upper corners cut off. Proceeding in the same way, areas should be cut off from the main rectangle to create the fin shown in Fig. 6.87.

From the ANSYS Main Menu, select **Preprocessor** → **Modelling** → **Create** → **Areas** → **Rectangle** → **By Dimensions**. Fig. 6.97 shows the frame in which appropriate inputs should be made.

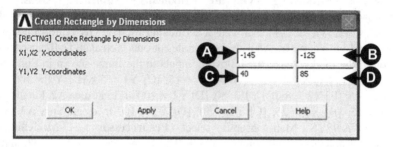

Fig. 6.97 Create rectangle by four coordinates.

To create area A1, input the following: [A] X1 = − 145; [B] X2 = − 125; [C] Y1 = 40; [D] Y2 = 85. In order to create area A2, input the following: [A] X1 = 125; [B] X2 = 145; [C] Y1 = 40; [D] Y2 = 85. In order to create area A3, input the following: [A] X1 = − 105; [B] X2 = −95; [C] Y1 = 25; [D] Y2 = 100. In order to create area A5, input the following: [A] X1 = 95; [B] X2 = 105; [C] Y1 = 25; [D] Y2 = 100.

From the ANSYS Main Menu, select **Preprocessor** → **Modelling** → **Operate** → **Booleans** → **Subtract** → **Areas**. The frame shown in Fig. 6.96 appears. Select the first area A4 (large rectangle) and click the [A] **OK** button. Next, select areas A1, A2, A3 and A5, and click the [A] **OK** button. Area A6 with appropriate cut-outs is created, and is shown in Fig. 6.98.

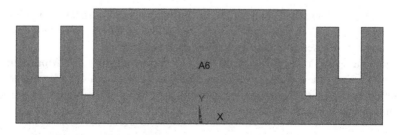

Fig. 6.98 Image of the fin after some areas were subtracted.

To finish construction of the fin's model, use the frame shown in Fig. 6.97 and make the following inputs: [A] X1 = −85; [B] X2 = −75; [C] Y1 = 25; [D] Y2 = 100. Area A1 is created. Next, input [A] X1 = −65; [B] X2 = −55; [C] Y1 = 25; [D] Y2 = 100 to create area A2. Then input [A] X1 = −45; [B] X2 = −35; [C] Y1 = 25; [D] Y2 = 100 to create area A3. Appropriate inputs should be made to create areas, to be cut out later, on the right-hand side of the fin. Thus input [A] X1 = 85; [B] X2 = 75; [C] Y1 = 25; [D] Y2 = 100 to create area A4. Input [A] X1 = 65; [B] X2 = 55; [C] Y1 = 25; [D] Y2 = 100 to create area A5. Input [A] X1 = 45; [B] X2 = 35; [C] Y1 = 25; [D] Y2 = 100 to create area A7. Next, from the ANSYS Main Menu, select **Preprocessor → Modelling → Operate → Booleans → Subtract → Areas.** The frame shown in Fig. 6.96 appears. First select area A6 and click the [A] **OK** button. Then select areas A1, A2, A3, A4, A5, and A7. Clicking the [A] **OK** button implements the command and a new area A8 with appropriate cut-outs is created. To finalise the construction of the model, make the following inputs to the frame shown in Fig. 6.97 to create area A1: [A] X1 = −25; [B] X2 = −15; [C] Y1 = 50; [D] Y2 = 100. Input [A] X1 = −5; [B] X2 = 5; [C] Y1 = 50; [D] Y2 = 100 to create area A2. Finally, input [A] X1 = 15; [B] X2 = 25; [C] Y1 = 50; [D] Y2 = 100 to create area A3. Again, from the ANSYS Main Menu, select **Preprocessor → Modelling → Operate → Booleans → Subtract → Areas**. The frame shown in Fig. 6.96 appears. Select first area A8 and click the [A] **OK** button. Next, select areas A1, A2 and A3. Clicking the [A] **OK** button produces area A4, as shown in Fig. 6.99.

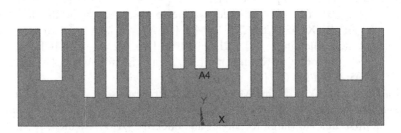

Fig. 6.99 Two-dimensional image of the fin.

Fig. 6.99 shows the final shape of the fin with dimensions as specified in Fig. 6.87. It is, however, a two-dimensional model. The width of the fin is 100 mm and this dimension can be used to create a three-dimensional model.

From the ANSYS Main Menu, select **Preprocessor → Modelling → Operate → Extrude → Areas → Along Normal**. Select Area 4 (to be extruded in the direction normal to the screen, i.e. Z-axis) and click the **OK** button. In response, a frame shown in Fig. 6.100 appears.

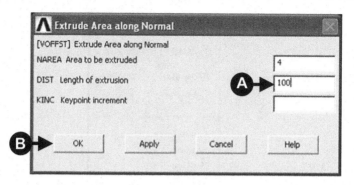

Fig. 6.100 Extrude area.

Input [A] **Length of extrusion** = 100 mm and click the [B] **OK** button. A three-dimensional model of the fin is created, as shown in Fig. 6.101.

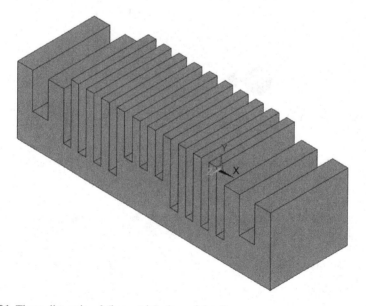

Fig. 6.101 Three-dimensional (isometric) view of the fin.

The fin is shown in isometric view without area numbers. To deselect numbering of areas, refer to Fig. 6.93, in which box **Area numbers** should be unchecked.

From the ANSYS Main Menu, select **Preprocessor** → **Meshing** → **Mesh Attributes** → **Picked Volumes**. The frame shown in Fig. 6.102 is created.

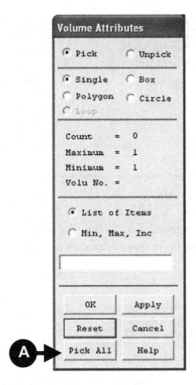

Fig. 6.102 Volume attributes.

Select [A] **Pick All** and the next frame, shown in Fig. 6.103, appears.

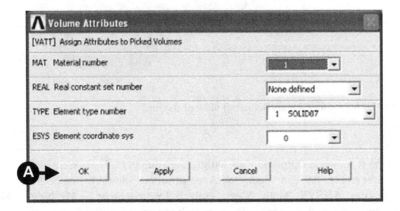

Fig. 6.103 Volume attributes with specified material and element type.

Material Number 1 and element type SOLID87 are as specified at the beginning of the analysis; to accept that, click the [A] **OK** button.

Now the meshing of the fin can be carried out. From the ANSYS Main Menu, select **Preprocessor** → **Meshing** → **Mesh** → **Volumes** → **Free**. The frame shown in Fig. 6.104 appears.

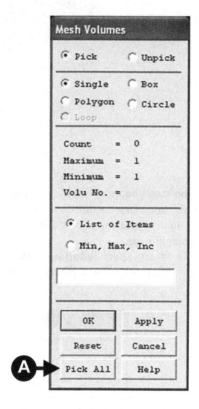

Fig. 6.104 Mesh volume.

Select the [A] **Pick All** option, as shown in Fig. 6.104, to mesh the fin. Fig. 6.105 shows a meshed fin.

Fig. 6.105 View of the fin with
mesh network.

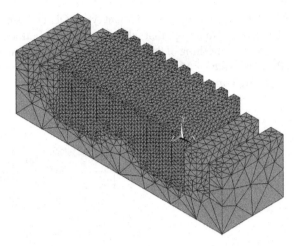

6.4.3   Solution

Prior to running the solution stage, boundary conditions must be applied properly. In
the case considered here, the boundary conditions are expressed by the heat transfer
coefficient, which is a quantitative measure of how efficiently heat is transferred from
the fin surface to the surrounding air.

From the ANSYS Main Menu, select **Solution** → **Define Loads** → **Apply** →
Thermal → **Convection** → **On Areas**. Fig. 6.106 shows the resulting frame.

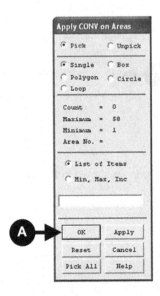

Fig. 6.106 Apply boundary conditions to the fin areas.

Select all areas of the fin except the bottom area and click the [A] **OK** button. The frame created as a result of that action is shown in Fig. 6.107.

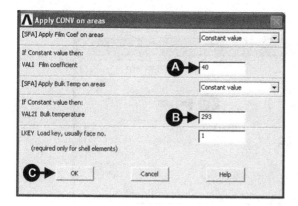

Fig. 6.107 Apply heat transfer coefficient and surrounding temperature.

Input [A] **Film coefficient** = 40 W/m^2 K and [B] **Bulk temperature** = 293 K, and click the [C] **OK** button. Next, the heat flux of intensity 1000 W/m must be applied to the base of the fin. Therefore, from the ANSYS Main Menu, select **Solution → Define Loads → Apply → Thermal → Heat Flux → On Areas**. The resulting frame is shown in Fig. 6.108.

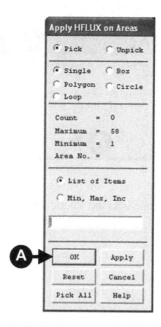

Fig. 6.108 Apply heat flux on the fin base.

Select the bottom surface (base) of the fin and click the [A] **OK** button. A new
frame appears (see Fig. 6.109) and the input made is as follows: [A] **Load HFLUX
value** = 1000 W/m. Clicking the [B] **OK** button implements the input.

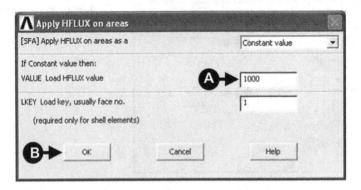

Fig. 6.109 Apply heat flux value on the fin base.

All required preparations have been made and the model is ready for its solution.
From the ANSYS Main Menu, select **Solution → Solve → Current LS**. Two frames
appear. One gives a summary of solution options. After checking the correctness of the
options, this frame should be closed using the menu at the top of the frame. The other
frame is shown in Fig. 6.110.

Fig. 6.110 Solve the
problem.

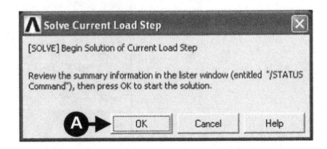

Clicking the [A] **OK** button starts the solution process.

6.4.4 Postprocessing

A successful solution is signalled by the message 'Solution is done'. The post-processing phase can now be initiated to view the results. The problem asks for temperature distribution within the developed area.

From the ANSYS Main Menu, select **General Postproc** → **Plot Results** → **Contour Plot** → **Nodal Solution**. The frame shown in Fig. 6.111 appears.

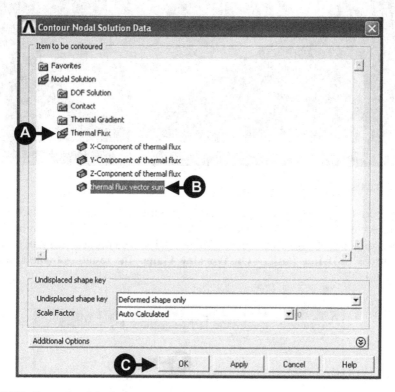

Fig. 6.111 Contour nodal solution.

Select [A] **Thermal Flux** and [B] **thermal flux vector sum**, and click [C] **OK** to produce the graph shown in Fig. 6.112.

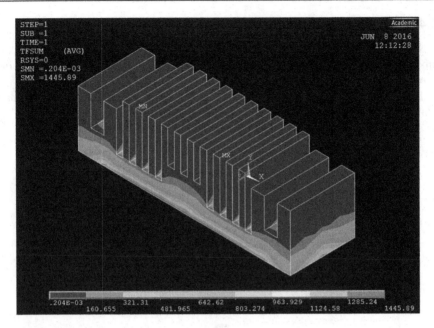

Fig. 6.112 Heat flux distribution.

In order to observe how the temperature changes from the base surface to the top surface of the fin, a path along which the variations take place has to be determined. From the ANSYS Main Menu, select **General Postproc → Path Operations → Define Path → On Working Plane**. The resulting frame is shown in Fig. 6.113.

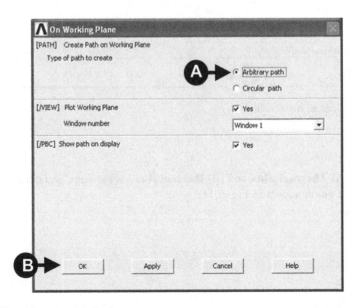

Fig. 6.113 Arbitrary path selection.

By activating the [A] **Arbitrary path** button and clicking the [B] **OK** button, another frame, shown in Fig. 6.114, is produced.

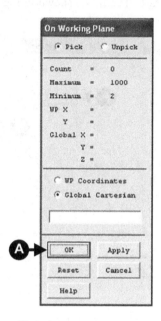

Fig. 6.114 Arbitrary path on working plane.

Two points should be picked: one that is on the bottom line at the middle of the fin and one, moving vertically upwards, on the top line of the fin. After that the [A] **OK** button should be clicked. A new frame appears and is shown in Fig. 6.115.

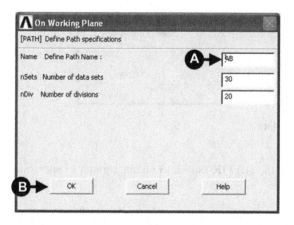

Fig. 6.115 Path name definition.

In the [A] **Define Path Name** box, write AB and click the [B] **OK** button.
From the ANSYS Main Menu, select **General Postproc** → **Path Operations** → **Map onto Path**. The frame shown in Fig. 6.116 appears.

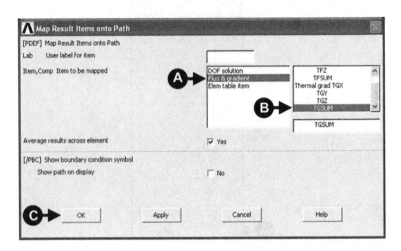

Fig. 6.116 Map results on the path.

Select [A] **Flux & gradient** and [B] **TGSUM**, and click the [C] **OK** button. Next, from the ANSYS Main Menu, select **General Postproc** → **Path Operations** → **Plot Path Item** → **On Graph**. Fig. 6.117 shows the resulting frame.

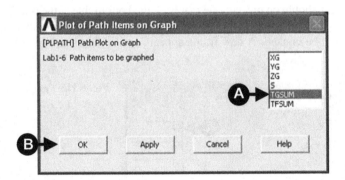

Fig. 6.117 Selection of items to be plotted.

Select [A] **TGSUM** and click the [B] **OK** button to obtain a graph as shown in Fig. 6.118.

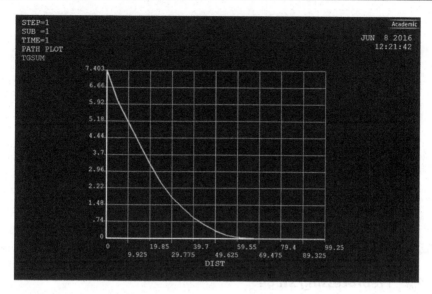

Fig. 6.118 Temperature gradient plot as a function of distance from the fin base.

The graph shows temperature gradient variation as a function of distance from the base of the fin.

6.5 Heat conduction

6.5.1 Problem description

A block of material with thermal conductivity $k = 10$ W/m°C is assumed to be infinitely long. On three sides of it, the surrounding temperature is 100°C and at its top end, it is 500°C. Determine the temperature distribution in the block.

The block is schematically shown in Fig. 6.119. A steady-state (transient thermal) analysis is adopted. The problem will be solved using the main approach of GUI (graphic user interface). However, the command approach will also be given at the end of the solution procedure to illustrate the features of this methodology.

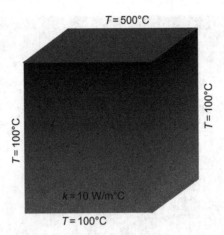

Fig. 6.119 Illustration of the problem to be solved.

6.5.2 Preprocessing stage

6.5.2.1 Model construction

From the ANSYS Main Menu, select **Preferences** → **Thermal**, as shown in Fig. 6.120.

Fig. 6.120 Selection of the analysis type.

In order to create geometry of the block from the ANSYS Main Menu, select **Preprocessor** → **Modelling** → **Create** → **Areas** → **Rectangle** → **By 2 Corners**. The frame shown in Fig. 6.121 is created.

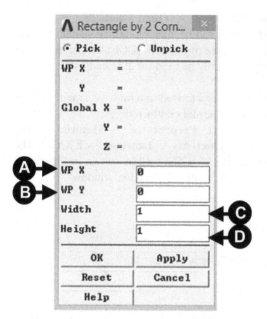

Fig. 6.121 Create rectangle by two corners.

Inputs into the frame are shown in Fig. 6.121: [A] **WP X** = 0, [B] **WP Y** = 0, [C] **Width** = 1, [D] **Height** = 1.

Next, the type of the element to be used in the analysis must be defined. From the ANSYS Main Menu, select **Preprocessor** → **Element Type** → **Add/Edit/Delete**. The frame shown in Fig. 6.122 is generated.

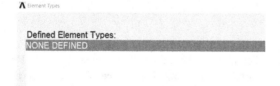

Fig. 6.122 Selection of element type.

Click [A] **Add** and select **Thermal Mass Solid → Quad 4Node 55**.

For this example, the PLANE55 (Thermal Solid, Quad 4node 55) element is used. This element has four nodes and a single degree of freedom (DOF) temperature-wise at each node. Please note that PLANE55 can only be used for 2D steady-state or transient thermal analysis.

The material of the block now has to be characterised in terms of its main thermal property required for the analysis, which is thermal conductivity, k.

From the ANSYS Main Menu, select **Preprocessor → Material Props → Material Models → Thermal → Conductivity → Isotropic → KXX**. The frame shown in Fig. 6.123 is created. Input [A] **KXX** = 10 as shown in Fig. 6.123 and click [B] **OK**. It is important to close the material selection window.

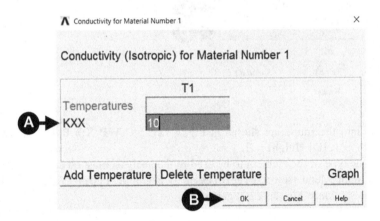

Fig. 6.123 Selection of material model.

Having selected the element type, the user now needs to embark on meshing. First, the mesh size is determined. From the ANSYS Main Menu, select **Preprocessor → Meshing → Size Cntrls → ManualSize → Areas → All Areas**. Fig. 6.124 shows the frame generated. Input [A] **SIZE** = 0.05 and click [B] **OK**, as shown in Fig. 6.124.

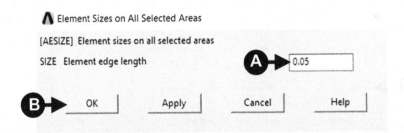

Fig. 6.124 Element size choice.

The model is now ready for meshing. From the ANSYS Main Menu, select **Preprocessor → Mesh → Areas → Free**. The frame in Fig. 6.125 is created and [A] **Pick All** should be selected to have the block meshed.

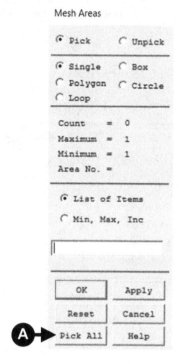

Fig. 6.125 Selection of all areas involved for meshing.

6.5.3 Solution stage

First, the analysis type must be defined. From the ANSYS Main Menu, select **Solution → Analysis Type → New Analysis → Steady-State**. The frame shown in Fig. 6.126 is created.

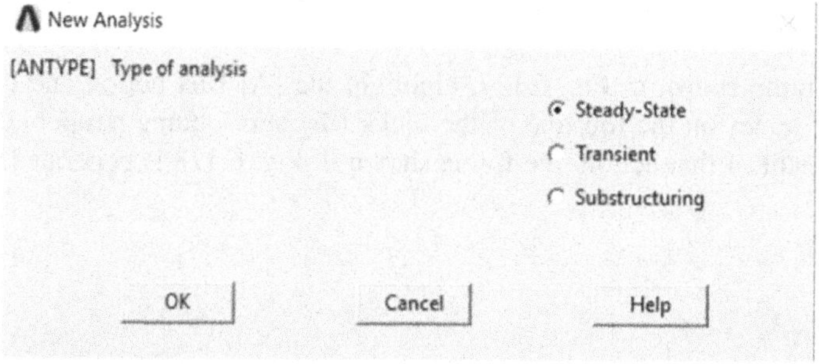

Fig. 6.126 Analysis type choice.

As always, the model must be constrained. For thermal problems, constraints can be in the form of temperature, heat flow, convection, heat flux, heat generation and radiation. In this particular example, all four sides of the block have fixed temperatures. Thus, from the ANSYS Main Menu, select **Solution → Define Loads → Apply → Thermal → Temperature → On Nodes**. The frame shown in Fig. 6.127 appears.

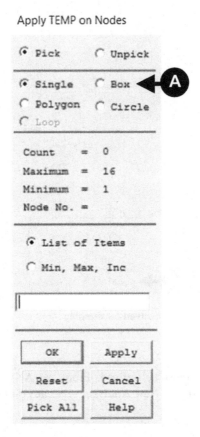

Fig. 6.127 Temperature constraint on nodes.

In the frame shown in Fig. 6.127, highlight the [A] **Box** button and draw a box around the nodes on the top line of the block (the surrounding temperature is 500° C). As a result of that action, the frame shown in Fig. 6.128 is generated.

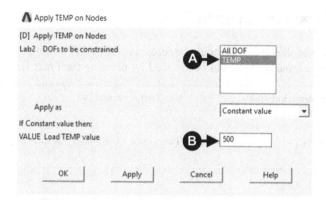

Fig. 6.128 Specific temperature constraint on selected nodes.

Please note that the DOF to be constrained is [A] **temperature** (highlighted in Fig. 6.128), and [B] **VALUE** Load TEMP = 500°C should be entered (also shown in Fig. 6.128).

Using the same approach, constraints (temperature) to the remaining three sides should be applied, ensuring that **VALUE** Load TEMP = 100°C in all three cases.

The last step in the solution stage is to solve the problem. Therefore, from the ANSYS Main Menu, select **Solution → Solve → Current LS**.

6.5.4 Postprocessing stage

6.5.4.1 Temperature plot

From the ANSYS Main Menu, select **General Postproc → Plot Results → Contour Plot → Nodal Solution → DOF Solution → Nodal Temperature**. The resulting plot is shown in Fig. 6.129.

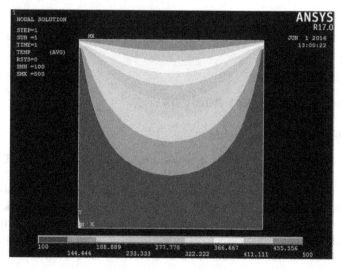

Fig. 6.129 Temperature map within the block.

6.6   Solution in command mode

All the command shown below should be entered in the ANSYS command area (top of
the main screen) and each should be finished by pressing the Enter key.

Input into the command area	Explanatory remarks
/title, Heat conduction	
/PREP7	
! define geometry	This is not a command but an explanatory remark
length = 1.0	
height = 1.0	
blc4, 0, 0, length, height	
! mesh 2D areas	This is not a command but an explanatory remark
ET, 1, PLANE55	Selection of element type
MP, KXX, 1, 10	Input of thermal conductivity
ESIZE, length/20	Selection of element size
AMESH, ALL	Meshing all areas involved
FINISH	
/SOLU	
ANTYPE, 0	Steady-state thermal analysis
! fixed temp BC's	
NSEL, S, LOC, Y, height	Select nodes on the top line with y = height
D, ALL, TEMP, 500	Boundary condition—temp fixed at 500°C
NSEL, ALL	
NSEL, S, LOC, X, 0	Select nodes on three sides of the block
NSEL, A, LOC, X, length	
NSEL, A, LOC, Y, 0	
D, ALL, TEMP, 100	Boundary condition—temp fixed at 100°C
NSEL, ALL	
SOLVE	
FINISH	
/POST1	
PLNSOL, TEMP, , 0,	Produces contour plot of temperature

6.7   Thermal stresses analysis

6.7.1   Problem formulation

This solved example will outline a simple analysis of coupled thermal/structural
problem.

A steel link, with no internal stresses, is pinned between two solid columns at a
reference temperature of 0°C (273 K), as shown in Fig. 6.130. One of the columns
is heated to a temperature of 75°C (348 K). As heat is transferred from the column
into the link, the link will attempt to expand. However, since it is pinned, this cannot
occur and stress is therefore created in the link. A steady-state solution of the resulting
stress will be arrived at to simplify the analysis.

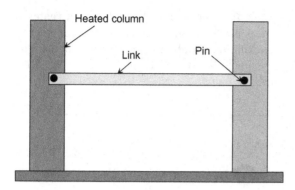

Fig. 6.130 Schematic presentation of the problem.

Loads will not be applied to the link, only a temperature change from 0°C to 75°C. The link is made of steel with a modulus of elasticity of 210 GPa, a thermal conductivity of 60.5 W/m°K and a thermal expansion coefficient of 12×10^{-6}/K.

6.7.2 Thermal settings

6.7.2.1 Geometry and thermal properties

The first thing to do is to allocate a name for the project. Therefore, after loading ANSYS software, from the Utility Menu, select **File** and next **Change jobname**. In the frame created, enter an appropriate name for the project and click **OK** to implement it. Next, from the ANSYS Main Menu, select **Preferences** and activate the [A] **Thermal** button as shown in Fig. 6.131.

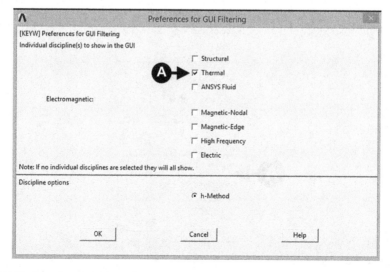

Fig. 6.131 Selecting analysis preference.

From the ANSYS Main Menu, select **Preprocessor** → **Modelling** → **Create** → **Keypoints** → **In Active CS**. In the response frame shown in Fig. 6.132, coordinates for keypoint 1 are given [B]. By clicking the **Apply** button, the same frame will reappear and then coordinates for keypoint 2, $X = 1$, $Y = 0$, should be entered. After that, press the [C] **OK** button to close the frame.

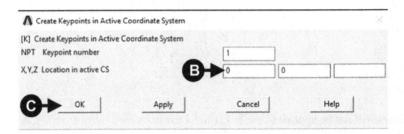

Fig. 6.132 Create keypoints.

From the ANSYS Main Menu, select **Preprocessor** → **Modelling** → **Create** → **Lines** → **Lines** → **In Active CS**. The frame shown in Fig. 6.133 appears.

Fig. 6.133 Create lines.

Pick keypoints 1 and 2, and click the [A] **OK** button to create a line representing the link 1 m long.

Next, the element type has to be defined. Therefore, from the ANSYS Main Menu, select **Preprocessor** → **Element Type** → **Add/Edit/Delete**. The frame shown in Fig. 6.134 is generated.

Fig. 6.134 Define element type.

Click the [A] **Add** button to generate the frame showing element types (see Fig. 6.135). From the list of elements, select [B] **Thermal mass link 3D conduction 33** and click the [C] **OK** button. This element is a uniaxial element with the ability to conduct heat between its nodes.

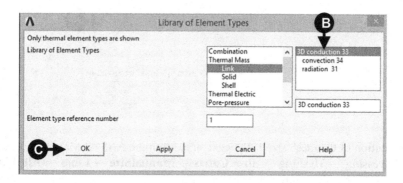

Fig. 6.135 Selecting element type for analysis.

Next, real constants for the chosen element type must be defined. Thus, from the ANSYS Main Menu, select **Preprocessor → Real Constants → Add/Edit/Delete**. In the frame that appears, press the **Add** button to activate the next frame listing the element type **LINK33**. Press the **OK** button and then the frame shown in Fig. 6.136 appears. In the real constants frame for LINK33, [A] cross-sectional **AREA** $= 4 \times 10^{-4}$ should be entered. This defines the link with a cross-section of 2 cm $\times$ 2 cm.

Fig. 6.136 Specify real constant.

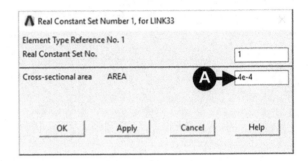

The next step in the analysis is to determine properties of a material from which the link is made. From the ANSYS Main Menu, select **Preprocessor → Material Props → Material Models → Thermal → Conductivity → Isotropic**. In the frame that appears (see Fig. 6.137), enter the following: [A] **KXX** $= 60.5$.

Fig. 6.137 Define thermal material.

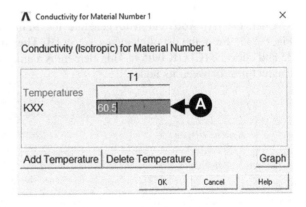

Definition of the mesh size is the next step. From the ANSYS Main Menu, select **Preprocessing → Meshing → Size Cntrls → ManualSize → Lines → All Lines**. Fig. 6.138 shows the frame that is generated.

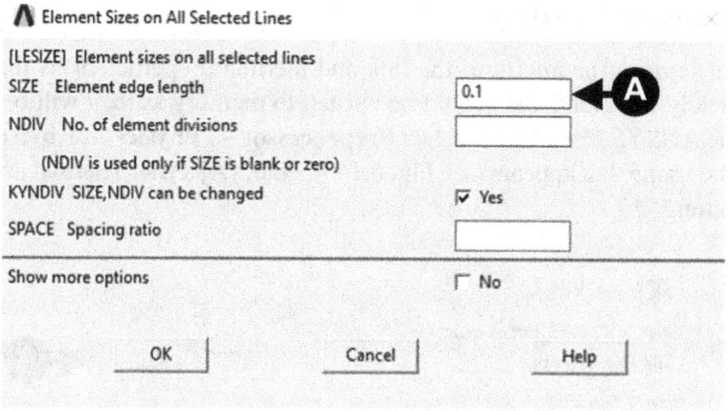

Fig. 6.138 Set element size.

An element with [A] edge length = 0.1 will be used, as shown in Fig. 6.138.

After deciding on the mesh size, meshing of the link can be attempted. Thus, from the ANSYS Main Menu, select **Preprocessor** → **Meshing** → **Mesh** → **Lines**. In the frame that appears (see Fig. 6.139), click the [A] **Pick All** button to mesh the link.

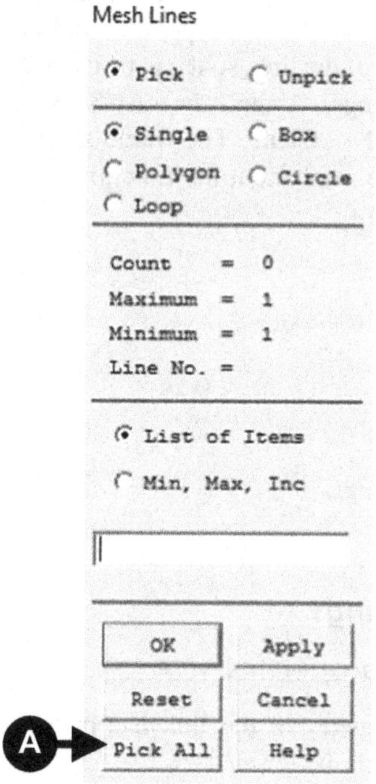

Fig. 6.139 Mesh lines.

6.7.2.2 Thermal settings

The thermal settings (geometry of the link and thermal properties of its material) are now completely described and should be written to memory as they will be used later on. From the ANSYS Main Menu, select **Preprocessor → Physics → Environment → Write**. In the frame that appears (see Fig. 6.140), enter [A] **Title**: Thermal and click the [B] **OK** button.

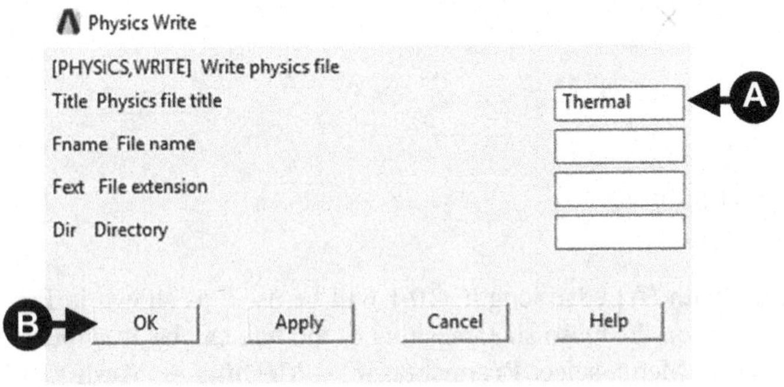

Fig. 6.140 Define thermal settings.

The next practical step in the analysis is to clear the environment. This action clears all the information such as the element type, material properties, etc. However, this does not clear the geometry of the link, so it can be used in the next stage when dealing with the structural settings. From the ANSYS Main Menu, select **Pre-processor → Environment → Clear**. The frame shown in Fig. 6.141 is generated. Click the [A] **OK** button to implement the function.

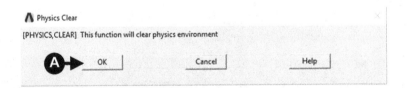

Fig. 6.141 Clear thermal settings.

6.7.3 Structural settings

6.7.3.1 Mechanical properties

In the previous steps, geometry of the link has been established, so now only its mechanical properties should be defined.

From the ANSYS Main Menu, select **Preprocessor → Element Type → Switch Elem Type**. Fig. 6.142 shows the frame generated.

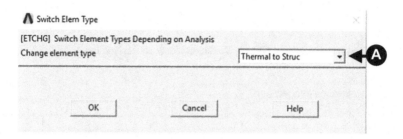

Fig. 6.142 Switch element type.

Choose [A] **Thermal to Struc** as shown in Fig. 6.142. This will automatically switch to the equivalent structural element. A warning saying that the new element has to be modified as necessary might appear. In this case, only the material properties of the link should be modified, as the geometry is unchanged.

From the ANSYS Main Menu, select **Preprocessor → Material Props → Material Models → Structural → Linear → Elastic → Isotropic**. In the frame generated (see Fig. 6.143), the following mechanical properties of the steel (link's material) should be entered: [A] Young's Modulus **EX** $= 210 \times 10^9$ and [B] Poisson's Ratio **PRXY** $= 0.3$.

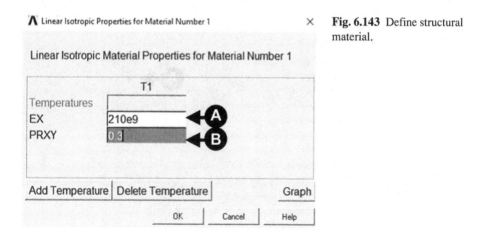

Fig. 6.143 Define structural material.

Furthermore, from the ANSYS Main Menu, select **Preprocessor → Material Models → Structural → Thermal Expansion → Secant Coefficient → Isotropic**. In the generated frame (see Fig. 6.144), enter [A] **ALPX** $= 12 \times 10^{-6}$.

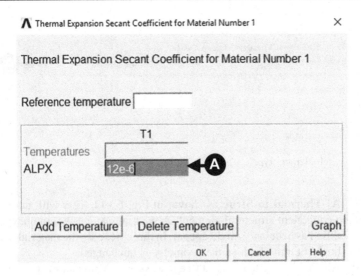

Fig. 6.144 Define thermal expansion coefficient.

Finally, the structural settings should be written. From the ANSYS Main Menu, select **Preprocessor → Physics → Environment → Write**. In the frame that appears (Fig. 6.145), enter [A] TITLE **Structural** and click the [B] **OK** button.

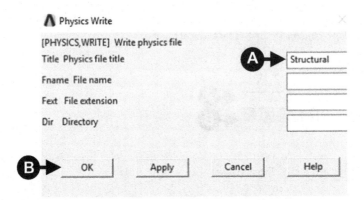

Fig. 6.145 Define structural settings.

6.7.4 Solution stage

Firstly, the analysis type needs to be defined. From the ANSYS Main Menu, select **Solution → Analysis Type → New Analysis**. In the generated frame, shown in Fig. 6.146, activate the [A] **Static** button.

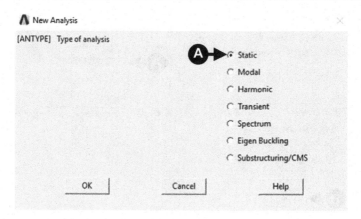

Fig. 6.146 Select analysis type.

After this, the thermal settings have to be recalled into the analysis. From the ANSYS Main Menu, select **Solution → Physics → Environment → Read**. In the generated frame shown in Fig. 6.147, select [A] **Thermal** and then click the [B] **OK** button.

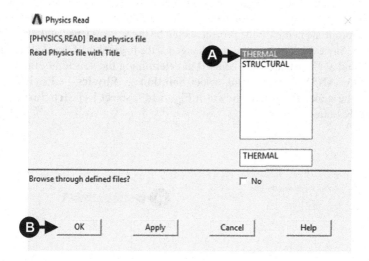

Fig. 6.147 Read thermal settings.

As usual, constraints to the model must be applied. Therefore, from the ANSYS Main Menu, select **Solution → Define Loads → Apply → Thermal → Temperature → On Keypoints**. In the resulting frame (Fig. 6.148), set the temperature of keypoint 1 (the left-most point) to [A] 348 K and click the [B] **OK** button.

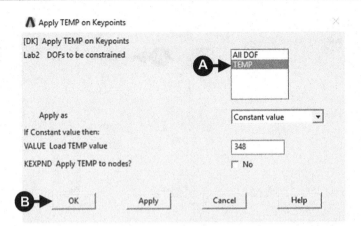

Fig. 6.148 Apply temperature constraints of keypoints.

Finally, from the ANSYS Main Menu, select **Solution** → **Solve** → **Current LS**. It is important to close the currently used environment and enable other environments (structural) to be opened without contamination, otherwise error messages might appear.

The thermal part of the solution has now been obtained. If the steady-state temperature within the link is plotted, then it can be seen as being a uniform 384 K as expected. This information is saved in a file named Jobname.rth, where rth is the thermal results file. An appropriate name for the project should be used, as assigned at the beginning of the project. Since the jobname was not changed at the beginning of the analysis, this data can be found at file.rth and will be used in determining the structural effects.

From the ANSYS Main Menu, select **Solution** → **Physics** → **Environment** → **Read**. In the generated frame shown in Fig. 6.149, select [A] **Structural** and click the [B] **OK** button.

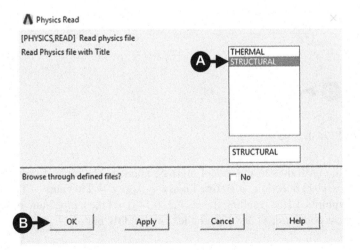

Fig. 6.149 Read structural settings.

Next, from the ANSYS Main Menu, select **Solution → Define Loads → Apply → Structural → Displacement → On Keypoints**. Constrain keypoint 1 in all [A] **DOF** (see Fig. 6.150) and keypoint 2 in the [B] **UX** direction as shown in Fig. 6.151.

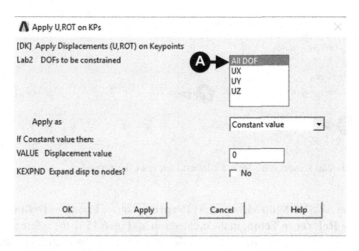

Fig. 6.150 Apply movement constraints (all DOF) on keypoints.

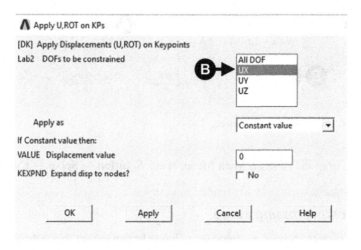

Fig. 6.151 Apply movement constraints (UX) on keypoints.

Stresses in the link are created by thermal expansion; therefore thermal effect must be recalled into the solution.

From the ANSYS Main Menu, select **Solution → Define Loads → Apply → Structural → Temperature → From Therm Analy**. In the resulting frame shown in Fig. 6.152, enter the file name (given to the project) [A] **file.rth** and click the [B]

OK button. This couples the results from the solution of the thermal settings to the information defining the structural settings.

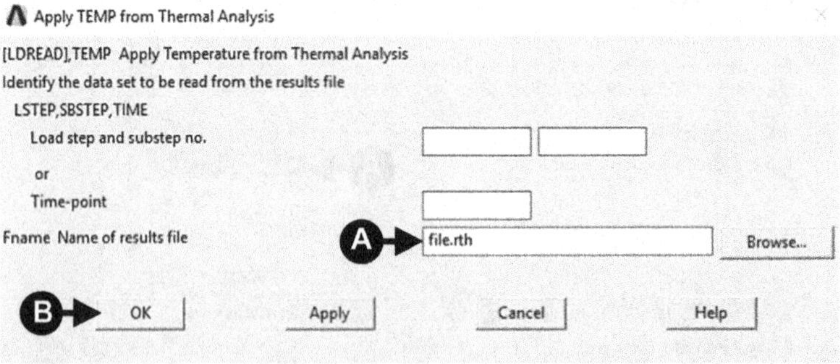

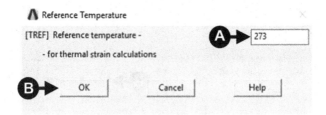

Fig. 6.152 Recall temperature from a file with name of the project.

From the ANSYS Main Menu, select **Preprocessor → Loads → Define Loads → Settings → Reference Temp**. In the created frame (Fig. 6.153), the reference temperature is set to [A] 273 K; click the [B] **OK** button to implement the selection.

Fig. 6.153 Input reference temperature.

Finally, from the ANSYS Main Menu, select **Solution → Solve → Current LS**.

6.7.5 Postprocessing stage

In order to view the results, postprocessing has to be carried out. Since the link is modelled as a single line, the stress cannot be listed in the usual way. Instead, an element table must be created first.

From the ANSYS Main Menu, select **General Postproc → Element Table → Define Table**. In the resulting frame, select **Add**. A new frame then opens (see Fig. 6.154), in which the following entries should be made: [A] **EQV** strain = 0, [B] **Lab.** = **CompStr**. From the results data, select [C] **by sequence num** and [D] **LS** in the adjacent pull-down menu, and ensure that the entry [E] **LS, 1** is made.

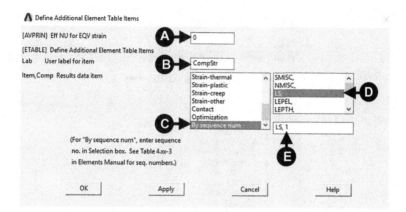

Fig. 6.154 Define element table.

To list stress data from the ANSYS Main Menu, select **General Postproc → Element Table → List Elem Table**. In the resulting frame (Fig. 6.155), select [A] **COMPSTR** and click the [B] **OK** button.

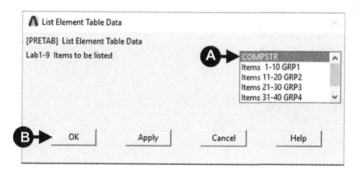

Fig. 6.155 List element table.

Because of this action, a frame containing a list of compressive stresses in the link's elements is displayed (see Fig. 6.156). As expected, the stress of compressive nature in each element is −189 MPa.

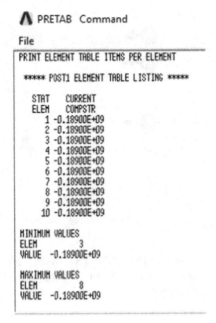

Fig. 6.156 List of stresses by element.

Application of ANSYS to contact between machine elements

7

7.1 General characteristic of contact problems

In almost every mechanical device, its constituent components are in either rolling or sliding contact. In most cases, contacting surfaces are nonconforming so that the area through which the load is transmitted is very small, even after some surface deformation, and the pressures and local stresses are very high. Unless the component is purposefully designed for the load and life expected, it may fail due to early general wear or local fatigue failure. The magnitude of the damage is a function of the materials and the intensity of the applied load as well as the surface finish, lubrication, and relative motion.

The intensity of the load can usually be determined from equations, which are functions of the geometry of the contacting surfaces, essentially the radii of curvature, and the elastic constants of the materials. Large radii and smaller modules of elasticity give larger contact areas and lower pressures.

A contact is said to be conforming (concave) if the surfaces of the two elements fit exactly or even closely together without deformation. Journal bearings are an example of concave contact. Elements that have dissimilar profiles are considered to be nonconforming (convex). When brought into contact without deformation, they first touch at a point (hence point contact) or along a line (line contact). In a ball bearing, the ball makes point contact with the inner and outer races, whereas in a roller bearing the roller makes line contact with both races. Line contact arises when the profiles of the elements are conforming in one direction and nonconforming in the perpendicular direction. The contact area between convex elements is very small compared with the overall dimensions of the elements themselves. Therefore, the stresses are high and concentrated in the region close to the contact zone and are not substantially influenced by the shape of the elements at a distance from the contact area.

Contact problem analyses are based on the Hertz theory, which is an approximation on two counts. First, the geometry of general curved surfaces is described by quadratic terms only and second, the two bodies, at least one of which must have a curved surface, are taken to deform as though they were elastic half-spaces. The accuracy of the Hertz theory is in doubt if the ratio a/R (a = radius of the contact area; R = radius of curvature of contacting elements) becomes too large. With metallic elements this restriction is ensured by the small strains at which the elastic limit is reached. However, a different situation arises with compliant elastic solids like rubber. A different problem is encountered with conforming (concave) surfaces in contact, for example, a pin in a closely fitting hole or by a ball and socket joint. Here, the arc of contact may be large compared with the radius of the hole or socket without incurring large strains.

Engineering Analysis with ANSYS Software. https://doi.org/10.1016/B978-0-08-102164-4.00007-8

Modern developments in computing have stimulated research into numerical methods to solve problems in which the contact geometry cannot be described adequately by the quadratic expressions used originally by Hertz. The contact of worn wheels and rails or the contact of conforming gear teeth with Novikov profile are typical examples. In the numerical methods, the contact area is subdivided into a grid and the pressure distribution represented by discrete boundary elements acting on the elemental areas of the grid. Usually, elements of uniform pressure are employed, but overlapping triangular elements offer some advantages. They sum to approximately linear pressure distribution and the fact that the pressure falls to zero at the edge of the contact ensures that the surfaces do not interfere outside the contact area. The three-dimensional equivalent of overlapping triangular elements is overlapping hexagonal pyramids on an equilateral triangular grid.

An authoritative treatment of contact problems can be found in the monograph by Johnson [1].

7.2   Example problems

7.2.1   Pin-in-hole interference fit

7.2.1.1   Problem description

One end of a steel pin is rigidly fixed into the solid plate while its other end is force fitted into the steel arm. The configuration is shown in Fig. 7.1.

Fig. 7.1 Illustration of the problem.

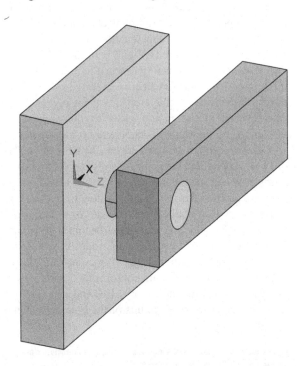

This is a three-dimensional analysis but because of the inherent symmetry of the model, analysis will be carried out for a quarter symmetry model only. There are two objectives of the analysis. The first objective is to observe the force fit stresses of the pin, which is pushed into the arm's hole with geometric interference. The second objective of the analysis is to find out stresses, contact pressures, and reaction forces due to a torque applied to the arm (force acting at the arm's end) and causing rotation of the arm. Stresses resulting from shearing of the pin and bending of the pin will be neglected purposefully.

The dimensions of the model are as follows:

Pin radius = 1 cm, pin length = 3 cm. Arm width = 4 cm, arm length = 12 cm, arm thickness = 2 cm. Hole in the arm: radius = 0.99 cm, depth = 2 cm (through thickness hole).

Both elements are made of steel with Young's modulus = 2.1×10^9 N/m^2, Poisson's ratio = 0.3 and are assumed to be elastic.

7.2.1.2 Construction of the model

In order to analyse the contact between the pin and the hole, a quarter symmetry model is appropriate. This is shown in Fig. 7.2.

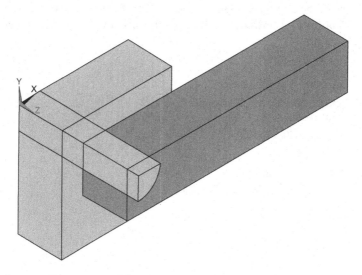

Fig. 7.2 A quarter symmetry model.

In order to create a model shown in Fig. 7.2, two 3D (three-dimensional) primitives are used: block and cylinder. The model is constructed using GUI (graphic user interface) only. For carrying out Boolean operations, it is convenient to have volumes numbered. This can be done by selecting from the Utility Menu **PlotCtrls → Numbering** and checking the appropriate box to activate **VOLU** (volume numbers) option.

From the ANSYS Main Menu, select **Preprocessor** → **Modelling** → **Create** → **Volumes** → **Block** → **By Dimensions**. In response, a frame shown in Fig. 7.3 appears.

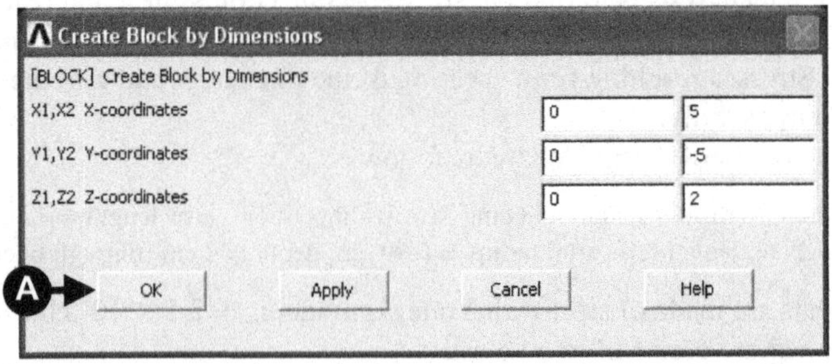

Fig. 7.3 Create block by dimensions.

It can be seen from Fig. 7.3 that appropriate X, Y, Z coordinates were entered. Clicking the [A] **OK** button implements the entries. A block with the length 5, width 5, and thickness 2 (vol. 1) is created.

Next, from the ANSYS Main Menu, select **Preprocessor** → **Modelling** → **Create** → **Volumes** → **Cylinder** → **By Dimensions**. In response, a frame shown in Fig. 7.4 appears.

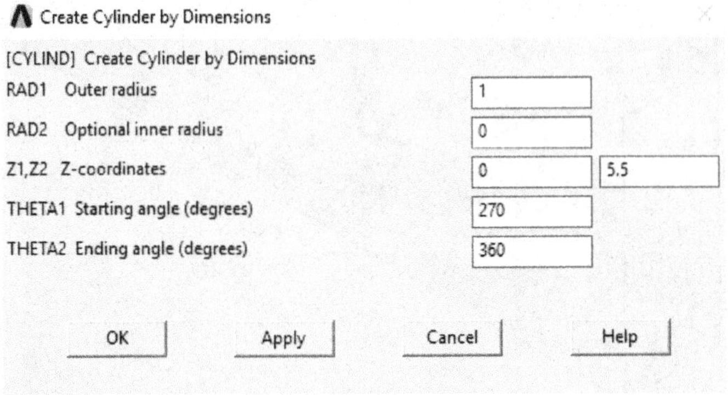

Fig. 7.4 Create cylinder by dimensions.

The inputs are shown in Fig. 7.4. Clicking the [A] **OK** button implements the entries and creates a solid cylinder sector with a radius 1 cm, length 5.5 cm, starting angle 270 degrees, and ending angle 360 degrees (vol. 2).

From the ANSYS Main Menu, select **Preprocessor** → **Modelling** → **Operate** → **Booleans** → **Overlap** → **Volumes**. A frame shown in Fig. 7.5 appears.

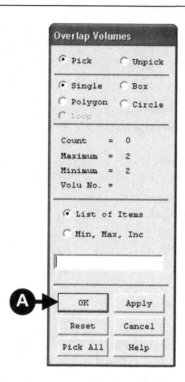

Fig. 7.5 Overlap volumes.

Block (vol. 1) and cylinder (vol.2) should be picked and [A] **OK** button pressed. As a result of that block and cylinder are overlapped.

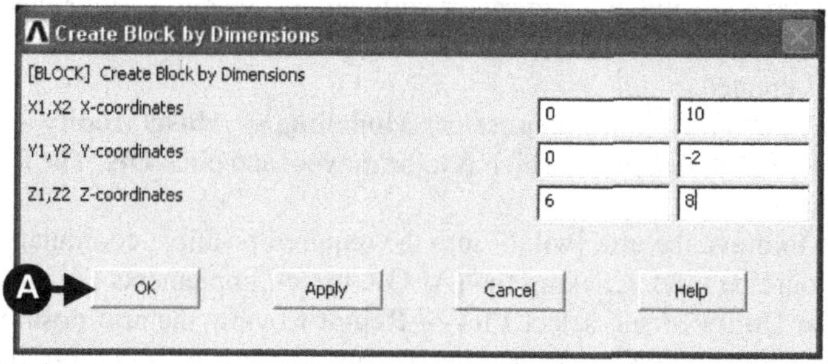

Fig. 7.6 Create block by dimensions.

From the ANSYS Main Menu, select **Preprocessor** → **Modelling** → **Create** → **Volumes** → **Block** → **By Dimensions**. The frame shown in Fig. 7.6 appears.

Coordinates X, Y, Z were used as shown in Fig. 7.6. Clicking the [A] **OK** button implements the entries and, as a result, a block volume was created with the length 10 cm, width 2 cm and thickness 2 cm (vol. 2).

From the ANSYS Main Menu, select **Preprocessor → Modelling →
Create → Volumes → Cylinder → By Dimensions**. The frame shown in Fig. 7.7
appears.

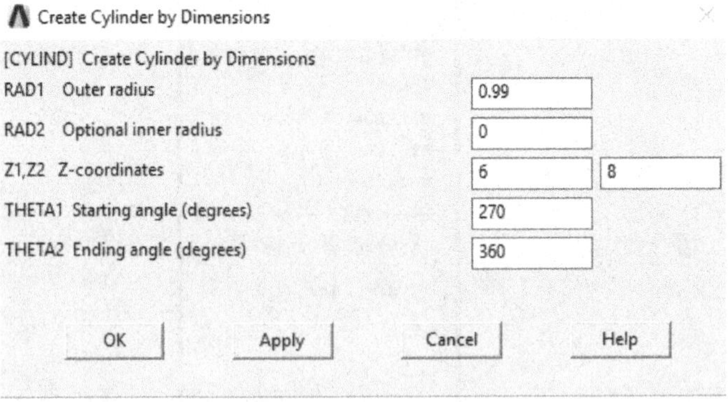

Fig. 7.7 Create cylinder by dimensions.

Input data entered are shown in Fig. 7.7. Clicking the [A] **OK** button implements
the entries. As a result, a solid cylinder sector with radius 0.99 cm, length 2 cm,
starting angle 270 degrees, and ending angle 360 degrees (vol. 2) is produced. Next,
volume 2 must be subtracted from volume 1 in order to produce a hole in the arm with
the radius of 0.99 cm, which is smaller than the radius of the pin. In this way, an interference fit between the pin and the arm is created.

From the ANSYS Main Menu, select **Preprocessor → Modelling → Operate→
Booleans → Subtract → Volumes**. The frame shown in Fig. 7.8 appears.

Volume 2 (short solid cylinder sector with radius 0.99 cm) is subtracted from volume 1 (the arm) by picking them in turn and pressing the [A] **OK** button. As a result,
volume 6 is created.

From the ANSYS Main Menu, select **Modelling → Move/Modify → Volumes**.
Then pick volume 6 (the arm), which is to be moved, and click **OK**. The frame shown
in Fig. 7.9 appears.

In order to move the arm (vol. 6) into the required position, coordinates shown in
Fig. 7.9 should be used. Clicking the [A] **OK** button implements the move action.

From the Utility Menu, select **Plot → Replot** to view the arm positioned in the
required location. Finally, from the Utility Menu, select **PlotCtrls → View
Settings → Viewing Direction**. The frame shown in Fig. 7.10 appears.

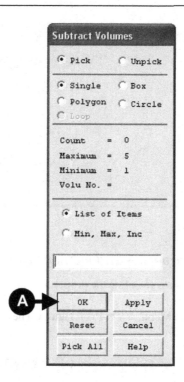

Fig. 7.8 Subtract volumes.

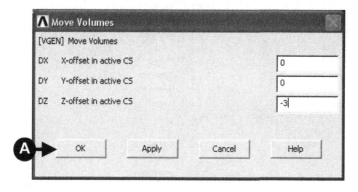

Fig. 7.9 Move volumes.

By selecting coordinates X, Y, Z as shown in Fig. 7.10 and activating [A] **Plot → Replot** command (Utility Menu), a quarter symmetry model, shown in Fig. 7.2, is finally created.

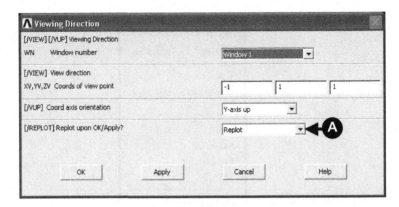

Fig. 7.10 Viewing direction.

7.2.1.3 Material properties and element type

The next step in the analysis is to define the properties of the material used to make the pin and the arm.

From the ANSYS Main Menu, select **Preferences**. The frame shown in Fig. 7.11 is produced.

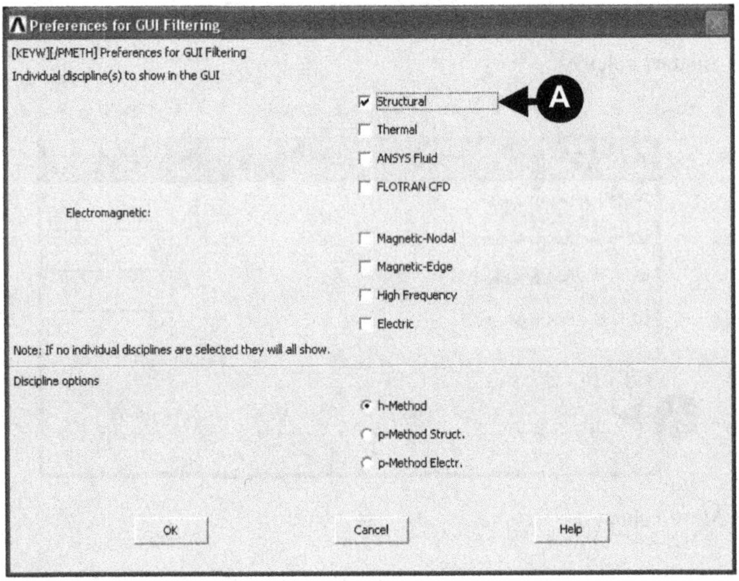

Fig. 7.11 Preferences.

From the Preferences list, [A] **Structural** option was selected as shown in Fig. 7.11.

From the ANSYS Main Menu, select **Preprocessor** → **Material Props** → **Material Models**.

Click in turn: **Structural: Linear: Elastic: Isotropic**. A frame shown in Fig. 7.12 appears.

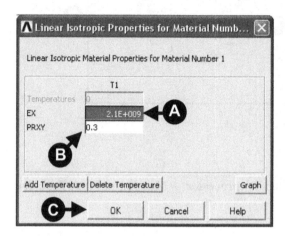

Fig. 7.12 Material properties.

Enter [A] $EX = 2.1 \times 10^9$ for Young's modulus and [B] **PRXY** $= 0.3$ for Poisson's ratio. Then click [C] **OK** and afterwards **Material: Exit**.

After defining properties of the material, the next step is to select the element type appropriate for the analysis.

From the ANSYS Main Menu, select **Preprocessor: Element Type: Add/Edit/ Delete**. The frame shown in Fig. 7.13 appears.

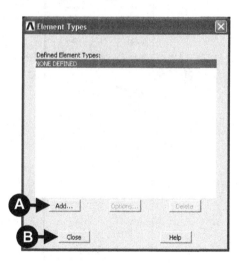

Fig. 7.13 Element types.

Click [A] **Add** in order to pull down another frame, shown in Fig. 7.14.

In the left column click [A] **Structural Solid** and in the right column click [B] **Brick 8node 185**. After that click [C] **OK** and [B] **Close** in the frame shown in Fig. 7.13. This completes the element type selection.

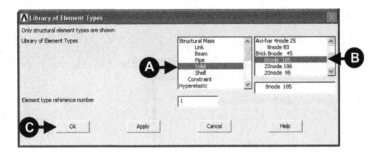

Fig. 7.14 Library of element types.

7.2.1.4 Meshing

From the ANSYS Main Menu, select **Preprocessor** → **Meshing** → **Mesh Tool**. The frame shown in Fig. 7.15 appears.

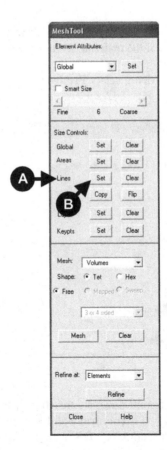

Fig. 7.15 Mesh tool options.

There are a number of options available. The first step is to go to [A] **Size Control** → **Lines** option and click the [B] **Set** button. This opens another frame (shown in Fig. 7.16) prompting to pick lines on which element size is going to be controlled.

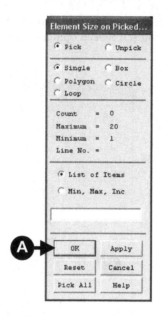

Fig. 7.16 Element size on picked lines.

Pick the horizontal and vertical lines on the front edge of the pin and click [A] **OK**. The frame shown in Fig. 7.17 appears. In the box [A] **No. of element divisions**, type 3 and change selection [B] **SIZE, NDIV can be changed** to **No** by checking the box and, finally, click [C] **OK**.

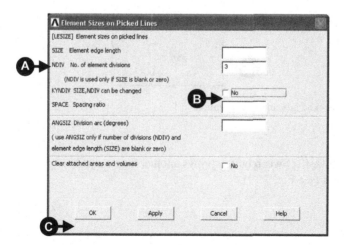

Fig. 7.17 Element sizes on picked lines.

Using the **MeshTool** frame again, shown in Fig. 7.18, click button [A] **Set** in the **Size Controls** → **Lines** option and pick the curved line on the front of the arm. Click **OK** afterwards. The frame shown in Fig. 7.17 appears. In the box **No. of element divisions,** type 4 this time and press the [C] **OK** button.

In the frame **MeshTool** (see Fig. 7.18), pull down [B] **Volumes** in the option **Mesh**. Check [C] **Hex/Wedge** and [D] **Sweep** options. This is shown in Fig. 7.18.

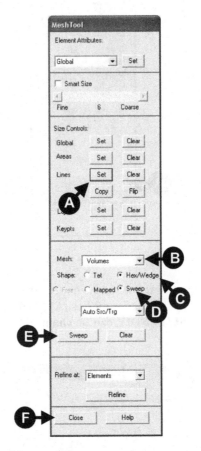

Fig. 7.18 Checked hex and sweep options.

Pressing the [E] **Sweep** button brings another frame asking to pick the pin and the arm volumes (see Fig. 7.19).

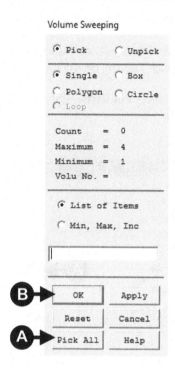

Fig. 7.19 Volume sweeping.

Selecting first [A] **Pick All** and then pressing the [B] **OK** button initiates the meshing process. The model after meshing looks like the image in Fig. 7.20.

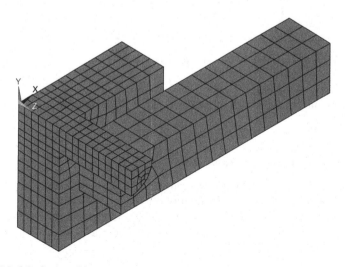

Fig. 7.20 Model after meshing process.

Pressing the [F] **Close** button on **MeshTool** frame (see Fig. 7.18) ends the mesh generation stage.

After meshing is complete, it is usually necessary to smooth element edges in order to improve the graphic display. This can be accomplished using the **PlotCtrls** facility in the Utility Menu.

From the Utility Menu, select **PlotCtrls** → **Style** → **Size and Shape**. The frame shown in Fig. 7.21 appears.

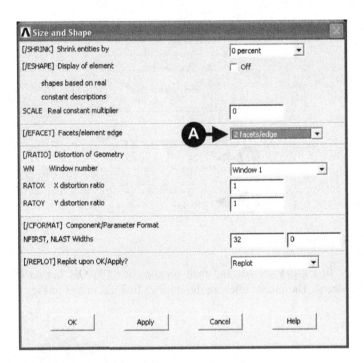

Fig. 7.21 Size and shape control of element edges.

In the option [A] **Facets/element edge**, select **2 facets/edge**, which is shown in Fig. 7.21.

7.2.1.5 Creation of contact pair

In solving the problem of contact between two elements, it is necessary to create a contact pair. Contact Wizard is the facility offered by ANSYS.

From the ANSYS Main Menu, select **Preprocessor** → **Modelling** → **Create** → **Contact Pair**. As a result of this selection, a frame shown in Fig. 7.22 appears.

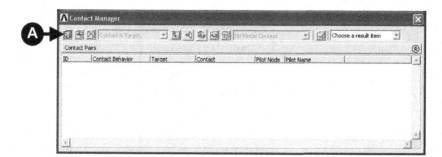

Fig. 7.22 Contact Manager.

The location of [A] Contact Wizard button is in the upper left-hand corner of the frame. By clicking on this button a new frame, shown in Fig. 7.23, is produced.

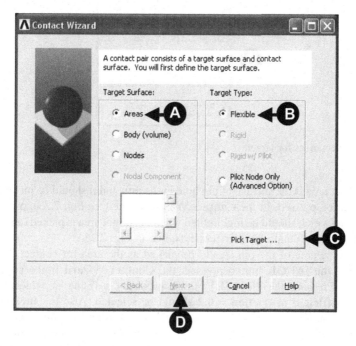

Fig. 7.23 Contact Wizard.

In the frame shown in Fig. 7.23, select [A] **Areas**, [B] **Flexible** and press the [C] **Pick Target** button. As a result of that selection, the frame shown in Fig. 7.24 is produced.

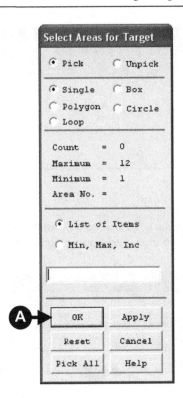

Fig. 7.24 Select areas for target.

The target area is the surface of pin hole in the arm and it should be picked and the
[A] **OK** button pressed. In the **Contact Wizard** frame (see Fig. 7.23), press the [D]
Next button (which should be highlighted when the target area is picked) and the **Pick
Contact** button to obtain another frame, shown in Fig. 7.25.

The surface area of the pin should be picked as the area for contact. When this
is done and the [A] **OK** button pressed, the **Contact Wizard** frame appears (see
Fig. 7.23). Pressing the **Next [D]** button produces a frame in which **Material
ID = 1**, **Coefficient of friction = 0.2** should be selected. Also the **Include Initial
penetration** option should be checked. Next, the **Optional settings** button should
be pressed in order to refine contact parameters further. In the new frame, **Normal
penalty stiffness = 0.1** should be selected. Also, the **Friction** tab located in the
top of the frame menu should be activated and **Stiffness matrix = Unsymmetric**
selected. Afterwards, pressing the **OK** button and then **Create** results in the image
shown in Fig. 7.26.

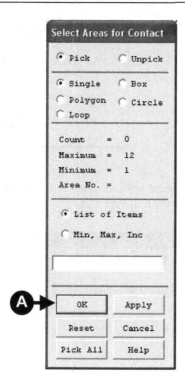

Fig. 7.25 Select areas for contact.

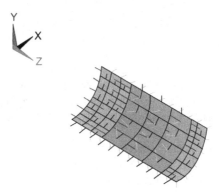

Fig. 7.26 Pin contact area.

Also, the **Contact Wizard** frame appears in the form shown in Fig. 7.27.

The message is that the contact pair has been created. Pressing the [A] **Finish** button closes the **Contact Wizard** tool.

Fig. 7.27 Contact Wizard final message.

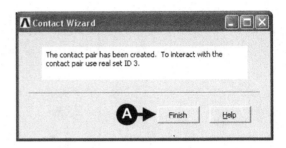

The Contact Manager frame appears again with the information pertinent to the problem considered. It is shown in Fig. 7.28.

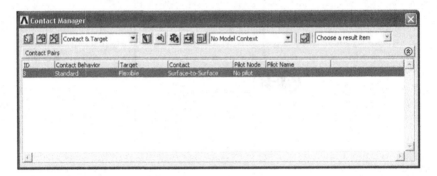

Fig. 7.28 Contact Manager summary information.

7.2.1.6 Solution

In the solution stage, solution criteria have to be specified first. As a first step in that process, symmetry constraints are applied on the quarter symmetry model.

From the ANSYS Main Menu, select **Solution** → **Define Loads** → **Apply** → **Structural** → **Displacement** → **Symmetry BC** → **On Areas**. A frame shown in Fig. 7.29 appears.

Four areas which were created when the full configuration model was sectioned to produce quarter symmetry model should be picked. When that is done, click the [A] **OK** button in the frame of Fig. 7.29.

The next step is to apply boundary constraints on the block of which the pin is an integral part.

From the ANSYS Main Menu, select **Solution** → **Define Loads** → **Apply** → **Structural** → **Displacement** → **On Areas**. Fig. 7.30 showing quarter symmetry model appears.

The back side of the block should be picked and then the **OK** button pressed. The frame shown in Fig. 7.31 appears.

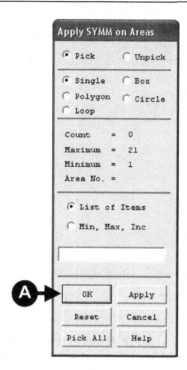

Fig. 7.29 Apply SYMM on areas.

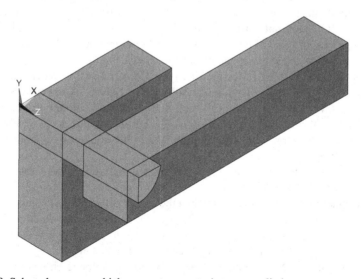

Fig. 7.30 Selected areas on which symmetry constraints are applied.

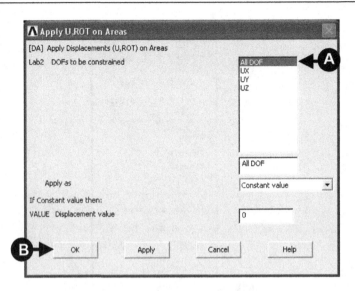

Fig. 7.31 Apply U, ROT on areas.

All degrees of freedom [A] **All DOF** should be constrained with the displacement value equal to zero (see Fig. 7.31). Clicking the [B] **OK** button applies the constraints.

Because the original problem formulation asks for stress analysis when the arm is pulled out of the pin, the analysis involves large displacement effects. The first type of load results from the interference fit between the pin and arm.

From the ANSYS Main Menu, select **Solution → Analysis Type → Sol'n Controls**. A frame shown in Fig. 7.32 appears. In the pull-down menu, select [A] **Large Displacement Static**. Further selected options should be: [B] **Time at end of load step** = 100; [C] **Automatic time stepping** (pull-down menu) = Off; [D] **Number of substeps** = 1. All specified selections are shown in Fig. 7.32. Pressing the [E] **OK** button applies the settings and closes the frame.

The next action is to solve for the first type of load, which is interference fit.

From the ANSYS Main Menu, select **Solution → Solve → Current LS**. A frame showing review of information pertaining to the planned solution action appears. After checking that everything is correct, select **File → Close** to close that frame. Pressing the **OK** button starts the solution. When the solution is completed, press the **Close** button.

In order to return to the previous image of the model, select **Utility Menu → Plot → Replot**.

The second type of load is created when the arm is pulled out of the pin. A number of actions have to be taken in order to prepare the model for the solution. The first action is to apply a displacement along Z axis equal to 2 cm (thickness of the arm) to all nodes on the front of the pin in order to observe this effect.

From the Utility Menu, choose **Select → Entities**. A frame shown in Fig. 7.33 appears.

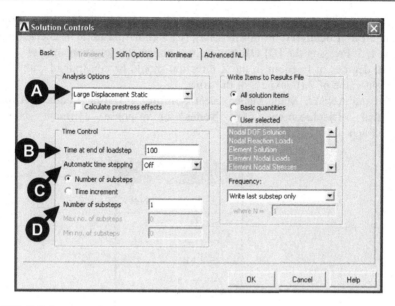

Fig. 7.32 Solution controls.

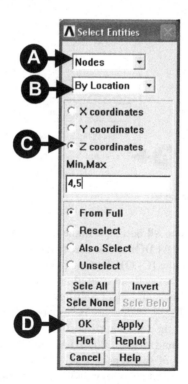

Fig. 7.33 Select entities.

In this frame, the following selections should be made: [A] **Nodes** (from pull-down menu); [B] **By Location** (pull-down menu); [C] **Z coordinates** (to be checked); **Min, Max** = 4, 5. Pressing the [D] **OK** button implements the selections made.

Next, degrees of freedom in the Z direction should be constrained with the displacement value of 2 (thickness of the arm).

From the ANSYS Main Menu, select **Solution → Define Loads → Apply → Structural → Displacement → On Nodes**. In response, a frame shown in Fig. 7.34 appears.

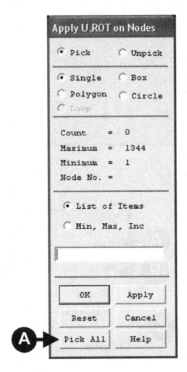

Fig. 7.34 Apply U, ROT on nodes.

By pressing the [A] **Pick All** button, a frame shown in Fig. 7.35 is called up.

As shown in Fig. 7.35, [A] **DOF to be constrained** = **UZ** and the [B] **Displacement value** = 2. Pressing the [C] **OK** button applies selected constraints.

Options for the analysis of pull-out operation have to be defined now.

From the ANSYS Main Menu, select **Solution → Analysis Type → Sol'n Controls**. A frame shown in Fig. 7.36 appears in response.

As shown in Fig. 7.36, the following selections defining solution controls were made: [A] **Time at end of load step** = 200; [B] **Automatic time stepping** (pull-down menu) = On; [C] **Number of substeps** = 100; [D] **Max no. of substeps** = 10,000; [E] **Min no. of substeps** = 10; **Frequency** (pull-down menu) = Write every Nth substep; where $N = -10$. Pressing the [F] **OK** button applies selected controls.

Now the model is ready to be solved for the load resulting from pulling the arm out.

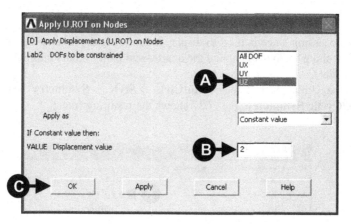

Fig. 7.35 Apply U, ROT on nodes.

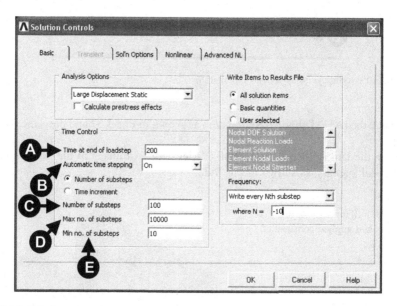

Fig. 7.36 Solution controls.

From the ANSYS Main Menu, select **Solution: Solve: Current LS**. A frame giving summary information pertinent to the solution appears. After reviewing the information select **File: Close** to close the frame. After that, pressing the **OK** button starts the solution. When the solution is done, press the **Close** button.

During solution process warning messages could appear. In order to make sure that the solution is done, it is practical to issue the command in the input box: **/NERR, 100, 100, 0FF**. This ensures that the ANSYS programme does not abort if it encounters a considerable number of errors.

7.2.1.7 Postprocessing

The postprocessing stage is used to display solution results in a variety of forms.

The first thing to do is to expand the quarter symmetry model into the full configuration model.

From the Utility Menu, select **PlotCtrls** → **Style** → **Symmetry Expansion** → **Periodic/Cyclic Symmetry**. Fig. 7.37 shows the resulting frame.

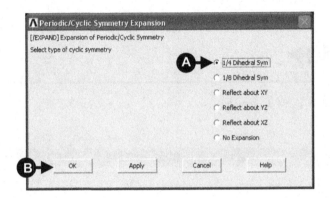

Fig. 7.37 Periodic/cyclic symmetry expansion.

Check the [A] ¼ **Dihedral Sym** button as shown in Fig. 7.37 and press the [B] **OK** button.

From the Utility Menu, select **Plot** → **Elements**. An image of the full configuration model appears, as shown in Fig. 7.38.

Fig. 7.38 Full model with mesh of elements and applied constraints.

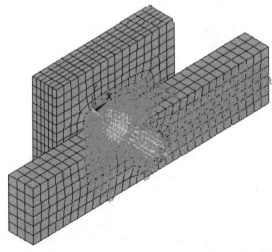

The first set of results to observe in the postprocessing stage is the state of stress due to interference fit between the pin and the hole in the arm.

From the ANSYS Main Menu, select **General Postproc** → **Read results** → **By Load Step**. The frame shown in Fig. 7.39 is produced.

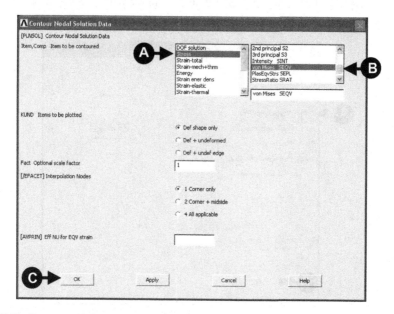

Fig. 7.39 Read results by load step number.

The selection [A] **Load step number** = 1 is shown in Fig. 7.39. By clicking the [B] **OK** button, the selection is implemented.

From the ANSYS Main Menu, select **General Postproc** → **Plot Results** → **Contour Plot** → **Nodal Solu**. In the resulting frame (see Fig. 7.40), the following selections are made: [A] **Item to be contoured** = Stress; [B] **Item to be contoured** = von Mises (SEQV). Pressing the [C] **OK** button implements the selections made.

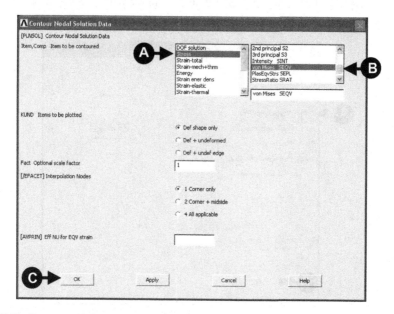

Fig. 7.40 Contour nodal solution data.

A contour plot of von Mises stress (nodal solution) is shown in Fig. 7.41.

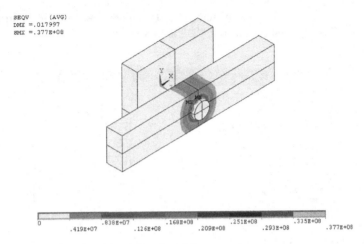

Fig. 7.41 Contour plot of nodal solution (von Mises stress).

Fig. 7.41 shows a stress contour plot for the assembly of the pin in the hole. In order to observe contact pressure on the pin resulting from the interference fit, it is necessary to read results by time/frequency.

From the ANSYS Main Menu, select **General Postproc → Read Results → By Time/Freq**.

In the resulting frame, shown in Fig. 7.42, the selection to be made is: [A] **Value of time or freq** = 120. Pressing the [B] **OK** button implements the selection.

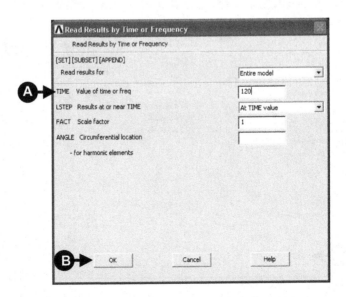

Fig. 7.42 Read results by time or frequency.

From the Utility Menu, choose **Select: Entities**. The frame shown in Fig. 7.43 appears.

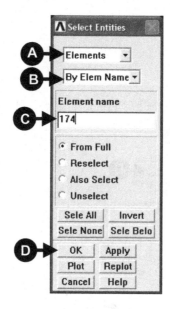

Fig. 7.43 Select entities.

In the frame shown in Fig. 7.43, the following selections are made: [A] **Elements** (from pull-down menu); [B] **By Elem Name** (from pull-down menu); [C] **Element Name** = 174. The element with the number 174 was introduced automatically during the process of creation of contact pairs described earlier. It is listed in the **Preprocessor → Element Type → Add/Edit/Delete** option. Selections are implemented by pressing the [D] **OK** button.

From the Utility Menu, select **Plot → Elements**. An image of the pin with surface elements is produced (see Fig. 7.44).

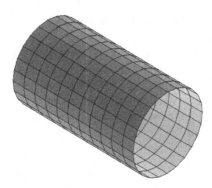

Fig. 7.44 Pin with surface elements.

From the ANSYS Main Menu, select **General Postproc** → **Plot Results** → **Contour Plot** → **Nodal Solu**. The frame shown in Fig. 7.45 appears.

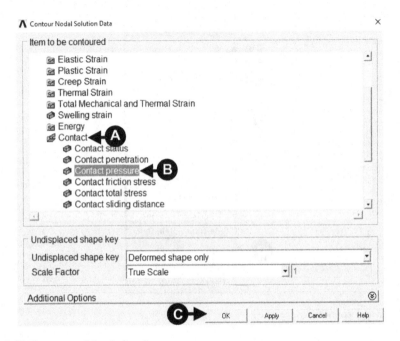

Fig. 7.45 Contour nodal solution data.

In the frame shown in Fig. 7.45, the following selections are made: [A] **Contact** and [B] **Pressure**. These are items to be contoured. Pressing the [C] **OK** button implements the selections made. Fig. 7.46 shows an image of the pin with pressure contours.

The last action to be taken is to observe the state of stress resulting from pulling out the arm from the pin.

From the Utility Menu, choose **Select** → **Everything**. Next, from the ANSYS Main Menu, select **General Postproc: Read Results: By Load Step**. The frame shown in Fig. 7.47 appears.

As shown in Fig. 7.47, [A] **Load step number** = 2″ was selected. Pressing the [B] **OK** button implements the selection.

From the ANSYS Main Menu, select **General Postproc** → **Plot Results** → **Contour Plot** → **Nodal Solu**. In the appearing frame (see Fig. 7.40), the following are selected as items to be contoured: [A] **Stress**; [B] **von Mises (SEQV)**. Pressing the [C] **OK** button implements the selections made. Fig. 7.48 shows stress contours on the pin resulting from pulling out the arm.

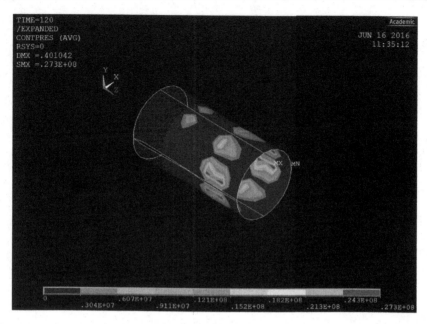

Fig. 7.46 Contact pressure contours on surface of the pin.

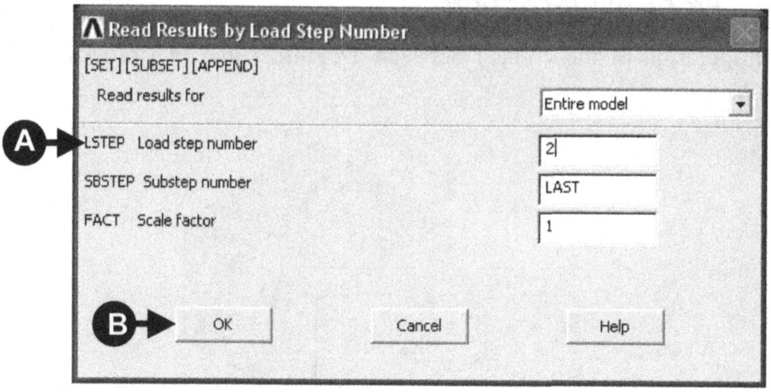

Fig. 7.47 Read results by load step number.

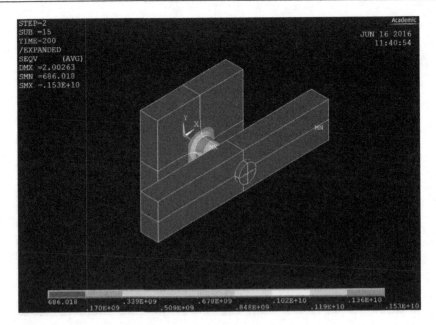

STEP=2
SUB =15
TIME=200
/EXPANDED
SEQV   (AVG)
DMX =2.00263
SMN =686.018
SMX =.153E+10

Academic
JUN 16 2016
11:40:54

MN

686.018   .339E+09   .678E+09   .102E+10   .136E+10
       .170E+09   .509E+09   .848E+09   .119E+10   .153E+10

Fig. 7.48 Pull-out stress contours on the pin.

7.2.2 Concave contact between cylinder and two blocks

7.2.2.1 Problem description

The configuration of the contact between a cylinder and two blocks is shown in Fig. 7.49.

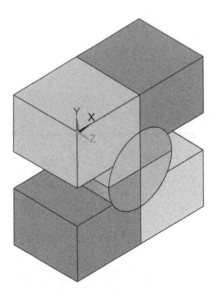

Fig. 7.49 Configuration of the contact between a cylinder and two blocks.

This is a typical contact problem, which in engineering applications is represented by a cylindrical rolling contact bearing. Also, the characteristic feature of the contact is that, nominally, surface contact takes place between elements. In reality, this is never the case due to surface roughness and unavoidable machining errors and dimensional tolerance. There is no geometrical interference when the cylinder and two blocks are assembled.

This is a 3D analysis and advantage could be taken of the inherent symmetry of the model. Therefore, the analysis will be carried out on a half symmetry model only. The objective of the analysis is to observe the stresses in the cylinder when the initial gap between two blocks is decreased by 0.05 cm.

The dimensions of the model are as follows:

Cylinder radius = 0.5 cm; cylinder length = 1 cm; block length = 2 cm; block width = 1 cm; block thickness = 0.75 cm. Both blocks are geometrically identical. All elements are made of steel with Young's modulus = 2.1×10^9 N/m^2, Poisson's ratio = 0.3 and are assumed elastic. The friction coefficient at the interface between cylinder and the block is 0.2.

7.2.2.2 Model construction

For the intended analysis, a half symmetry model is appropriate. This is shown in Fig. 7.50.

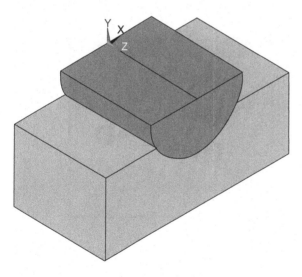

Fig. 7.50 A half symmetry model.

In order to create a model shown in Fig. 7.50, the use of two 3D (three-dimensional) primitives, namely block and cylinder, is made. The model is constructed using GUI facilities only. When carrying out Boolean operations on volumes, it is quite convenient to number them. This is done by selecting from the **Utility Menu → PlotCtrls → Numbering** and checking the appropriate box to activate the **VOLU** (volume numbers) option.

From the ANSYS Main Menu, select **Preprocessor** → **Modelling** → **Create** → **Volumes** → **Block** → **By Dimensions**. In response, a frame shown in Fig. 7.51 appears.

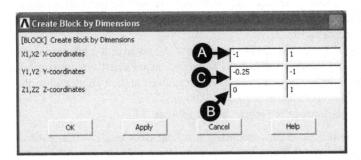

Fig. 7.51 Create block by dimensions.

It can be seen from Fig. 7.51 that appropriate X, Y, Z coordinates were entered to create a block (vol. 1) with the length 2 cm ([A] X1 = − 1; X2 = 1), width 1 cm ([B] Z1 = 0; Z2 = 1), and thickness 0.75 cm ([C] Y1 = − 0.25; Y2 = − 1). Next, from the ANSYS Main Menu, select **Preprocessor** → **Modelling** → **Create** → **Volumes** → **Cylinder** → **By Dimensions**. In response, a frame shown in Fig. 7.52 appears.

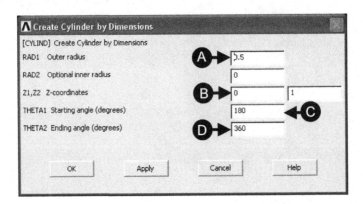

Fig. 7.52 Create volume by dimensions.

The input into the frame, shown in Fig. 7.52, created a solid cylinder sector with the [A] radius 0.5 cm, [B] length 1 cm, [C] starting angle 180 degrees, and [D] ending angle 360 degrees (vol. 2).

From the ANSYS Main Menu, select **Preprocessor** → **Modelling** → **Operate** → **Booleans** → **Overlap** → **Volumes**. A frame shown in Fig. 7.53 appears.

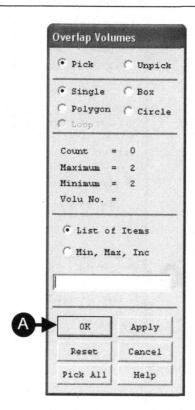

Fig. 7.53 Overlap volumes.

Block (vol. 1) and cylinder (vol.2) should be picked and the [A] **OK** button pressed. As a result of that, the block and cylinder are overlapped and three volumes created: vol. 5 (block), vol. 3 (section of the cylinder within the block), and vol. 4 (remaining of the cylinder after a section of it has been subtracted).

From the ANSYS Main Menu, select **Modelling** → **Delete** → **Volume and Below**. The frame shown in Fig. 7.54 appears.

Picking vol. 4 and clicking the [A] **OK** button deletes it. The same operation should be repeated to delete vol. 3. As a result of that, a block should be produced. The front view of the block is shown in Fig. 7.55.

From the ANSYS Main Menu, select **Preprocessor** → **Modelling** → **Create** → **Volumes** → **Cylinder** → **By Dimensions**. In response, a frame shown in Fig. 7.56 appears.

A cylinder created earlier (see Fig. 7.52 for inputs) is reproduced (vol. 1).

In order to create loading conditions at the contact interface, the cylinder is moved towards the block by 0.05 units.

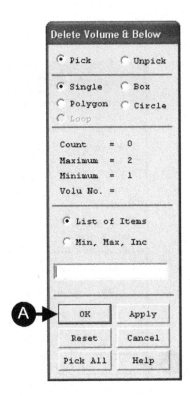

Fig. 7.54 Delete volume and below.

Fig. 7.55 Block with a cylindrical cut-out.

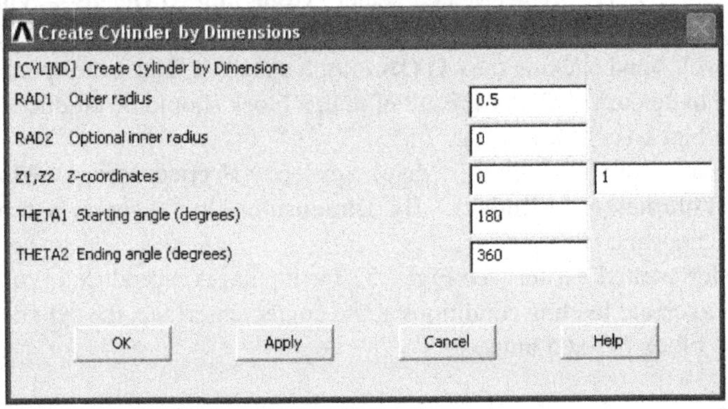

Fig. 7.56 Create cylinder by dimensions.

From the ANSYS Main Menu, select **Modelling** → **Move/Modify** → **Volumes**. A frame shown in Fig. 7.57 appears.

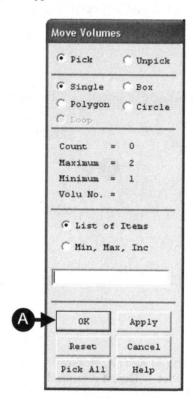

Fig. 7.57 Move volumes.

Selecting vol. 1 and clicking the [A] **OK** button calls up another frame shown in Fig. 7.58.

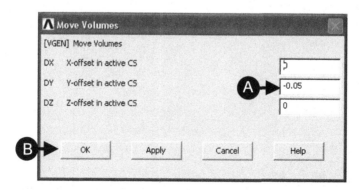

Fig. 7.58 Move volumes.

Fig. 7.58 shows that the cylinder was moved by [A] 0.05 cm downwards—that is, towards the block, after clicking the [B] **OK** button.

From the Utility Menu, select **Plot** → **Replot** to view the cylinder positioned in the required location. Finally, from the Utility Menu, select **PlotCtrls** → **View Settings** → **Viewing Direction**. The frame shown in Fig. 7.59 appears.

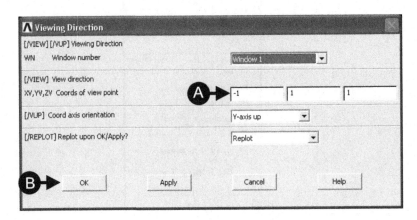

Fig. 7.59 Viewing direction.

By selecting coordinates [A] X, Y, Z, as shown in Fig. 7.59, clicking the [B] **OK** button, and activating the **Plot** → **Replot** command (Utility Menu), a half symmetry model, shown in Fig. 7.50, is finally created.

7.2.2.3 Material properties

Before any analysis is attempted, it is necessary to define the properties of the material to be used.

From the ANSYS Main Menu, select **Preferences**. The frame shown in Fig. 7.60 is produced.

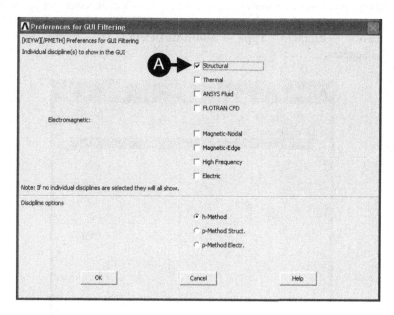

Fig. 7.60 Preferences: structural.

From the Preferences list, [A] **Structural** option was selected, as shown in Fig. 7.60. From the ANSYS Main Menu, select **Preprocessor → Material Props → Material Models**. Click in turn: **Structural → Linear → Elastic → Isotropic**. A frame shown in Fig. 7.61 appears.

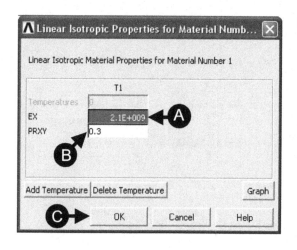

Fig. 7.61 Material properties.

Enter [A] $EX = 2.1 \times 10^9$ for Young's modulus and [B] **PRXY** $= 0.3$ for Poisson's ratio. Then click the [C] **OK** button and afterwards **Material** → **Exit**. After defining properties of the material, the next step is to select element type appropriate for the analysis performed. From the ANSYS Main Menu, select **Preprocessor** → **Element Type** → **Add/Edit/Delete**. The frame shown in Fig. 7.62 appears.

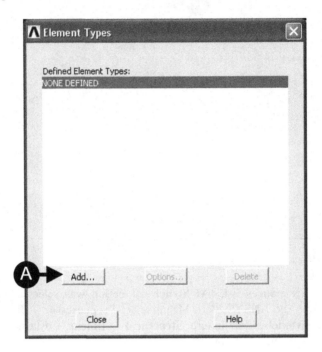

Fig. 7.62 Element types.

Click [A] **Add** in order to pull down another frame, shown in Fig. 7.63.

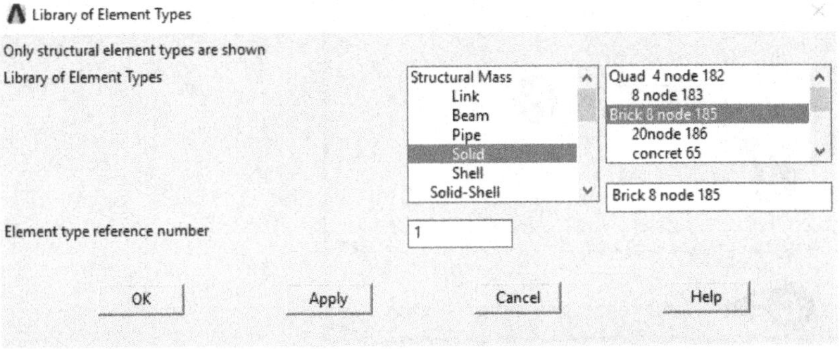

Fig. 7.63 Library of element types.

In the left column click [A] **Structural Solid** and in the right column click [B] **Brick 8node 185**. After that click [C] **OK** and **Close** in order to finish the element type selection.

7.2.2.4 Meshing

From the ANSYS Main Menu, select **Preprocessor: → Meshing → Mesh Tool**. The frame shown in Fig. 7.64 appears.

Fig. 7.64 Mesh tool options.

There are a number of options available. The first step is to go to **Size Control** → **Lines** option and click the [A] **Set** button. This opens another frame (shown in Fig. 7.65), prompting you to pick lines on which element size is going to be controlled.

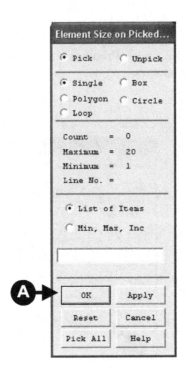

Fig. 7.65 Element size on picked lines.

Pick two horizontal lines on the front edge of the cylinder and click the [A] **OK** button. The frame shown in Fig. 7.66 appears.

In the box [A] **No. of element divisions**, type 3 and change selection [B] **SIZE, NDIV can be changed** to **No** by checking the box and click the [C] **OK** button. Both selections are shown in Fig. 7.66.

Similarly, using **Mesh Tool** frame, click button **Set** in the **Size Controls: Lines** option and pick the curved line on the front of the block. Click **OK** afterwards. The frame shown in Fig. 7.66 appears. In the box [A] **No. of element divisions**, type 4 this time and press the [C] **OK** button.

In the frame **Mesh Tool** (see Fig. 7.67), pull down [A] **Volumes** in the option **Mesh**.

Check [B] **Hex/Wedge** and [C] **Sweep** options. This is shown in Fig. 7.67.

Pressing the [D] **Sweep** button brings another frame asking you to pick volumes to be swept (see Fig. 7.68).

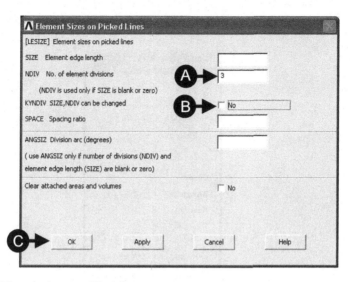

Fig. 7.66 Element sizes on picked lines.

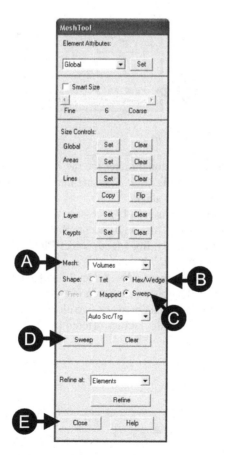

Fig. 7.67 Checked hex and sweep options.

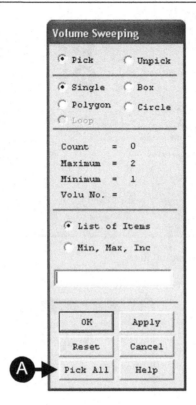

Fig. 7.68 Volume sweeping.

Pressing the [A] **Pick All** button initiates the meshing process. After meshing, the model looks like the image in Fig. 7.69. Pressing the [E] **Close** button in **Mesh Tool** frame ends the mesh generation stage.

After meshing is complete, it is usually necessary to smooth element edges in order to improve the graphic display. This can be accomplished using the **PlotCtrls** facility in the Utility Menu. From the Utility Menu, select **PlotCtrls** → **Style** → **Size and Shape**. The frame shown in Fig. 7.70 appears.

In the option [A] **Facets/element edge**, select **2 facets/edge** and click the [B] **OK** button to implement the selection, as shown in Fig. 7.70.

In solving the problem of contact between two elements, it is necessary to create a contact pair. Contact Wizard is the facility offered by ANSYS.

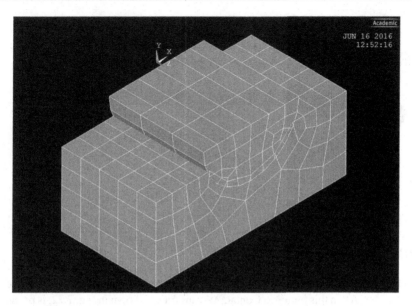

Fig. 7.69 Model after meshing process.

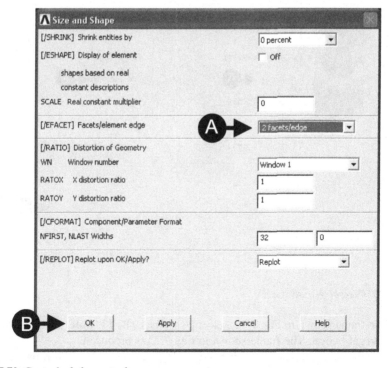

Fig. 7.70 Control of element edges.

7.2.2.5 Creation of a contact pair

From the ANSYS Main Menu, select **Preprocessor** → **Modelling** → **Create** → **Contact Pair**. As a result of this selection, a frame shown in Fig. 7.71 appears.

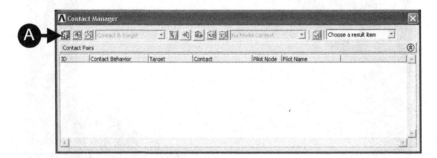

Fig. 7.71 Contact Manager.

The Contact Wizard button is located in the upper left-hand corner of the frame. By clicking the [A] on this button, a Contact Wizard frame, shown in Fig. 7.72, is produced.

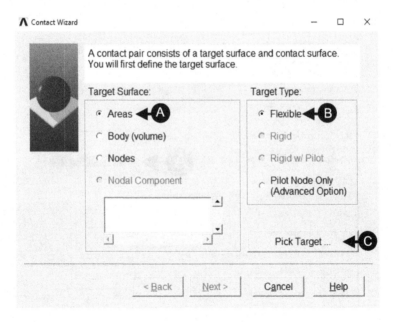

Fig. 7.72 Contact Wizard (target).

In the frame shown in Fig. 7.72, select [A] **Areas**, [B] **Flexible**, and press the [C] **Pick Target** button. The frame shown in Fig. 7.73 is produced.

Select the curved surface in the block as a target and press the [A] **OK** button in the frame of Fig. 7.73. Again the Contact Wizard frame appears, and this time the **Next** button should be pressed to obtain the frame shown in Fig. 7.74.

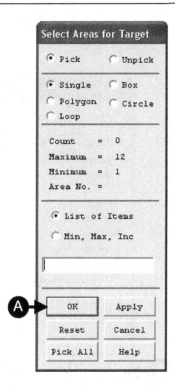

Fig. 7.73 Select body for target.

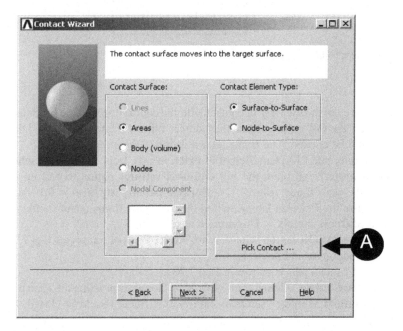

Fig. 7.74 Contact Wizard (contact).

Press the [A] **Pick Contact** button to create the frame shown in Fig. 7.75.

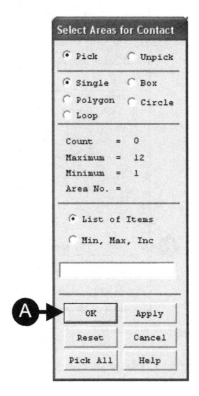

Fig. 7.75 Select bodies for contact.

Pick the area of the cylinder in contact with the concave area in the block as a contact and press the **OK** button. Again the Contact Wizard frame appears, and the **Next** button should be clicked. The frame shown in Fig. 7.76 appears.

In this frame, select [A] **Coefficient of Friction** = 0.2, check box [B] **Include initial penetration**. Next, press the [C] **Optional settings** button to call up another frame. In the new frame (Fig. 7.77), **Normal penalty stiffness = 0.1** should be selected. Also, the **Friction** tab located in the top of the frame menu should be activated and **Stiffness matrix = Unsymmetric** selected.

Pressing the [A] **OK** button brings back **Contact Wizard** frame (see Fig. 7.76); the [D] **Create** button should be pressed.

A created contact pair is shown in Fig. 7.78.

Finally, the **Contact Wizard** frame should be closed by pressing the **Finish** button. Also, the **Contact Manager** summary information frame should be closed.

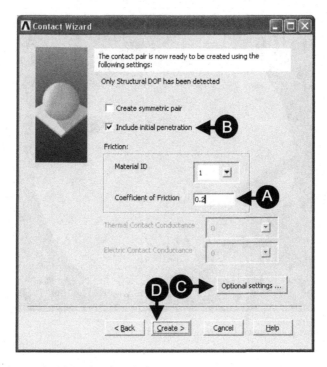

Fig. 7.76 Contact Wizard (optional settings).

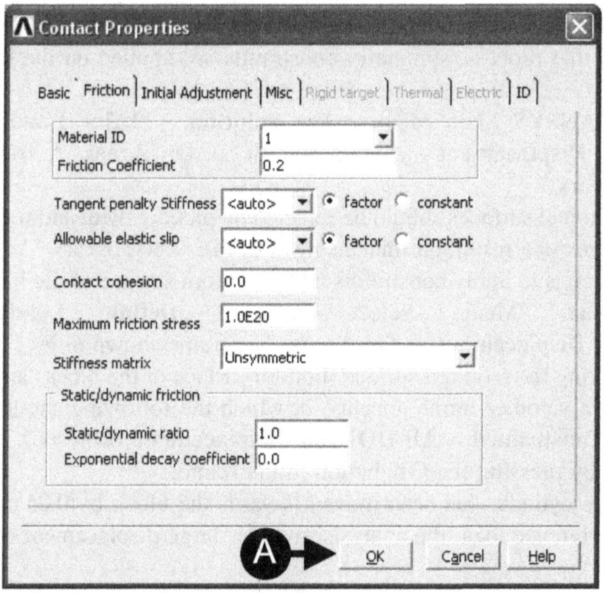

Fig. 7.77 Contact properties—optional settings.

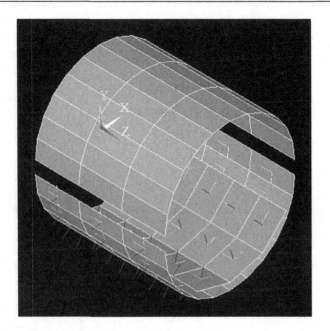

Fig. 7.78 Contact pair created by Contact Wizard.

7.2.2.6 Solution

Before the solution process can be attempted, solution criteria have to be specified. As a first step in that process, symmetry constraints are applied on the half symmetry model.

From the ANSYS Main Menu, select **Solution** → **Define Loads** → **Apply** → **Structural** → **Displacement** → **Symmetry BC** → **On Areas**. A frame shown in Fig. 7.79 appears.

Three horizontal surfaces should be selected by picking them and then clicking the [A] **OK** button. As a result, an image shown in Fig. 7.80 appears.

The next step is to apply constraints on the bottom surface of the block. From the ANSYS Main Menu, select **Solution** → **Define Loads** → **Apply** → **Structural** → **Displacement** → **On Areas**. The frame shown in Fig. 7.81 appears.

After selecting the required surface (bottom surface of the block) and pressing the [A] **OK** button, another frame appears, in which the following should be selected: **DOFs to be constrained** = **All DOF** and **Displacement value** = 0. Selections are implemented by pressing the **OK** button in the frame.

Because the cylinder has been moved towards the block by 0.05 cm, in order to create an interference load, the analysis involves large displacement effects.

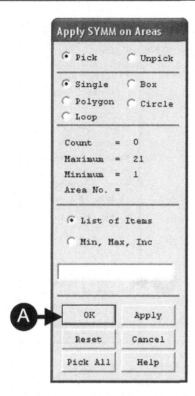

Fig. 7.79 Apply SYMM on areas.

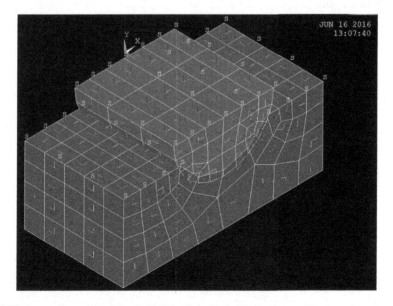

Fig. 7.80 Symmetry constraints applied on three horizontal areas.

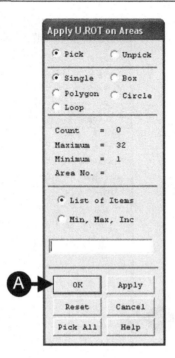

Fig. 7.81 Apply U, ROT on areas.

From the ANSYS Main Menu, select **Solution → Analysis Type → Sol'n Controls**. A frame shown in Fig. 7.82 appears.

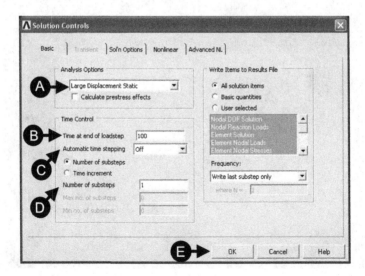

Fig. 7.82 Solution controls.

In the pull-down menu, select [A] **Large Displacement Static**. Further selected options should be: [B] **Time at end of load step** = 100; [C] **Automatic time stepping** (pull-down menu) = **Off**; [D] **Number of substeps** = 1. All specified selections are shown in Fig. 7.82. Pressing the [E] **OK** button implements the settings and closes the frame.

Now the modelling stage is completed and the solution can be attempted. From the ANSYS Main Menu, select **Solution** → **Solve** → **Current LS**. A frame showing a review of information pertaining to the planned solution action appears. After checking that everything is correct, select **File: Close** to close that frame. Pressing the **OK** button starts the solution. When the solution is completed, press the **Close** button.

In order to return to the previous image of the model, select **Utility Menu** → **Plot** → **Replot**.

7.2.2.7 Postprocessing

Solution results can be displayed in a variety of forms using postprocessing facility. For the results to be viewed for the full model, the half symmetry model used for analysis has to be expanded.

From the Utility Menu, select **PlotCtrls** → **Style** → **Symmetry Expansion** → **Periodic/Cyclic Symmetry**. Fig. 7.83 shows the resulting frame.

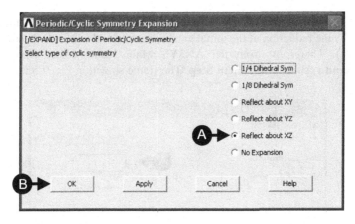

Fig. 7.83 Periodic/cyclic symmetry expansion.

In the frame shown in Fig. 7.83, [A] **Reflect about XZ** was selected. After clicking the [B] **OK** button in the frame and selecting **Plot: Elements** from the Utility Menu, an image of a full model, shown in Fig. 7.84, is produced.

Fig. 7.84 Full model with mesh of
elements and applied constraints.

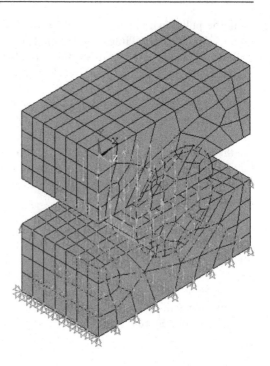

The objective of the analysis presented here was to observe stresses in the cylinder
produced by the reduction of the initial gap between two blocks by 0.05 cm (an interference fit). Therefore, from the ANSYS Main Menu, select **General Postproc → Read results → By Load Step**. The frame shown in Fig. 7.85 is produced.

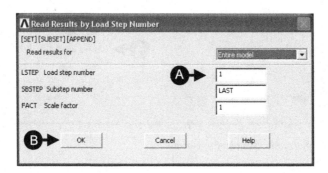

Fig. 7.85 Read results by load step number.

The selection [A] **Load step number** = 1, shown in Fig. 7.85, is implemented by
clicking the [B] **OK** button.

From the ANSYS Main Menu, select **General Postproc → Plot Results → Contour
Plot → Nodal Solu**. In the resulting frame (see Fig. 7.86), the following selections are
made: [A] **Item to be contoured = Stress**; [B] **Item to be contoured = von Mises
(SEQV)**. Pressing the [C] **OK** button implements the selections made.

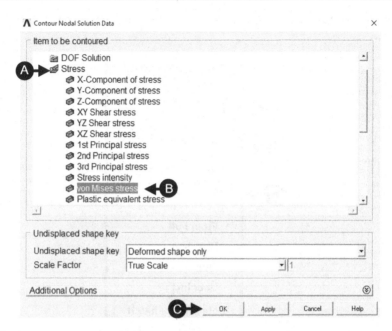

Fig. 7.86 Contour nodal solution data.

A contour plot of von Mises stress (nodal solution) is shown in Fig. 7.87.

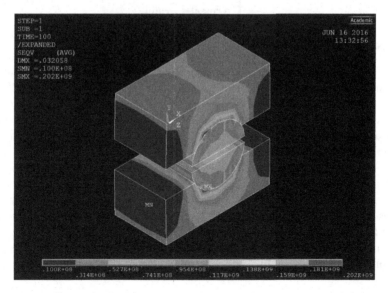

Fig. 7.87 Contour plot of nodal solution (von Mises stress).

Fig. 7.87 shows a von Mises stress contour for the whole assembly. If one is interested in observing contact pressure on the cylinder alone, then a different presentation of solution results is required.

From the Utility Menu, choose **Select** → **Entities**. The frame shown in Fig. 7.88 appears.

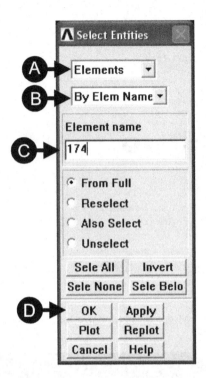

Fig. 7.88 Select entities.

In the frame shown in Fig. 7.88, the following selections are made: [A] **Elements** (first pull-down menu); [B] **By Elem Name** (second pull-down menu); [C] **Element Name** = 174. The element with the number 174 was introduced automatically during the process of creation of contact pairs described earlier. It is listed in the **Preprocessor** → **Element Type** → **Add/Edit/Delete** option. Selections are implemented by pressing the [D] **OK** button.

From the Utility Menu, select **Plot** → **Elements**. An image of the cylinder surface with a mesh of elements is produced (see Fig. 7.89).

It is seen that the gap equal to 0.05 units exists between two half of the cylinder. It is the result of moving half of the cylinder towards the block (by 0.05 cm) in order to create loading at the interface.

From the ANSYS Main Menu, select **General Postproc** → **Plot Results** → **Contour Plot** → **Nodal Solu**. The frame shown in Fig. 7.90 appears.

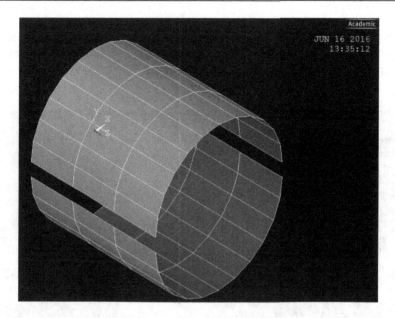

Fig. 7.89 Cylinder with surface elements (174).

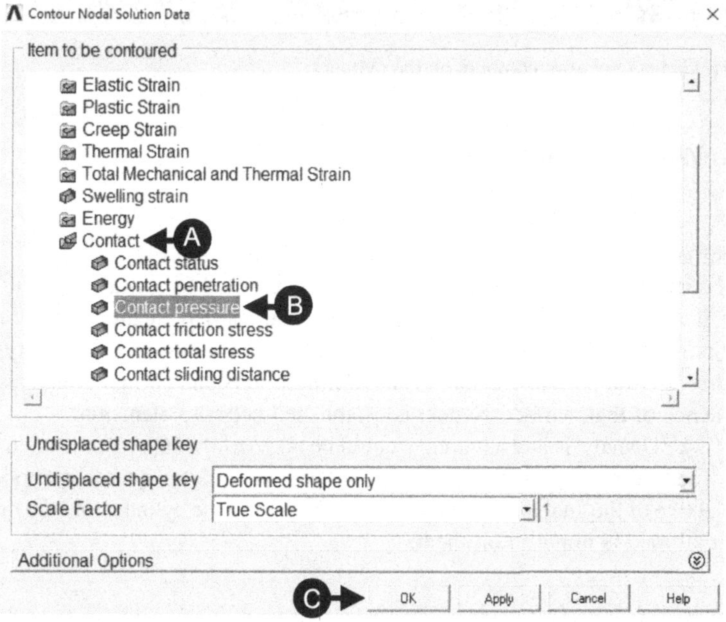

Fig. 7.90 Contour nodal solution data.

In the frame, shown in Fig. 7.90, the following selections are made: [A] **Contact** and [B] **Pressure**. These are items to be contoured. Pressing the [C] **OK** button implements the selections made. In response to that, an image of the cylinder surface with pressure contours on it is produced, as shown in Fig. 7.91.

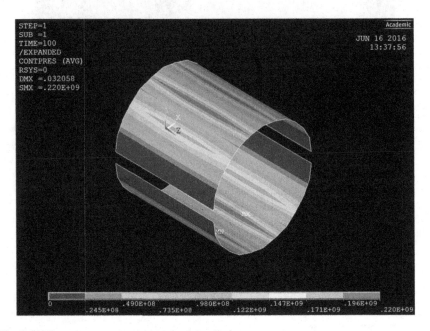

Fig. 7.91 Contact pressure contours on the cylinder.

7.2.3   Wheel-on-rail line contact

7.2.3.1   Problem description

The configuration of the contact to be analysed is shown in Fig. 7.92.

This contact problem, which in practice is represented by a wheel-on-rail configuration, is well known in engineering. Also, the characteristic feature of the contact is that, nominally, contact between elements takes place along line. In reality, this is never the case due to unavoidable elastic deformations and surface roughness. As a consequence of that, surface contact is established between elements.

This is a 3D analysis and advantage could be taken of the inherent symmetry of the model. Therefore, the analysis will be carried out on a quarter symmetry model only. The objective of the analysis is to observe the stresses in the cylinder and the rail when an external load is imposed on them.

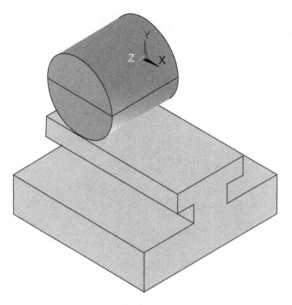

Fig. 7.92 Configuration of the contact.

The dimensions of the model are as follows:

Diameter of the cylinder = 1 cm; cylinder length = 2 cm. Rail dimensions: base width = 4 cm; head width = 2 cm; head thickness = 0.5 cm; rail height = 2 cm.

Both elements are made of steel with Young's modulus = 2.1×10^9 N/m^2, Poisson's ratio = 0.3 and are assumed elastic. The friction coefficient at the interface between cylinder and the rail is 0.1.

7.2.3.2 Model construction

For the intended analysis, a half symmetry model is appropriate. This is shown in Fig. 7.93.

The model is constructed using GUI facilities only. First, a 2D model is created (using rectangles and circles as primitives). This is shown in Fig. 7.94.

Next, using the 'extrude' facility, areas are converted into volumes and a 3D model constructed. When carrying out Boolean operations on areas or volumes, it is convenient to number them. This is done by selecting from the **Utility Menu** → **PlotCtrls** → **Numbering** and checking the appropriate box to activate the **AREA** (area numbers) or **VOLU** (volume numbers) option.

From the ANSYS Main Menu, select **Preprocessor** → **Modelling** → **Create** → **Areas** → **Rectangle** → **By Dimensions**. A frame shown in Fig. 7.95 appears.

Entered coordinates, [A] (X1 = 0; X2 = 2) and [B] (Y1 = −0.5; Y2 = −2.5), are shown in Fig. 7.95. This creates area A1.

Fig. 7.93 A quarter symmetry model.

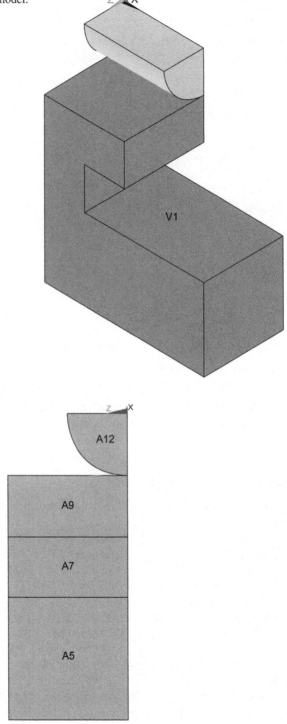

Fig. 7.94 Two-dimensional image (front) of the quarter symmetry model. Involved areas are numbered.

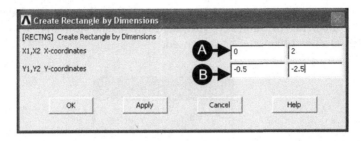

Fig. 7.95 Create rectangle by dimensions.

From the ANSYS Main Menu, select **Preprocessor → Modelling → Create → Areas → Rectangle → By Dimensions**. A frame with entered coordinates, [A] (X1 = 1; X2 = 2) and [B] (Y1 = −0.5; Y2 = −1.5), is shown in Fig. 7.96. This creates area A2.

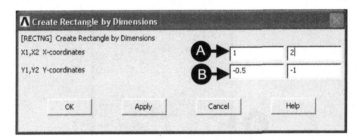

Fig. 7.96 Create rectangle by dimensions.

From the ANSYS Main Menu, select **Preprocessor → Modelling → Create → Areas → Rectangle → By Dimensions**. A frame with entered coordinates, [A] (X1 = 0.5; X2 = 2) and [B] (Y1 = −1; Y2 = −1.5), is shown in Fig. 7.97. This creates area A3.

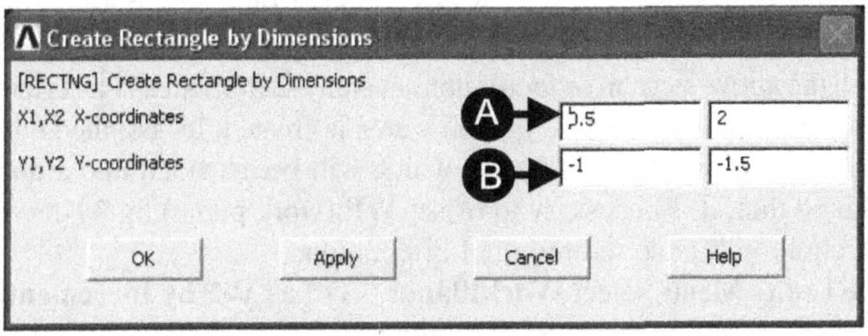

Fig. 7.97 Create rectangle by dimensions.

From the ANSYS Main Menu, select **Preprocessor** → **Modelling** →
Operate → **Booleans** → **Subtract** → **Areas**. A frame shown in Fig. 7.98 appears,
asking for selection of areas to be subtracted.

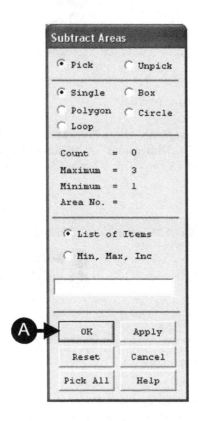

Fig. 7.98 Subtract areas.

Select the first area, A1, by clicking on it. Then click the [A] **OK** button in the
frame of Fig. 7.98. Next, select area A3 by clicking on it and pressing the [A] **OK**
button. Area A3 is subtracted from area A1. This operation creates area A4.

Repeat all the above steps in order to subtract area A2 from area A4. As a result of this
operation, a cross-section of half of the rail's area is created. Its assigned number is A1.

Next, a quarter of a circle is created, which will be extruded into a quarter of the
cylinder. To do that, it is necessary to offset **WP** (work plane) by 90 degrees, so that
the quarter circle will have the required orientation.

From the Utility Menu, select **WorkPlane** → **Offset WP by Increments**. Fig. 7.99
shows the frame resulting from the selection.

It can be seen from Fig. 7.99 that the ZX plane was offset by [A] 90 degrees clockwise (the clockwise direction is negative).

From the ANSYS Main Menu, select **Preprocessor** → **Modelling** → **Create** →
Areas → **Circle** → **By Dimensions**. The frame shown in Fig. 7.100 appears where

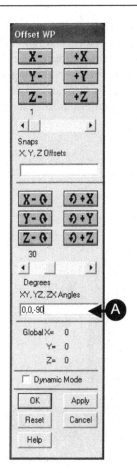

Fig. 7.99 Offset WP.

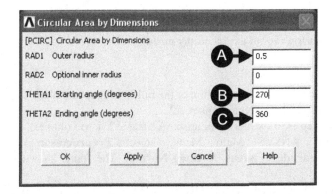

Fig. 7.100 Circular area by dimensions.

appropriate data were entered to create a solid quarter of a circle with [A] outer radius
of 0.5 cm, [B] starting angle 270 degrees, and [C] ending angle 360 degrees.

When the solid quarter circular area is created, it is important to restore **WP** offset
to its original setting. This can be done by following steps associated with Fig. 7.99
and selecting, XY, YZ, ZX angles as 0, 0, and 90 degrees, respectively.

In an isometric view, the 2D quarter model is shown in Fig. 7.101.

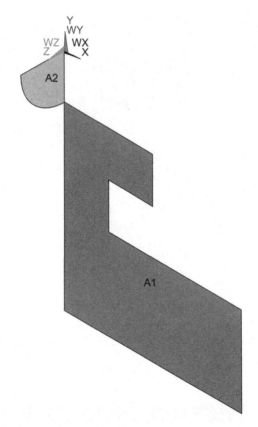

Fig. 7.101 Isometric view of the 2D quarter model.

It can be seen that the cross-section of the rail is assigned number A1 and the quarter of the solid circle is given number A2.

The next step is to extrude both areas, A1 and A2, into volumes.

From the ANSYS Main Menu, select **Preprocessor → Modelling →
Operate → Extrude → Areas → Along Normal**. The frame shown in Fig. 7.102
appears.

Pick area A1 and click the [A] **OK** button to pull down another frame shown in
Fig. 7.103.

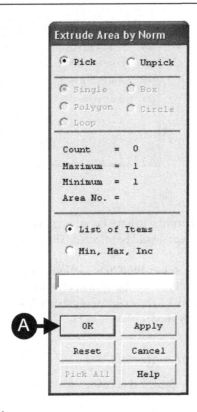

Fig. 7.102 Extrude areas by norm.

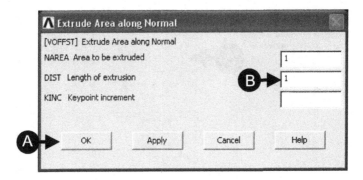

Fig. 7.103 Extrude area along normal.

Selections made are shown in Fig. 7.103. By clicking the [A] **OK** button, a 2D cross-section of the rail is extruded by [B] 1 cm (length of extrusion) into a volume. The direction of extrusion is normal to the rail's cross-section.

In a similar way, a quarter solid circle can be extruded to create a quarter of the cylinder with 1 cm length. Fig. 7.104 shows the frame and selections made.

Fig. 7.104 Extrude area along normal.

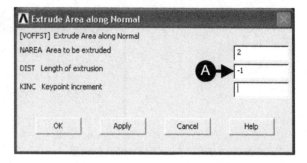

It should be noted that in order to have the quarter cylinder oriented as required, [A] length of extrusion $= -1$ cm should be selected. The minus sign denotes the direction of extrusion.

From the Utility Menu, select **PlotCtrls** → **Numbering** and check in **VOLU** and check out **AREA**. This will change the system of numbering from areas to volumes. The rail is allocated number V1 and cylinder number V2. This completes the construction of a quarter symmetry model, shown in Fig. 7.93.

7.2.3.3 Properties of material

Before any analysis is attempted, it is necessary to define the properties of the material to be used.

From the ANSYS Main Menu, select **Preferences**. The frame in Fig. 7.105 is produced.

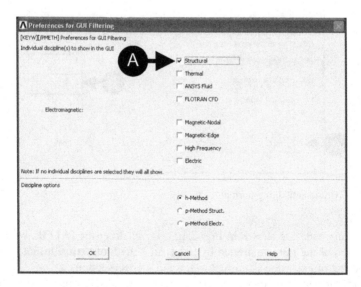

Fig. 7.105 Preferences: structural.

As shown in Fig. 7.105, [A] **Structural** was the option selected.

From the ANSYS Main Menu, select **Preprocessor** → **Material Props** → **Material Models**. Double-click **Structural** → **Linear** → **Elastic** → **Isotropic**. A frame shown in Fig. 7.106 appears.

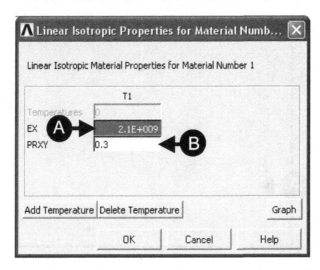

Fig. 7.106 Material properties.

Enter [A] **EX** = 2.1 × 10^9 for Young's modulus and [B] **PRXY** = 0.3 for Poisson's ratio. Then click the [C] **OK** button and afterwards **Material: Exit**.

After defining properties of the material, the next step is to select element type appropriate for the analysis performed.

From the ANSYS Main Menu, select **Preprocessor** → **Element Type** → **Add/Edit/Delete**. The frame shown in Fig. 7.107 appears.

Click the [A] **Add** button in order to pull down another frame, shown in Fig. 7.108.

In the left column click [A] **Structural Solid** and in the right column click [B] **Brick 8node 185**. After that click [C] **OK** and **Close** in order to finish the element type selection.

7.2.3.4 Meshing

From the ANSYS Main Menu, select **Preprocessor** → **Meshing** → **Mesh Tool**. The frame shown in Fig. 7.109 appears.

There are a number of options available. The first step is to go to **Size Control** → **Lines** option and click the [A] **Set** button. This opens another frame (shown in Fig. 7.110) prompting you to pick lines on which element size is going to be controlled.

Pick two lines, one arcuate and the other horizontal in contact with the top surface of the rail, located on the front side of the cylinder, and click the [A] **OK** button. The frame shown in Fig. 7.111 appears.

Fig. 7.107 Element types.

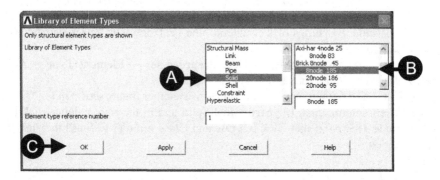

Fig. 7.108 Library of element types.

In the box [A] **No. of element divisions**, type 30 and change selection [B] **SIZE, NDIV can be changed** to **No** by checking the box out and then click [C] **OK**. Both selections are shown in Fig. 7.111.

Similarly, using the **MeshTool** frame (see Fig. 7.109), click the [A] **Set** button in the **Size Controls → Lines** option and pick two lines located on the top surface of the rail: one coinciding with the line previously picked and the other at the right angle to

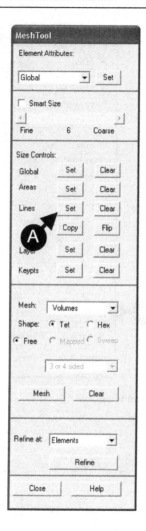

Fig. 7.109 Mesh tool.

the first one. Click [A] **OK** as shown in Fig. 7.110. The frame shown in Fig. 7.111 appears again. In the box [A] **No. of element divisions**, type 30 and press the [C] **OK** button.

In the frame **MeshTool** (see Fig. 7.112), pull down [A] **Volumes** in the option **Mesh**. Check the [B] **Hex** and [C] **Sweep** options.

Pressing the [D] **Sweep** button brings up another frame, shown in Fig. 7.113, asking you to pick volumes to sweep.

Pressing the [A] **Pick All** button initiates the meshing process. The model after meshing looks like the image in Fig. 7.114.

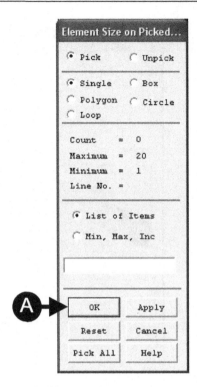

Fig. 7.110 Element size on picked lines.

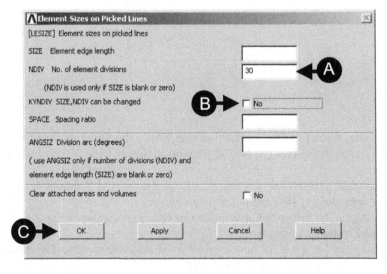

Fig. 7.111 Element sizes.

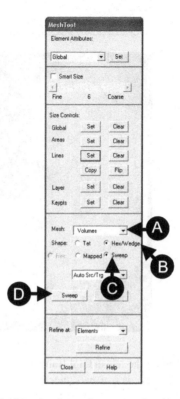

Fig. 7.112 Mesh tool (checked hex/wedge and sweep options).

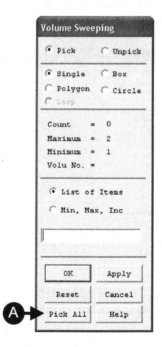

Fig. 7.113 Volume sweeping.

Fig. 7.114 Model after
meshing process.

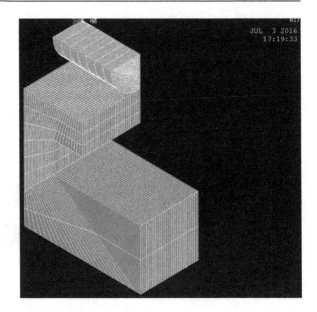

Pressing the **Close** button on the **MeshTool** frame ends the mesh generation stage.

After the meshing is complete, it is usually necessary to smooth element edges in order to improve the graphic display. This can be accomplished using the **PlotCtrls** facility in the Utility Menu.

From the Utility Menu, select **PlotCtrls** → **Style** → **Size and Shape**. The frame shown in Fig. 7.115 appears.

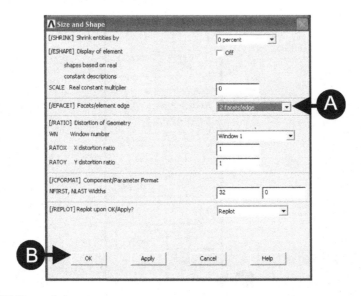

Fig. 7.115 Size and shape.

In the option [A] **Facets/element edge**, select **2 facets/edge** and click the [B] **OK** button as shown in Fig. 7.115.

7.2.3.5 Creation of a contact pair

From the ANSYS Main Menu, select **Preprocessor** → **Modelling** → **Create** → **Contact Pair**. As a result of this selection, a frame shown in Fig. 7.116 appears.

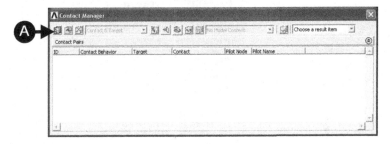

Fig. 7.116 Contact Manager.

The upper left-hand corner of the frame contains the [A] **Contact Wizard** button. By clicking on this button, a Contact Wizard frame, shown in Fig. 7.117, is produced.

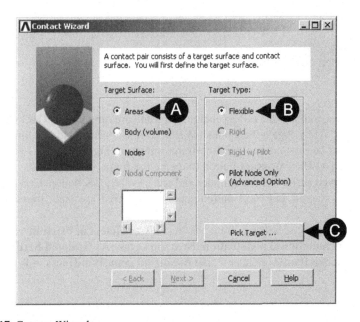

Fig. 7.117 Contact Wizard.

In the frame shown in Fig. 7.117, select [A] **Areas**, [B] **Flexible**, and press the [C] **Pick Target** button. The frame shown in Fig. 7.118 is produced.

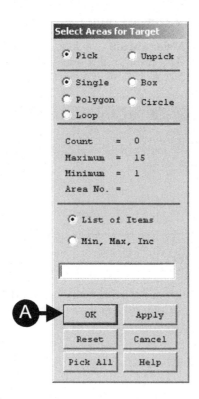

Fig. 7.118 Select areas for target.

Select the top area of the rail by clicking on it and pressing the [A] **OK** button in the frame of Fig. 7.118. Again the Contact Wizard frame appears, and this time the **Next** button should be pressed to obtain the frame shown in Fig. 7.119.

Press the [A] **Pick Contact** button to create the frame in Fig. 7.120.

Select curved surface of the cylinder and press the [A] **OK** button. Contact Wizard frame appears again and **Next** button should be pressed to obtain frame shown in Fig. 7.121.

In the [A] **Optional Settings** frame, select set **Coefficient of Friction** = 0.1, check out box—**Include initial penetration**. Next, press the [A] **Optional Settings** button to call up another frame. In the new frame, **Normal penalty stiffness** = 0.1 should be

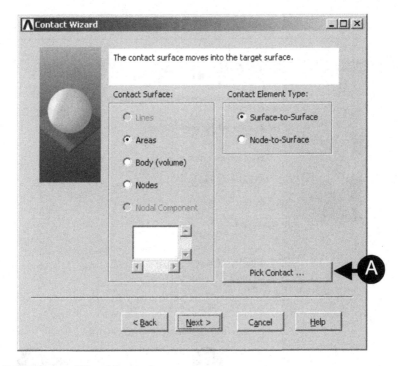

Fig. 7.119 Contact Wizard (contact).

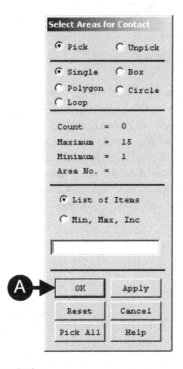

Fig. 7.120 Select areas for contact.

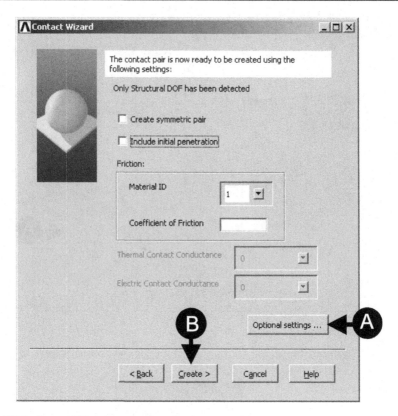

Fig. 7.121 Contact Wizard (optional settings).

selected. Also, the **Friction** tab located in the top of the frame menu should be activated and **Stiffness matrix = Unsymmetric** selected. The **Initial Adjustment** tab,
also located at the top of the menu, should be pressed and in the box [A] **Automatic
Contact Adjustment**, **Close gap** should be selected, as shown in Fig. 7.122.

This will prevent solution crushing due to the gap between target and contact surfaces being greater than allowed by the programme.

Pressing the [B] **OK** button brings back the Contact Wizard frame (see Fig. 7.121),
and the [B] **Create** button should then be pressed. The created contact pair is shown in
Fig. 7.123.

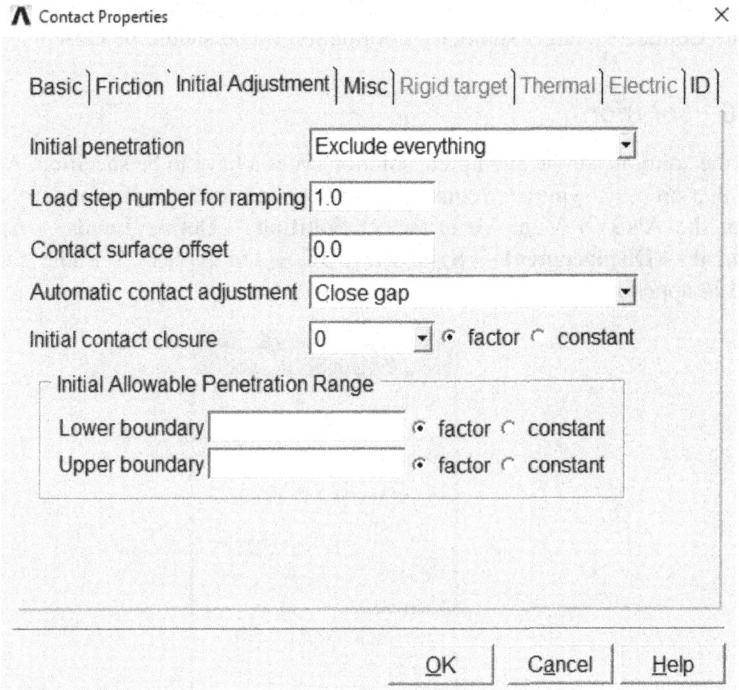

Λ Contact Properties   ✕

Basic | Friction ˙ Initial Adjustment | Misc | Rigid target | Thermal | Electric | ID |

Initial penetration   Exclude everything   ▾

Load step number for ramping 1.0

Contact surface offset   0.0

Automatic contact adjustment   Close gap   ▾

Initial contact closure   0   ▾ ⦿ factor ⦾ constant

┌─ Initial Allowable Penetration Range

   Lower boundary [   ] ⦿ factor ⦾ constant
   Upper boundary [   ] ⦿ factor ⦾ constant

OK | Cancel | Help

Fig. 7.122 Contact properties—optional settings.

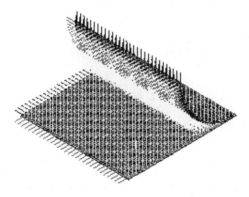

Fig. 7.123 Contact pair (surface).

Finally, the Contact Wizard frame should be closed by pressing the **Finish** button. Also, the Contact Manager summary information frame should be closed.

7.2.3.6 Solution

Before the solution can be attempted, solution criteria have to be specified. As a first step in that process, symmetry constraints are applied on the half symmetry model.

From the ANSYS Main Menu, select **Solution** → **Define Loads** → **Apply** → **Structural** → **Displacement** → **Symmetry BC** → **On Areas**. A frame shown in Fig. 7.124 appears.

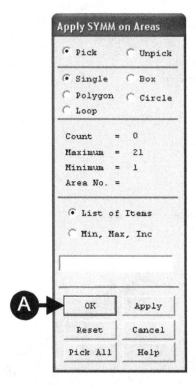

Fig. 7.124 Apply SYMM on areas.

Four surfaces at the back of the quarter symmetry model should be selected and the [A] **OK** button clicked. Symmetry constraints applied to the model are shown in Fig. 7.125.

The next step is to apply constraints on the bottom surface of the block. From the ANSYS Main Menu, select **Solution** → **Define Loads** → **Apply** → **Structural** → **Displacement** → **On Areas**. The frame shown in Fig. 7.126 appears.

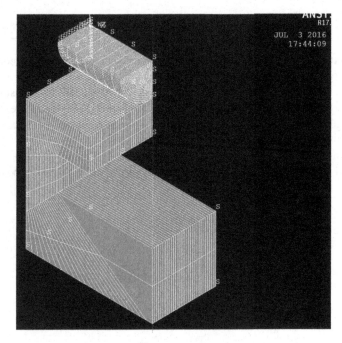

Fig. 7.125 Symmetry constraints applied to the model.

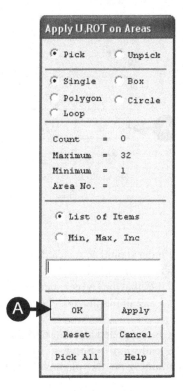

Fig. 7.126 Apply U, ROT on areas.

The bottom surface of the rail should be selected. After selecting the required surface and pressing the [A] **OK** button, another frame appears in which the following should be selected: **DOFs to be constrained** = All DOF and **Displacement value** = 0. Selections are implemented by pressing the **OK** button in the frame.

The final action is to apply external loads. In the case considered here, a pressure acting on the top surface of the cylinder will be used.

From the ANSYS Main Menu, select **Solution** → **Define Loads** → **Apply** → **Structural** → **Pressure** → **On Areas**. A frame shown in Fig. 7.127 appears.

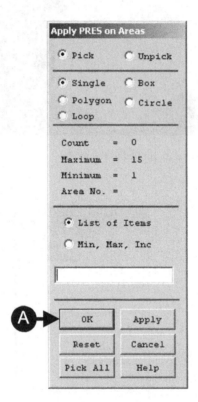

Fig. 7.127 Selection of area.

The top surface of the cylinder should be selected and the [A] **OK** button pressed to pull down another frame, shown in Fig. 7.128.

It can be seen from Fig. 7.128 that the [A] constant pressure of 0.5 MPa was applied to the selected surface. Fig. 7.129 shows the model ready for solution with constraints and applied load.

Now the modelling stage is completed and the solution can be attempted. From the ANSYS Main Menu, select **Solution** → **Solve** → **Current LS**. A frame showing a review of information pertaining to the planned solution action appears. After

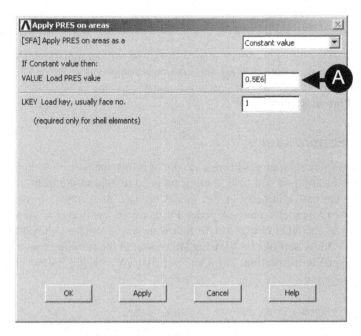

Fig. 7.128 Apply PRES on areas (magnitude).

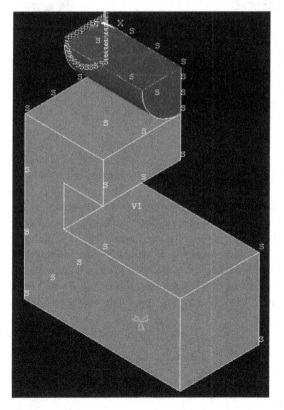

Fig. 7.129 Constraints and loads applied to the model.

checking that everything is correct, select **File: Close** to close that frame. Pressing the **OK** button starts the solution. When the solution is completed, press the **Close** button.

In order to return to the previous image of the model, select **Utility Menu → Plot → Replot**.

7.2.3.7 Postprocessing

In order to display solution results in a variety of forms, the postprocessing facility is used. In the example solved here, there is no need to expand the quarter symmetry model into the half symmetry or full model because the contact stresses are best observed from a quarter symmetry model. Furthermore, the isometric viewing direction used so far should be changed in the following way. From the Utility Menu, select **PlotCtrls → View Settings → Viewing Direction**. In the resulting frame, shown in Fig. 7.130, set **View direction**: [A] **XV** $= -1$, [B] **YV** $= 1$, [C] **ZV** $= -1$ and click the [D] **OK** button.

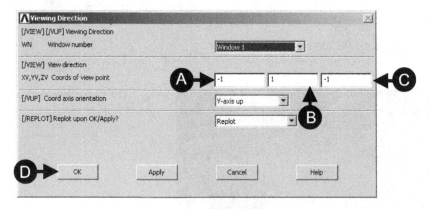

Fig. 7.130 Viewing direction.

The quarter symmetry model in the selected viewing direction is shown in Fig. 7.131.

From the ANSYS Main Menu, select **General Postproc → Read results → By Load Step**. The frame shown in Fig. 7.132 is produced. The selection [A] **Load step number** $= 1$, shown in Fig. 7.132, is implemented by clicking the [B] **OK** button.

From the ANSYS Main Menu, select **General Postproc → Plot Results → Contour Plot → Nodal Solu**. In the resulting frame, the following selections are made: [A] **Item to be contoured** = **Stress**; [B] **Item to be contoured** = **von Mises stress** (see Fig. 7.133). Pressing the [C] **OK** button implements the selections made.

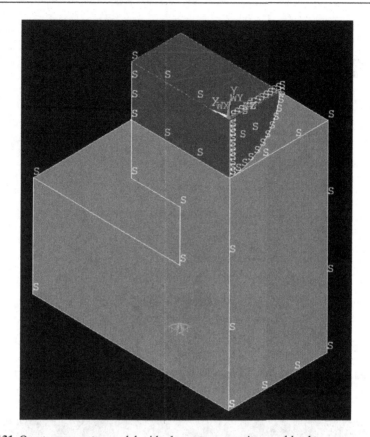

Fig. 7.131 Quarter symmetry model with elements, constraints, and loads.

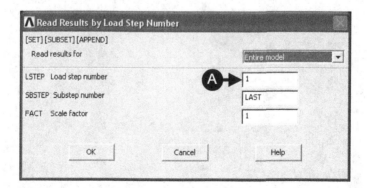

Fig. 7.132 Read results by load step number.

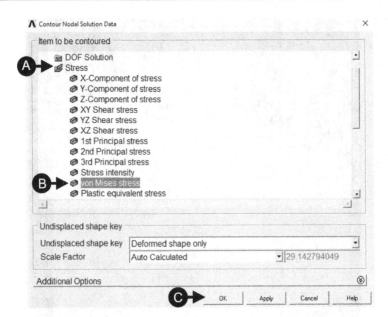

Fig. 7.133 Contour nodal solution data.

A contour plot of von Mises stress (nodal solution) is shown in Fig. 7.134.

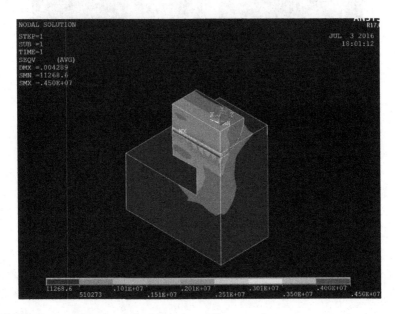

Fig. 7.134 Contour plot of nodal solution (von Mises stress).

Fig. 7.134 shows a von Mises stress contour for both the rail and cylinder. If one is interested in observing contact pressure on the cylinder surface alone, then a different presentation of solution results is required.

From the Utility Menu, choose **Select: Entities**. The frame shown in Fig. 7.135 appears.

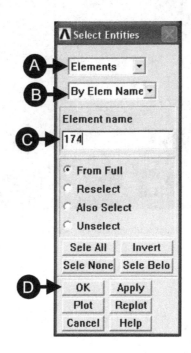

Fig. 7.135 Select entities.

In the frame shown in Fig. 7.135, the following selections are made: [A] **Elements** (first pull-down menu); [B] **By Elem Name** (second pull-down menu); [C] **Element Name** = 174. The element with the number 174 was introduced automatically during the process of creation of contact pairs described earlier. It is listed in the **Preprocessor → Element Type → Add/Edit/Delete** option. Pressing the [A] **OK** button in the frame shown in Fig. 7.135 implements the selections made.

From the Utility Menu, select **Plot → Elements**. An image of the cylinder with a mesh of elements is produced (see Fig. 7.136).

From the ANSYS Main Menu, select **General Postproc → Plot Results → Contour Plot → Nodal Solu**. The frame shown in Fig. 7.137 appears.

In the frame shown in Fig. 7.137, the following selections are made: [A] **Contact** and [B] **Contact pressure**. These are items to be contoured. Pressing the [C] **OK** button implements the selections made. In response to that, an image of the cylinder surface with pressure contours is produced, as shown in Fig. 7.138.

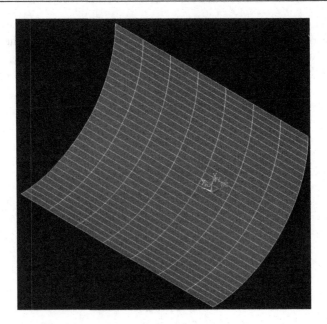

Fig. 7.136 Surface of the cylinder with contact elements.

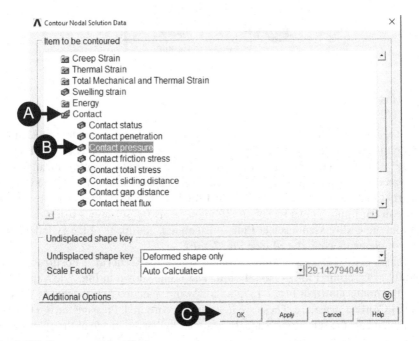

Fig. 7.137 Contour nodal solution.

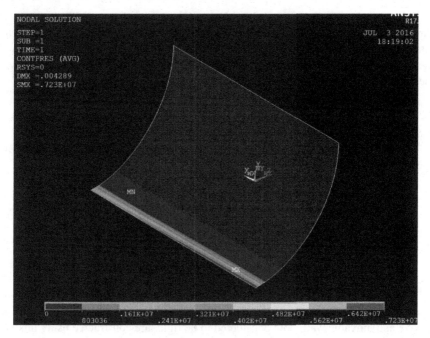

Fig. 7.138 Contact pressure on the cylinder surface.

7.2.4 Contact between elements made of elastic and viscoelastic materials

7.2.4.1 Problem description

The configuration of the contact between an O-ring made of rubber (hyperelastic material) and the groove is shown in Fig. 7.139.

An O-ring of solid circular cross-section is forced to conform to the shape of a rectangular groove by a moving wall as schematically shown in Fig. 7.139.

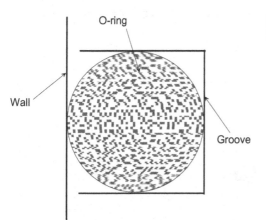

Wall

O-ring

Groove

Fig. 7.139 Configuration of the contact between an O-ring and the groove.

Following the initial squeeze of the O-ring (through movement of the wall), pressure is applied to the surface of the O-ring. Because of the sealing provided by the intrusion of the side walls, the pressure is effective over less than 180 degrees of the O-ring top surface. It is necessary to observe the conformity of the O-ring with the groove walls and stresses created by the pressure acting over its top surface.

The contact is characterised by the following data:

Young's modulus of the wall and groove material $= 2.1 \times 10^9$ Pa;
surface pressure applied to the O-ring $= 0.1 \times 10^3$ Pa;
material of the O-ring is modelled as hyperelastic material of Mooney-Rivlin (2 parameters) type with constants: C10 = 80, C01 = 20;
coefficient of friction between O-ring and wall = 0.1;
wall movement = 0.2 mm;
radius of the O-ring = 2.5 mm;
depth of the groove = 4.5 mm;
width of the groove = 5.5 mm; and
length of the wall = 10 mm.

7.2.4.2 Model construction

The O-ring is constructed using a hyperelastic element (Mooney-Rivlin), and the groove and movable wall, both considered to be rigid, are constructed using link elements. The contact elements are constructed using ANSYS facility—Contact Manager. The loads are applied by wall motion and groove cavity pressurisation. The pressure sealing on the O-ring is assumed to take place at 15 degrees off horizontal. The model is constructed using GUI facilities only.

From the ANSYS Main Menu, select **Preferences** and check the **Structural** option. Next, the elements to be used in the analysis are selected.

From the ANSYS Main Menu, select **Preprocessor** → **Element Type** → **Add/Edit/Delete**. The frame shown in Fig. 7.140 appears.

Fig. 7.140 Element types to be selected.

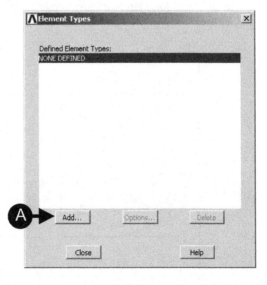

Pressing the [A] **Add** button calls up another frame as shown in Fig. 7.141.

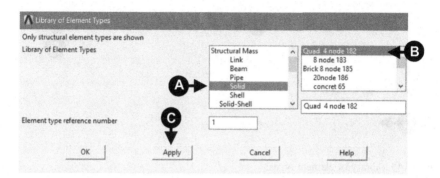

Fig. 7.141 Library of element types.

Select the [A] **Solid** and [B] **Quad 4 node 182** element and click the [C] **OK** button. This creates a frame shown in Fig. 7.142, where Type 1 **PLANE182** is shown.

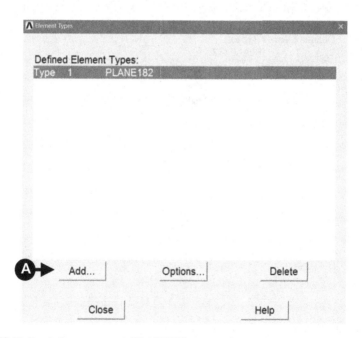

Fig. 7.142 Defined element types—PLANE182.

Click the [A] **Add** button as shown in Fig. 7.142, and select the element illustrated in Fig. 7.142.

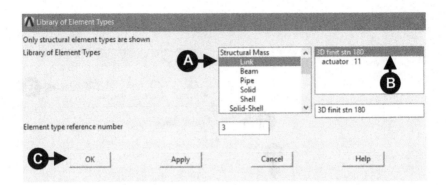

Fig. 7.143 Library of element types.

Selections of [A] **Link** and [B] **3D finit stn 180** were made as shown in Fig. 7.143. The selection of elements is shown in Fig. 7.144.

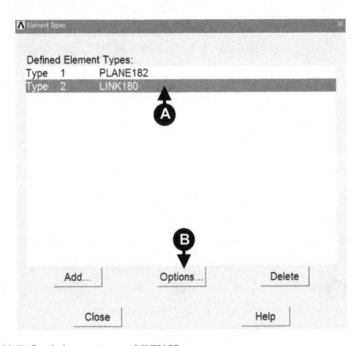

Fig. 7.144 Defined element types—LINK180.

In this figure, highlight [A] **LINK180** and click [B] **Options**. The frame shown in Fig. 7.145 is called up; [A] **Rigid (classic)** should be selected from the pull-down menu and the [B] **OK** button clicked to implement that choice.

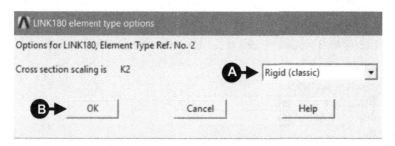

Fig. 7.145 LINK180 element options.

7.2.4.3 Selection of materials

The next step is to establish a database for materials used.

From the ANSYS Main Menu, select **Preprocessor** → **Material Props** → **Material Models** → **Structural** → **Nonlinear** → **Elastic** → **Hyperelastic** → **Mooney-Rivlin** (**2 parameters**).

As a result of the selection, a frame shown in Fig. 7.146 appears.

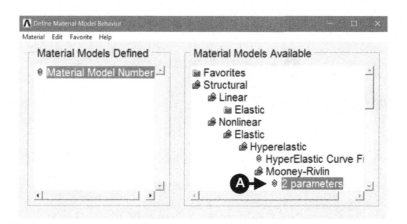

Fig. 7.146 Define material model behaviour.

Double-click on selection [A] to call up the frame shown in Fig. 7.147. Enter for C10 = 80, C01 = 20 and click [A] to implement the selections.

Now, having selected material for the O-ring, the next task is to define material for the wall and the groove. Thus, select **Material** from the menu at the top of the frame shown in Fig. 7.146 and click the **New Model** option. In response a frame shown in Fig. 7.148 appears.

By clicking the [A] **OK** button, a new material model number 2 is created, as shown in Fig. 7.149.

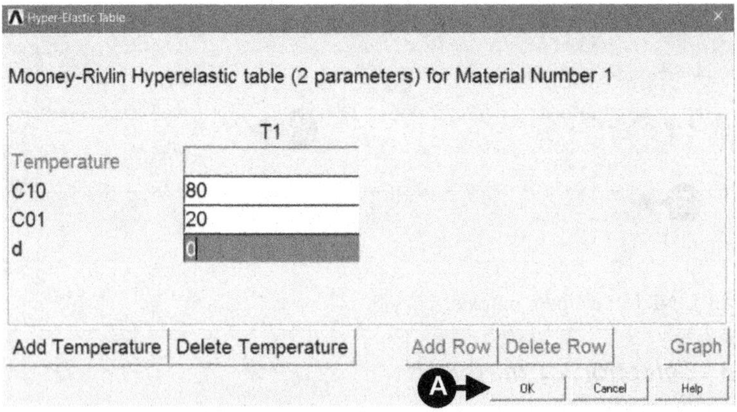

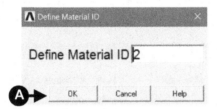

Fig. 7.147 Mooney-Rivlin hyperelastic table (two parameters) for material number 1.

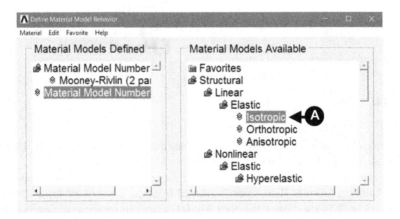

Fig. 7.148 Define material with ID = 2.

Fig. 7.149 Define material model behaviour.

Selections made are shown in Fig. 7.149. By double-clicking on [A] **Isotropic**, a frame shown in Fig. 7.150 appears.

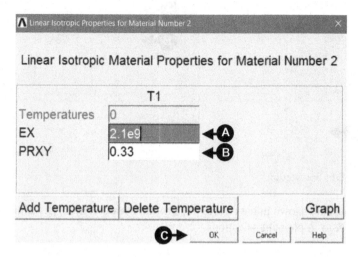

Fig. 7.150 Define material—linear isotropic.

The following selections are made: [A] $EX = 2.1 \times 10^9$ Pa, and [B] Poisson's coefficient **PRXY** = 0.33. By clicking the [C] **OK** button, the selections are implemented.

Having selected materials for the contact assembly, the frame should be closed by selecting **Material** and **Exit**.

The next step in the process of solving the problem is to characterise the link element which is used to mesh the wall and the groove.

From the ANSYS Main Menu, select **Preprocessor** → **Sections** → **Link** → **Add**. The frame shown in Fig. 7.151 appears.

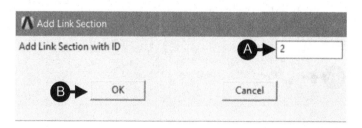

Fig. 7.151 Add link section.

Input [A] **ID** = 2 and [B] **OK** to implement the selection. Afterwards, the frame shown in Fig. 7.152 is called up.

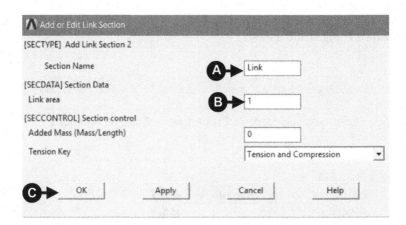

Fig. 7.152 Edit link section.

The inputs are shown in the figure: [A] **Section Name** = Link, [B] **Link area** = 1 and [C] **OK** to implement selections.

7.2.4.4 Geometry of the O-ring and meshing

The next phase in the model creation process is to draw the components involved.

From the ANSYS Main Menu, select **Preprocessor → Modelling → Create → Areas → Circle → By Dimensions**. As a result of the selection, a frame shown in Fig. 7.153 appears.

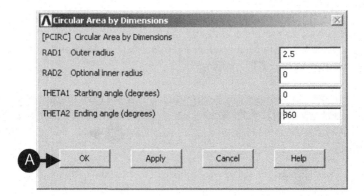

Fig. 7.153 Create circular area by dimensions.

By entering **RAD1** = 2.5, **RAD2** = 0, **THETA1** = 0, **THETA2** = 360, and clicking the [A] **OK** button, a solid circular area is created.

Next, the circular area, representing the O-ring, is meshed. From the ANSYS Main Menu, select **Preprocessor → Meshing → MeshTool**. Fig. 7.154 shows the resulting frame.

In the frame of Fig. 7.154, select [A] **Lines—Set** and [B] **Close**. A new frame is produced (see Fig. 7.155).

Pick all four arcuate segments on the circumference of the circular area and click the [A] **OK** button. The new frame shown in Fig. 7.156 appears.

In the frame of Fig. 7.156, enter number of element divisions, [A] **NDIV = 10** and uncheck box [B] **NDIV can be changed**. Clicking the [C] **OK** button implements the selections made.

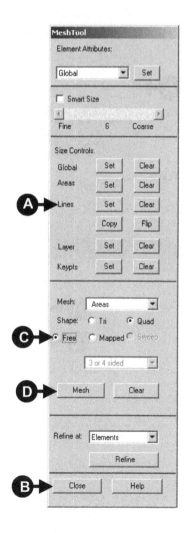

Fig. 7.154 Mesh tool.

Element Size on Picked Lines
Pick
Unpick
Single
Box
Polygon
Circle
Loop
Count = 0
Maximum = 4
Minimum = 1
Line No. =
List of Items
Min, Max, Inc
A
OK
Apply
Reset
Cancel
Pick All
Help

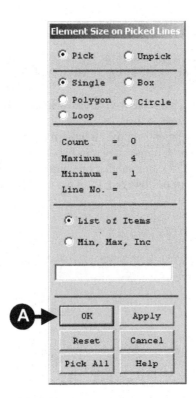

Fig. 7.155 Element size on picked lines.

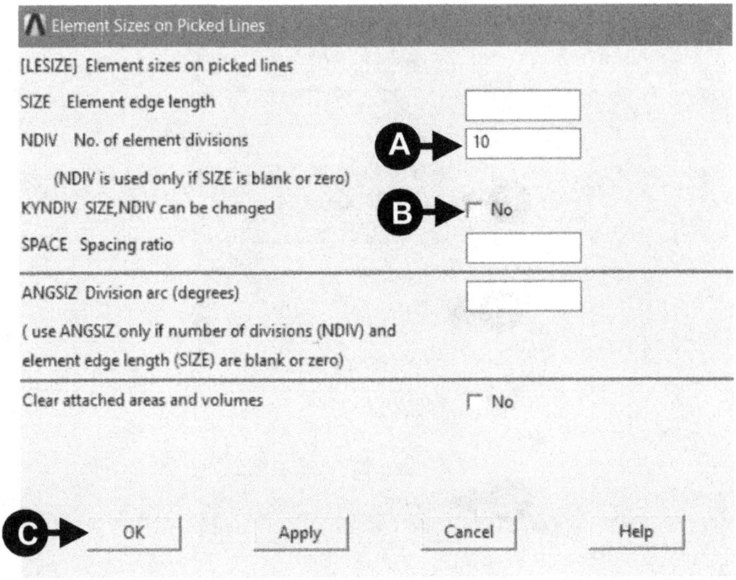

Element Sizes on Picked Lines
[LESIZE] Element sizes on picked lines
SIZE Element edge length
NDIV No. of element divisions
A
10
(NDIV is used only if SIZE is blank or zero)
KYNDIV SIZE,NDIV can be changed
B
No
SPACE Spacing ratio
ANGSIZ Division arc (degrees)
(use ANGSIZ only if number of divisions (NDIV) and
element edge length (SIZE) are blank or zero)
Clear attached areas and volumes
No
C
OK
Apply
Cancel
Help

Fig. 7.156 Set element size on selected lines.

In the frame shown in Fig. 7.154 (Mesh Tool), activate [C] **Free** button and click the [D] **Mesh** button. In the appearing frame (Fig. 7.157), click [A] **Pick All** to have the circular area meshed.

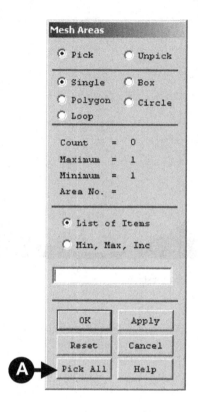

Fig. 7.157 Mesh areas.

Fig. 7.158 shows the circular area after the meshing process.

7.2.4.5 Creating wall and groove

Next, the wall and the groove are modelled as link components with an area equal to 1.

From the ANSYS Main Menu, select **Preprocessing** → **Modelling** → **Create** → **Keypoints** → **In Active CS**. The frame shown in Fig. 7.159 appears.

The input into the frame of Fig. 7.159 is as follows: **keypoint number** = 401 (it is an arbitrary selection, but it has to be greater than any number allocated to existing keypoints), $X = -2$; $Y = -2.5$. In a similar way, the other keypoints required for the groove and the wall as link components are created. The input coordinates are as follows:

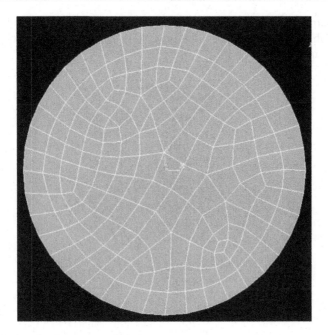

Fig. 7.158 O-ring circular area meshed.

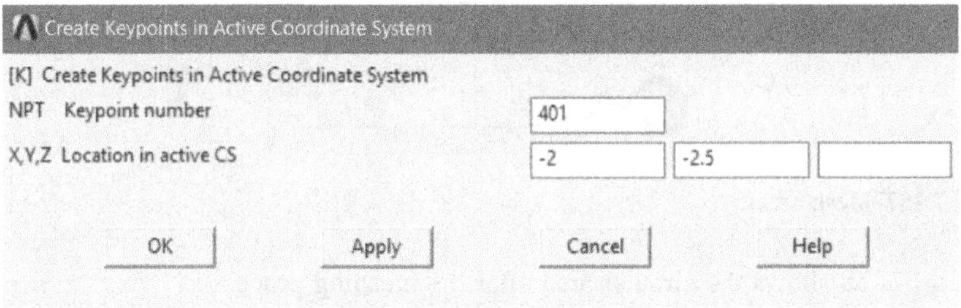

Fig. 7.159 Create keypoints in active coordinate system.

Keypoint number = 402, X = 2.5, Y = −2.5; keypoint number = 403, X = 2.5, Y = 3; keypoint number = 404, X = −2, Y = 3; keypoint number 405, X = −2.5, Y = 5; keypoint number = 406, X = −2.5, Y = −5.

The next lines will be created using defined keypoints. From the ANSYS Main Menu, select **Modelling: Create → Lines → Straight Line**. The frame shown in Fig. 7.160 appears.

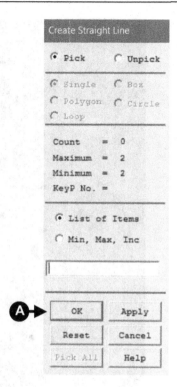

Fig. 7.160 Create straight line from keypoints.

Click in sequence on the keypoints defining the groove: 401, 402, 403, and 404 and then [A] **OK** as shown in Fig. 7.160. Afterwards, click in sequence on the keypoints defining the wall: 405 and 406 and then [A] **OK** as shown in Fig. 7.160. The resulting outline of the wall and the groove looks like that shown in Fig. 7.161.

The next step in the process of modelling is to mesh created lines using the LINK180 element selected earlier. From the ANSYS Main Menu, select **Meshing → Size Cntrls → ManualSize → Lines → Picked Lines**. In response, a frame shown in Fig. 7.162 is called up.

Pick all four lines defining the groove and the wall and click [A] **OK**. In response, a frame shown in Fig. 7.163 is created.

In the frame of Fig. 7.163, enter number of element divisions, [A] **NDIV = 10** and uncheck box [B] **NDIV can be changed**. Clicking the [C] **OK** button implements the selections made.

Fig. 7.161 View of the groove
and the wall.

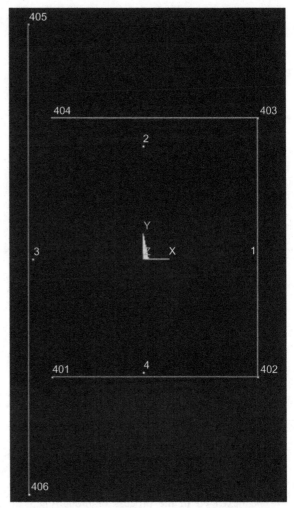

From the ANSYS Main Menu, select **Modelling** → **Create** → **Elements** → **Elem Attributes**. In response, the frame shown in Fig. 7.164 is created. Make entries as shown in Fig. 7.164, as they define element type and material to be used when meshing the groove and the wall.

From the ANSYS Main Menu, select **Meshing** → **Mesh** → **Lines**. This selection creates the frame shown in Fig. 7.165. Pick all four lines belonging to the groove-wall assembly and click [A] **OK**.

The O-ring assembly when fully meshed looks like that shown in Fig. 7.166.

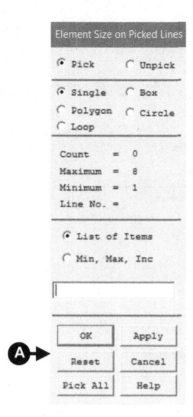

Fig. 7.162 Element size on picked lines.

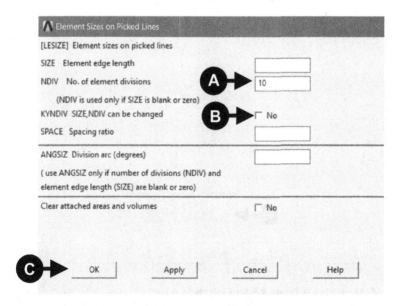

Fig. 7.163 Set element size on lines defining the wall and the groove.

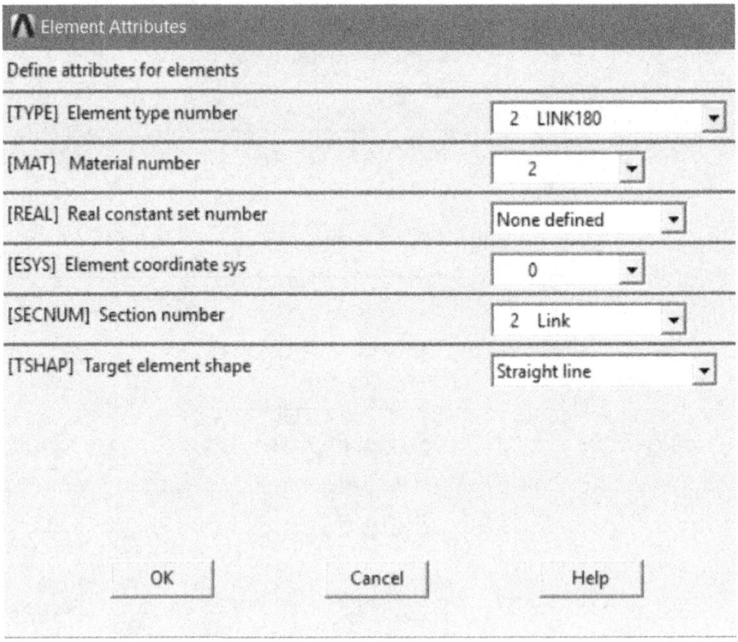

Fig. 7.164 Create element attributes.

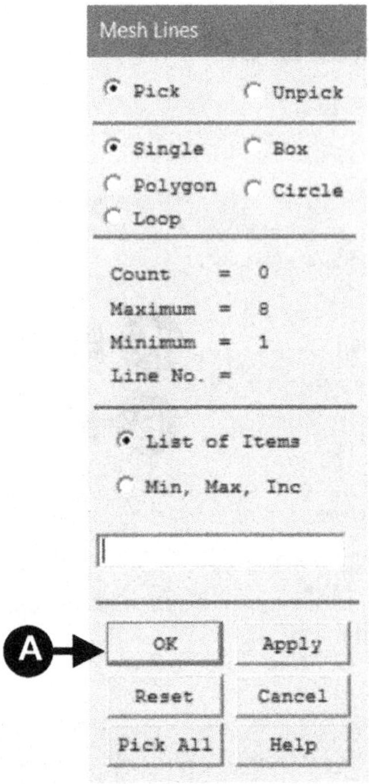

Fig. 7.165 Mesh lines defining the wall and the groove.

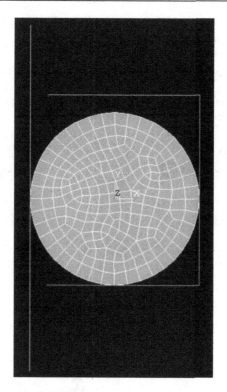

Fig. 7.166 View of the meshed O-ring assembly.

7.2.4.6   Creation of a contact pair

In solving contact problems, a contact pair has to be created. Contact Wizard is a useful facility provided by ANSYS.

From the ANSYS Main Menu, select **Preprocessor → Modelling → Create → Contact Pair**. As a result, a frame shown in Fig. 7.167 appears.

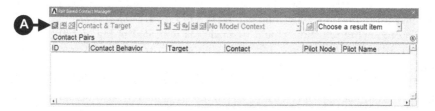

Fig. 7.167 Contact Manager frame.

Clicking on the [A] **Contact Wizard** button located in the upper left-hand corner of the frame calls up a new frame as shown in Fig. 7.168.

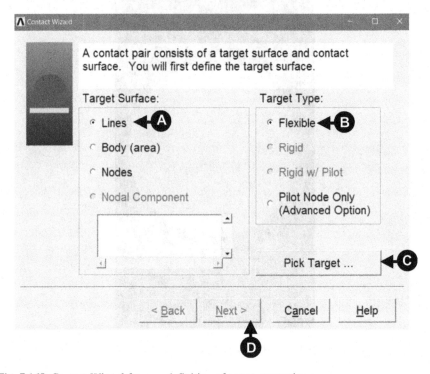

Fig. 7.168 Contact Wizard frame—definition of target properties.

In the frame shown in Fig. 7.168, select [A] **Lines**, [B] **Flexible**, and press the [C] **Pick Target** button. As a result of this selection, the frame shown in Fig. 7.169 is produced.

Target lines are: vertical line of the wall, lower horizontal line of the groove, and vertical line of the groove. After highlighting those lines, the [A] **OK** button should be pressed, as instructed by Fig. 7.169. Next, in the Contact Wizard frame (see Fig. 7.168), press the [D] **Next** button (this should be highlighted when the target lines are picked) to obtain another frame as shown in Fig. 7.170.

The following selection should be made (as indicated in Fig. 7.170): [A] **Body (area)**, [B] **Node-to-Surface**, [C] **Pick Contact**. In response, the frame shown in Fig. 7.171 is produced, prompting you to select a contact.

The surface of the O-ring (circle) should be selected and [A] **OK** pressed. The Contact Wizard frame appears (see Fig. 7.170). Pressing the [D] **Next** button (this should be highlighted) results in a frame in which the following entries should be made. **Material ID = 1, Coefficient of friction = 0.1**. Also, **Include Initial penetration** option ought to be unchecked. Next, the **Optional settings** button should be pressed to refine contact parameters further. In the new frame, the following entries are

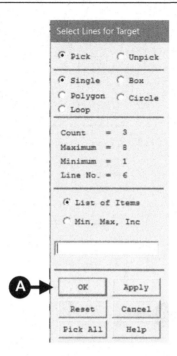

Fig. 7.169 Select lines for target.

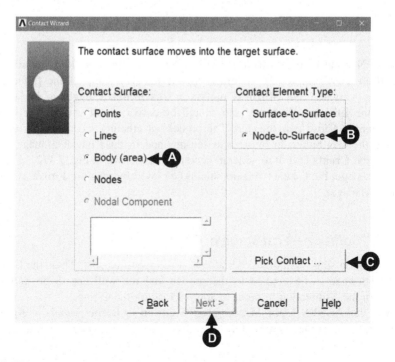

Fig. 7.170 Contact Wizard—definition of contact properties.

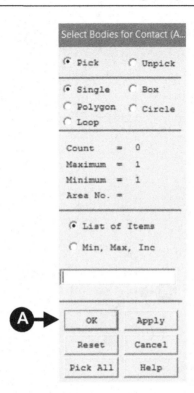

Fig. 7.171 Select circular area of O-ring as contact.

required: **Normal penalty stiffness = 0.1** and **Behaviour of contact surface = No separation**. Furthermore, the **Friction** tab located in the top of the frame menu should be activated and **Stiffness matrix = Unsymmetric** ought to be selected. Finally, the **Initial Adjustment** tab should be activated and **Automatic contact adjustment = Close gap** selected. The process of creating a contact pair ends by pressing the **OK** button to implement settings and in the Contact Manager frame pressing the **Create** button to generate a contact, as shown in Fig. 7.172.

At this stage, the Contact Wizard should be closed by pressing **Finish** as well as Contact Manager.

7.2.4.7 Solution—First stage of loading

In the solution stage, solution criteria have to be specified first. Thus, the first step constraints have to be applied to the groove and the wall has to be moved by 0.2 mm in the X-direction in order to apply a load on the O-ring.

From the ANSYS Main Menu, select **Solution → Define Loads → Apply → Structural → Displacement → On lines.** This selection generates the frame shown in Fig. 7.173.

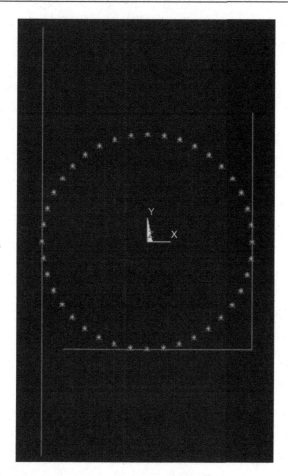

Fig. 7.172 View of the contact pair for the O-ring assembly.

All lines defining the groove should be highlighted and the [A] **OK** button pressed. In response, a frame shown in Fig. 7.174 appears where [A] **All DOF**, [B] **Value = 0**, and [C] **OK** should be selected.

Next, the wall should be moved by 0.2 mm. Recall the frame shown in Fig. 7.173 and select the vertical line defining the wall. Clicking [A] **OK** produces a frame as shown in Fig. 7.175.

Selections made are shown in Fig. 7.175: [A] **UX**, [B] **Value = 0.2**, [C] **OK**.

Before a solution is attempted, solution controls have to be selected. From the ANSYS Main Menu, select **Solution → Analysis Type → Sol'n Controls**. This selection produces the frame shown in Fig. 7.176. Selections made are shown in this figure: [A] **Large Displacement Static**, [B] **Time at end of loadstep = 1**, [C] **Write every substep**. Also ensure that **Number of substeps = 25, Max no. of substeps = 2000, Min no. of substeps = 5**.

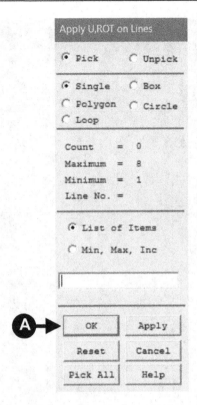

Fig. 7.173 Apply constraints on lines.

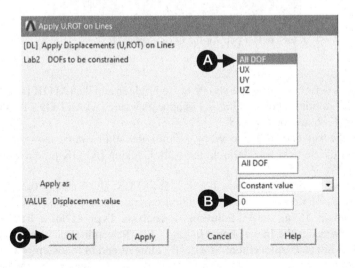

Fig. 7.174 Apply constraints (all DOF) on lines defining groove.

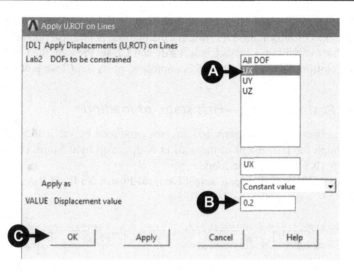

Fig. 7.175 Apply constraints (UX) on line defining wall.

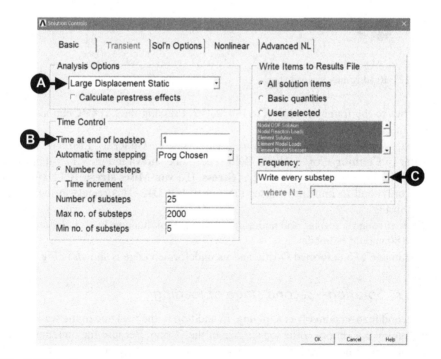

Fig. 7.176 Define solution controls.

From the ANSYS Main Menu, select **Solution** → **Solve** → **Current LS**. A frame showing review of information relevant for the planned solution action appears. After checking that everything is correct, select **File** and **Close**. Pressing the **OK** button initiates the solution. When the solution is complete, press the **Close** button.

7.2.4.8 Postprocessing—First stage of loading

In order to observe deformations and stresses produced by the load applied to the O-ring through the movement of the wall in X-direction by 0.2 mm, postprocessing facilities of ANSYS should be used.

From the ANSYS Main Menu, select **General Postproc: Read Results: By Load Step**. Fig. 7.177 shows the resulting frame.

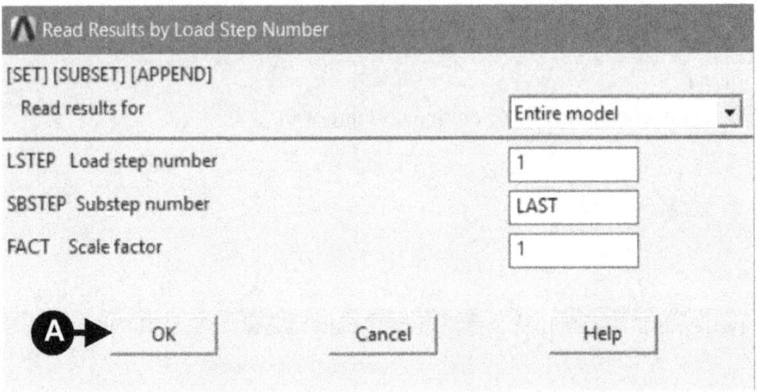

Fig. 7.177 Read results by load step.

Entries to the frame are shown in Fig. 7.177. Pressing the [A] **OK** button implements the selections made.

Next, from the ANSYS Main Menu, select **General Postproc** → **Plot Results** → **Contour Plot** → **Nodal Solu**. The frame shown in Fig. 7.178 appears.

Selections made are as follows: [A] **Stress**, [B] **von Mises stress**, [C] **Deformed shape only**, and to implement selections, click [D] **OK**. This results in the image shown in Fig. 7.179.

In order to see deformed and undeformed shapes simultaneously, a choice shown in Fig. 7.180 should be made.

The image of a deformed O-ring and its undeformed edge is shown in Fig. 7.181.

7.2.4.9 Solution—Second stage of loading

The second load step involves applying, in addition to the load due to the wall movement, pressure acting over the top surface of the O-ring. Because the pressure effectively acts over the angle from 14 to 166 degrees, in order to apply it properly it is convenient to change the coordinate system from Cartesian to Polar.

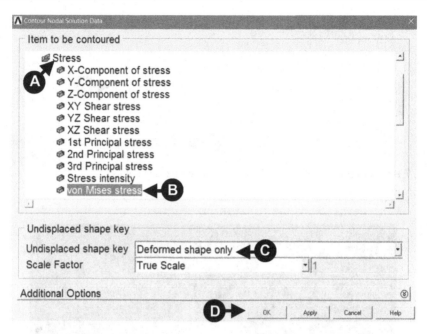

Fig. 7.178 Plot of von Mises stress as nodal plot.

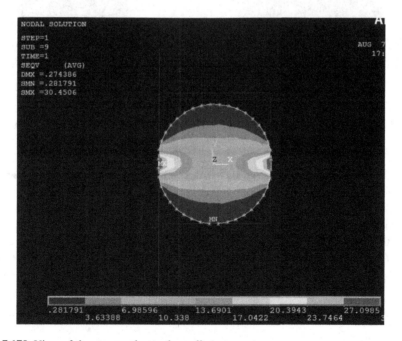

Fig. 7.179 View of the stresses due to the wall movement.

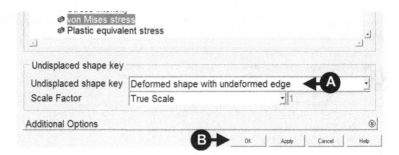

Fig. 7.180 Selection of deformed shape and undeformed edge plots.

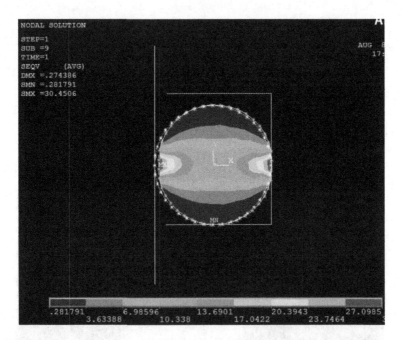

Fig. 7.181 Undeformed O-ring assembly together with deformed shape and resulting stresses.

From the ANSYS Utility Menu, select **WorkPlane → WP Settings**. The frame shown in Fig. 7.182 appears.

In the frame, select [A] **Polar** and [B] **OK** to implement the choices. Next, from the ANSYS Utility Menu, select **WorkPlane → Change Active CS to → Working Plane**. This selection ensures that the active coordinate system is identical with that of the WP coordinate system, which is the Polar system.

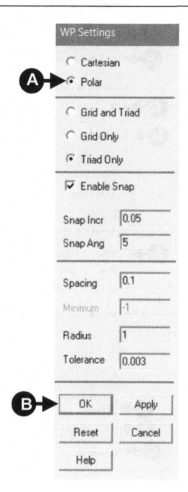

Fig. 7.182 Working plane settings.

From the ANSYS Utility Menu, select **Select** → **Entities**. This selection produces a frame shown in Fig. 7.183.

First, the elements belonging only to the O-ring should be selected. This is done by making the following selections shown in Fig. 7.183: [A] **Elements**, [B] **By Elem Name**, [C] **182** (element Plane 182 was used to mesh O-ring), [D] **From Full**, and [E] **OK**. Next, the nodes attached to the selected elements have to be chosen. Thus from the ANSYS Utility Menu, choose **Select: Entities**. The frame shown in Fig. 7.184 appears.

Fig. 7.184 shows that the following selections were made: [A] **Nodes**, [B] **Attached to**, [C] **Elements**, [D] **From Full**, and [E] **OK** to implement choices.

Again from the ANSYS Utility Menu, choose **Select** → **Entities**; this produces the frame shown in Fig. 7.185.

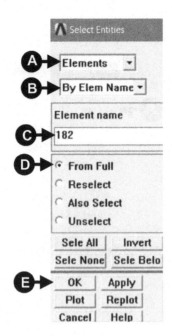

Fig. 7.183 Select entities—elements by element name.

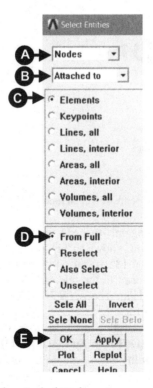

Fig. 7.184 Select entities—nodes attached to element.

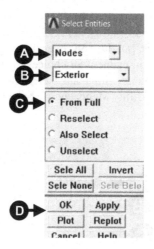

Fig. 7.185 Select entities—exterior nodes of the O-ring.

As shown in Fig. 7.185, the following entries were made: [A] **Nodes**, [B] **Exterior**, [C] **From Full**, and [D] **OK** to implement selections.

Finally, from the ANSYS Utility Menu, choose **Select** → **Entities**. Fig. 7.186 shows the resulting frame.

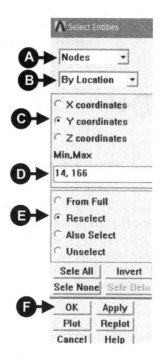

Fig. 7.186 Select entities—nodes by location.

In order to select nodes belonging to the O-ring on which pressure is applied (second stage loading), the selections shown in Fig. 7.186 were made: [A] **Nodes**, [B] **By Location**, [C] **Y coordinates**, [D] **14, 166** (over this angle pressure acts on O-ring), [E] **Reselect**, and [F] **OK** to implement all choices made.

Next, all associated elements with nodes selected above should be picked up. Therefore, from the ANSYS Utility Menu, choose **Select: Entities**. The frame shown in Fig. 7.187 appears.

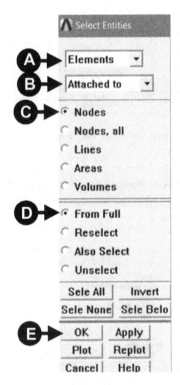

Fig. 7.187 Select entities—elements attached to previously selected nodes.

Selections made are as follows (see Fig. 7.187): [A] **Elements**, [B] **Attached to**, [C] **Nodes**, [D] **From Full**, and [E] **OK** to implement the selections.

Now is the right moment to deselect all contact elements, Type 4 Conta175 (this information can be found in the element type frame), from being involved in the load transmission. Thus, from the ANSYS Utility Menu, choose **Select → Entities**. The frame shown in Fig. 7.188 is produced.

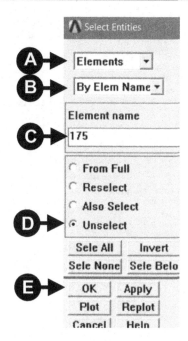

Fig. 7.188 Select entities—element by element name.

All selections made are shown in Fig. 7.188: [A] **Elements**, [B] **By Elem Name**, [C] **175** (this is name of contact element), [D] **Unselect**, and [E] **OK** to implement choices.

In order to apply pressure to selected external nodes belonging to the O-ring, the following steps should be taken.

From the ANSYS Main Menu, select **Preprocessor** → **Solution** → **Define Loads** → **Apply** → **Structural** → **Pressure** → **On Nodes**. The frame shown in Fig. 7.189 appears.

Pressing the [A] **Pick All** button creates another frame, shown in Fig. 7.190.

The load value [A] **VALUE = 0.1** × **10³** Pa is entered in the frame shown in Fig. 7.190. Pressing [B] **OK** implements the entry. Finally, from the ANSYS Utility Menu, choose **Select: Everything**.

This action recalls all elements and nodes belonging to the O-ring assembly. Usually, before the solution is attempted, solution options should be selected. However, in this case they are the same for those selected for the first stage of loading.

Now the solution of the problem ought to be attempted by selecting from the ANSYS Main Menu: **Solution** → **Solve** → **Current LS**. A frame showing a review of information relevant for the planned solution action appears. After checking that everything is correct, select **File** and **Close**. Pressing the **OK** button initiates the solution. When the solution is completed, press the **Close** button.

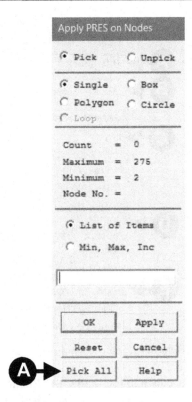

Fig. 7.189 Apply pressure on nodes.

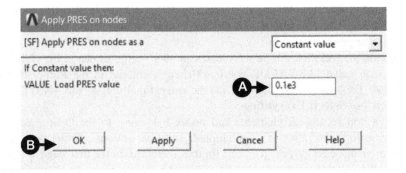

Fig. 7.190 Apply pressure—define its value.

7.2.4.10   Postprocessing—Second stage of loading

In order to see stresses and deformations of the O-ring, follow the steps outlined in Section 7.2.4.8. Fig. 7.191 shows the von Mises stress due to squeeze and external pressure acting on the O-ring (deformed shape only), while Fig. 7.192 shows the von Mises stress and external pressure acting on the O-ring, but this time the undeformed edge of the assembly is visible.

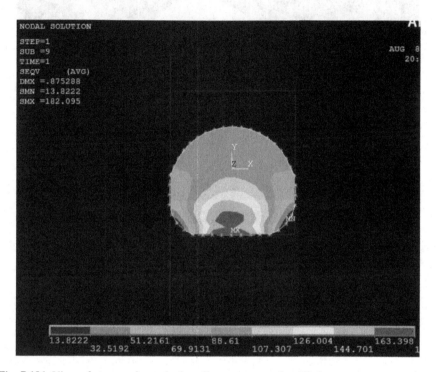

Fig. 7.191 View of stresses due to both wall movement and applied pressure.

7.2.5   Contact between two beams

7.2.5.1   Problem description

Two beams, as shown in Fig. 7.193, are having the following dimensions: length = 100 mm, cross-section = 10 mm × 10 mm. They are made of a material with a Young's modulus $E = 200,000$ Pa, Poisson's ratio = 0.3, and are rigidly constrained at the ends (see Fig. 7.193). A load of 10 kN is applied halfway through the length of the upper beam, resulting in that beam bending and subsequently making contact with the lower beam. There are two objectives of the analysis. The first is to learn how to utilise contact elements in order to simulate how the two beams react when they come into contact with each other. The second objective is to find out stresses and displacements associated with the contact and at the location of the contact.

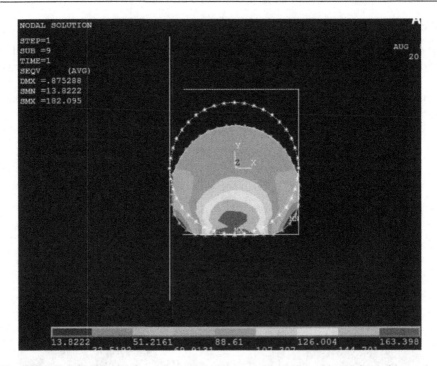

Fig. 7.192 Combined plot of undeformed O-ring assembly and resulting deformations and stresses due to applied load.

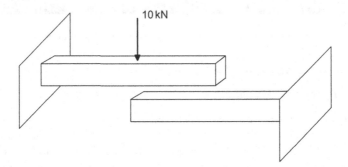

Fig. 7.193 Schematic of the problem.

7.2.5.2 Construction of the model

The model of the two beams to make a contact under a load is constructed using a GUI, which is the main approach. However, this is not the only approach used, and for the purposes of illustration, the command approach will be given as well. The model in this instance consists of two identical rectangles, which can be created as follows.

From the ANSYS Main Menu, select **Preprocessor** → **Modelling** → **Create** → **Areas** → **Rectangle** → **By 2 Corners**. In response, a frame shown in Fig. 7.194 appears.

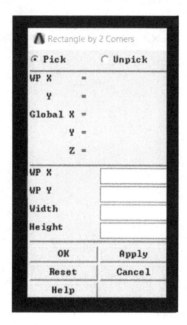

Fig. 7.194 Draw rectangle by two corners.

Input **WPX** = 0; **WPY** = 15; **Width** = 100; **Height** = 10, and press the **OK** button to create the first beam.

In order to create the second beam, the above procedure is repeated but the input data into the frame shown in Fig. 7.194 are: **WPX** = 50; **WPY** = 0; **Width** = 100; **Height** = 10. Again, the **OK** button should be pressed. The following geometry is created as a result (see Fig. 7.195).

Fig. 7.195 View of the rectangles representing beams.

7.2.5.3 Material properties and element type selection

The next step is to define the type of the analysis and the type of elements to be used.

From the ANSYS Main Menu, select **Preferences**, and in the appearing frame, tick **Structural** and then press the **OK** button.

Then, from the ANSYS Main Menu, select **Preprocessor** → **Element Type** → **Add/Edit/Delete**. The frame shown in Fig. 7.196 appears.

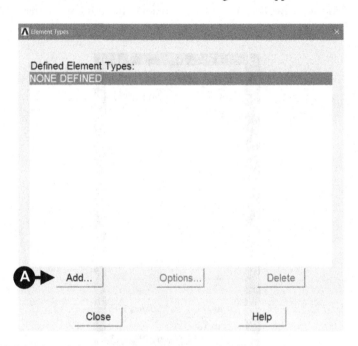

Fig. 7.196 Selection of element type.

Click [A] **Add** in order to access another frame as shown in Fig. 7.197.

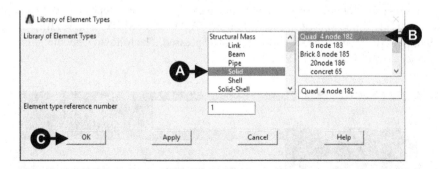

Fig. 7.197 Library of available element types.

In the left column, select [A] **Solid** and in the right column choose the [B] **Quad 4node 182** element. After that, click the [C] **OK** button.

While the Element Type window (see Fig. 7.196) is still open, click **Options**, and in the new appearing frame change [A] **Element Behaviour K3** to **Plane Strs w/thk** and click [B] **OK**, as shown in Fig. 7.198.

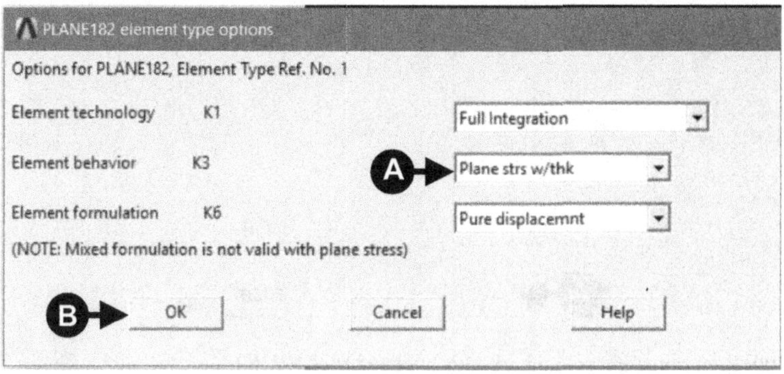

Fig. 7.198 Options for element PLANE182.

This allows a thickness of the beams to be input for the elements.

For the thickness of the beam to be used in the analysis, real constants must be defined. From the ANSYS Main Menu, select **Preprocessor** → **Real Constants** → **Add/Edit/Delete**. The frame shown in Fig. 7.199 appears. In this frame, press the [A] **Add** button to pull down another frame as shown in Fig. 7.200.

Fig. 7.199 Real constant definition.

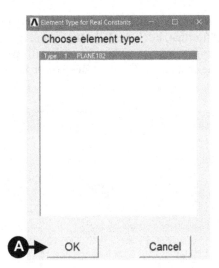

Fig. 7.200 Real constant associated with element PLANE182.

Click the [A] **OK** button to generate a third frame shown in Fig. 7.201. In this frame, input [A] **thickness THK** = 10 mm as shown. After that, click the [B] **OK** button and close the other frames.

Fig. 7.201 Input into real constant set linked to PLANE182.

Now, the materials for the two beams ought to be defined. From the ANSYS Main Menu, select **Preprocessor → Materials Props → Material Models**. Next, click **Structural → Linear → Elastic → Isotropic** and in the generated frame, input [A] **EX** = 200,000 (for Young's modulus) and [B] **PRXY** = 0.3 (for Poisson's ratio). After that, click the [C] **OK** button. Fig. 7.202 shows the inputs.

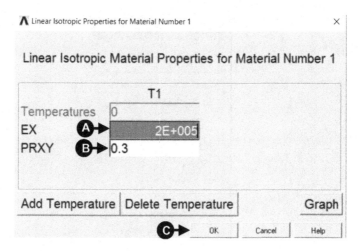

Fig. 7.202 Selection of material for the beams.

It is important to close the window showing the selection of a material.

7.2.5.4 Meshing

From the ANSYS Main Menu, select **Preprocessor → Meshing → Size Control → Manual Size → Areas → All Areas**. The frame shown in Fig. 7.203 is generated. For this example, element edge length is [A] 2 mm, as shown in Fig. 7.203.

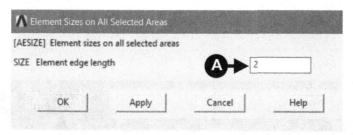

Fig. 7.203 Element size in terms of its edge length.

The next step is to mesh both beams. Therefore, from the ANSYS Main Menu, select **Preprocessor → Meshing → Mesh → Areas → Free → Pick All**. This is illustrated in Fig. 7.204.

Select [A] **Pick All** and press the [B] **OK** button to mesh the selected areas. The mesh resulting from this action in shown in Fig. 7.205.

7.2.5.5 Creation of a contact pair

It is very convenient, when solving contact problems, to use the **Contact Wizard** tool, available in ANSYS.

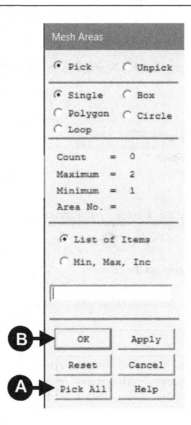

Fig. 7.204 Meshing areas.

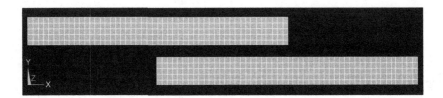

Fig. 7.205 View of meshed areas.

From the ANSYS Main Menu, select **Preprocessor** → **Modelling** → **Create** → **Contact Pair**. This action results in a frame shown in Fig. 7.206.

The location of the Contact Wizard button is in the upper left-hand corner of the frame. By clicking on button [A], a new frame, shown in Fig. 7.207, is generated.

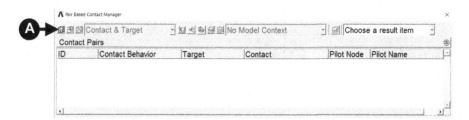

Fig. 7.206 Contact Manager.

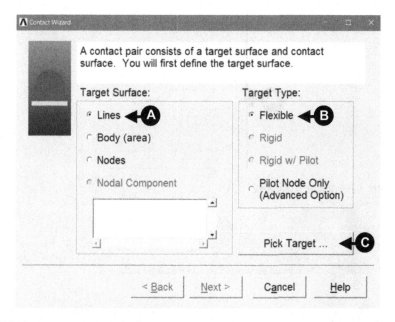

Fig. 7.207 Contact Wizard—pick target.

In this frame, select [A] **Lines** (upper line of the lower beam) and [B] **Flexible**, and press the [C] **Pick Target** button. As a result of this selection, the frame shown in Fig. 7.208 is produced.

The target area is the surface of the lower beam and should be picked and the **OK** button pressed. In the Contact Wizard (see Fig. 7.207), press the **Next** button (which should be highlighted once the target area is selected) to obtain the frame shown in Fig. 7.209. In this frame, the following selections should be made: [A] **contact surface—Nodes** and contact element type [B]—**Node-to-Surface**, as shown in Fig. 7.209.

Pressing the [C] **Pick Contact** button results in the frame shown in Fig. 7.210.

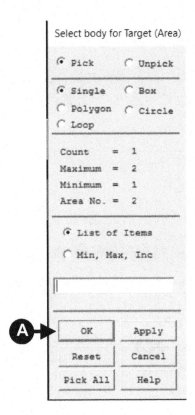

Fig. 7.208 Selection of target area.

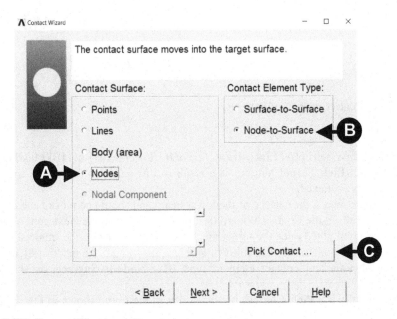

Fig. 7.209 Contact Wizard—pick contact.

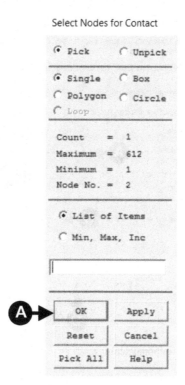

Fig. 7.210 Selection of contact nodes.

Responding to the prompt of this frame, select the node located at the lower right-hand corner of the upper beam and press the [A] **OK** button. When this is done, the Contact Wizard frame appears (see Fig. 7.209), and the **Next** button should be pressed in order to generate the frame in Fig. 7.211.

The entries into this frame are: [A] **material ID** = 1, [B] **coefficient of friction** = 0 (frictionless contact is assumed). The **Include initial penetration** option should be checked. Pressing the [C] **Optional settings** button enables the making of further adjustments to the model. Especially important is changing the **Stiffness matrix** to **Un-symmetric**, and that option can be accessed by pressing first the **Optional settings** button (see Fig. 7.211) and in the new frame the **Friction** tab located in the top of that frame menu. When all the above is done, pressing the [D] **Create** button (see frame in Fig. 7.211) produces an image of the contact pair, as shown in Fig. 7.212.

When a contact pair image is created, it signifies the end of the contact pair creation process and the **Contact Manager** frame should be closed.

However, the contact elements created by the Contact Manager have to have real constants defined. This is done by selecting **Preprocessor → Real Constants → Add/Edit/Delete**. As a result, the frame shown in Fig. 7.213 appears.

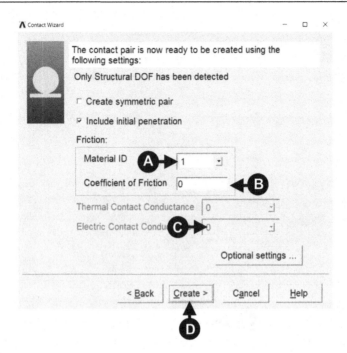

Fig. 7.211 Contact Wizard—optional settings.

Fig. 7.212 View of the contact pair.

Select [A] **Set 3** (containing contact elements) and press the [B] **Add** button. In response, the frame shown in Fig. 7.214 is generated.

Select [A] **Type 3 CONTACT175** and click [B] **OK**. In response, a frame shown in Fig. 7.215 is produced.

There are two important entries which have to be made. The first is to set normal penalty stiffness, [A] = **FKN** = 200,000 and target edge extension factor, and [B] = **TOLS** = 10 (this can be found by scrolling down the table shown in Fig. 7.215).

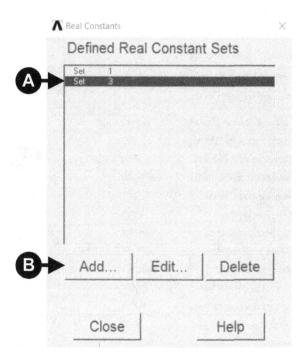

Fig. 7.213 Definition of real constants for contact pair.

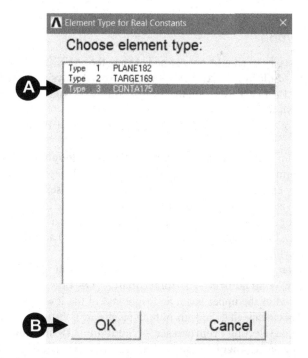

Fig. 7.214 Selection of element type used in real constant definition.

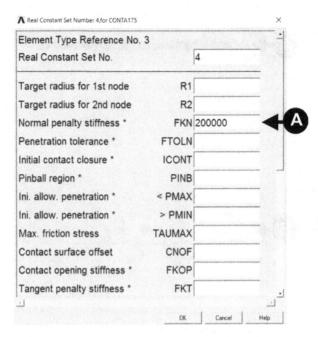

Fig. 7.215 Input into real constant table.

7.2.5.6 Solution

First, an analysis type should be defined. From the ANSYS Main Menu, select **Preprocessor → Solution → Analysis Type → New Analysis → Static**.

Next, select **Preprocessor → Solution → Analysis Type → Solution Control**. The image shown in Fig. 7.216 will appear.

Ensure the following selections are made under the [A] **Basic** tab as shown in Fig. 7.216.

Especially important is to set [B] **Automatic time stepping** = On. This allows ANSYS to determine appropriate sizes to break the load steps into. Decreasing the step size usually ensures better accuracy, although it increases computing time.

Additionally, activate the [C] **Nonlinear** tab located at the top of the frame shown in Fig. 7.216. A new frame of **Solution Controls** will appear (see Fig. 7.217).

Ensure that [A] **Maximum number of iterations** = 100 and [B] **Line search** is On, as shown in Fig. 7.217.

Next, from the ANSYS Main Menu, select **Preprocessor → Solution → Define Loads → Apply → Structural → Displacement → On Lines**.

Fix the left end of the upper beam and right end of the lower beam, choosing all **DOF** option. This means that the ends of both beams are constrained in all directions.

Furthermore, on selecting **Preprocessor → Solution → Define Loads → Apply → Structural → Force/Moment → On Nodes**, the frame shown in Fig. 7.218 appears.

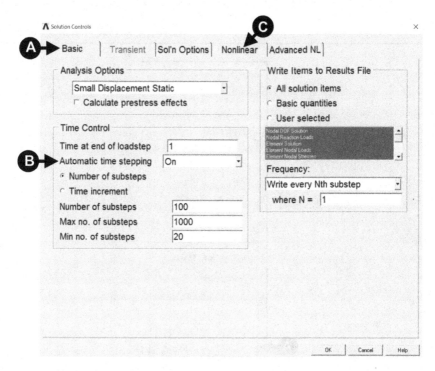

Fig. 7.216 Solution control adjustments—basic.

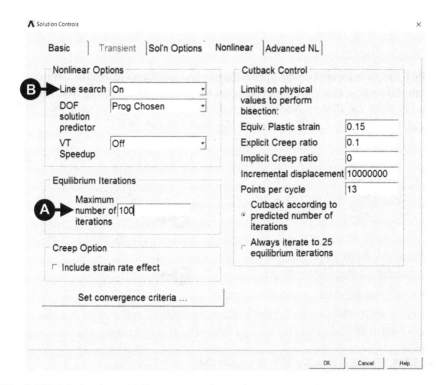

Fig. 7.217 Solution control adjustments—advanced.

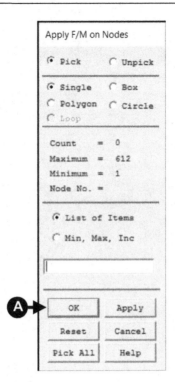

Fig. 7.218 Constraints applied to nodes.

Pick the node located on the upper line of the upper beam and halfway through its length. To do this, plot the images of the beam as nodes by selecting from the utility menu of ANSYS: **Plot** → **Nodes**. When the selection is done, pressing the [A] **OK** button in frame of Fig. 7.218 generates the frame shown in Fig. 7.219.

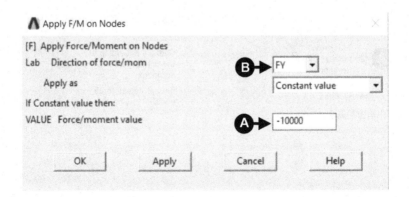

Fig. 7.219 Input into the table of node's constraints.

Apply [A] **VALUE Force/moment** = $-10,000$ N in [B] the **FY** direction to the centre of the top surface of the upper beam. Note that this is a point load on a 2D surface. The minus sign signifies the fact that the load is acting downwards.

Finally, select **Preprocessor** $\rightarrow$ **Solution** $\rightarrow$ **Solve** $\rightarrow$ **Current LS**. On pressing the **Solve** button, it takes some time to obtain the solution. When it is successful, the image shown in Fig. 7.220 appears, illustrating steps in consecutive iterations and convergence.

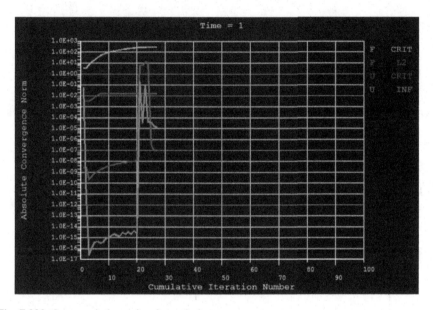

Fig. 7.220 Output windows showing solution process.

7.2.5.7 Postprocessing

In order to view the results and to present them in different formats, the ANSYS postprocessing stage is used.

From the ANSYS Main Menu, select **General PostProc**. Next, from the utility menu, select **PlotCtrls** $\rightarrow$ **Style** $\rightarrow$ **Displacement Scaling**. The frame shown in Fig. 7.221 will appear.

Click the [A]—1.0 (true scale) button (as shown in Fig. 7.221). This selection is very important.

In order to see the stress distribution in the beams, the following selections should be made:

From the Utility Menu, select **PlotCtrls** $\rightarrow$ **Style** $\rightarrow$ **Contours** $\rightarrow$ **Non-Uniform Contours**. This generates the frame shown in Fig. 7.222. Ensure that entries are as shown in that figure.

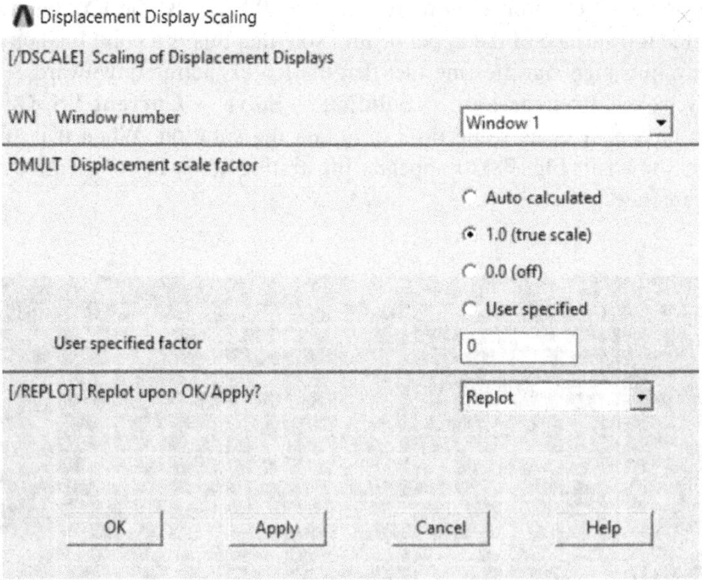

Fig. 7.221 Adjustment of the displacement display scale.

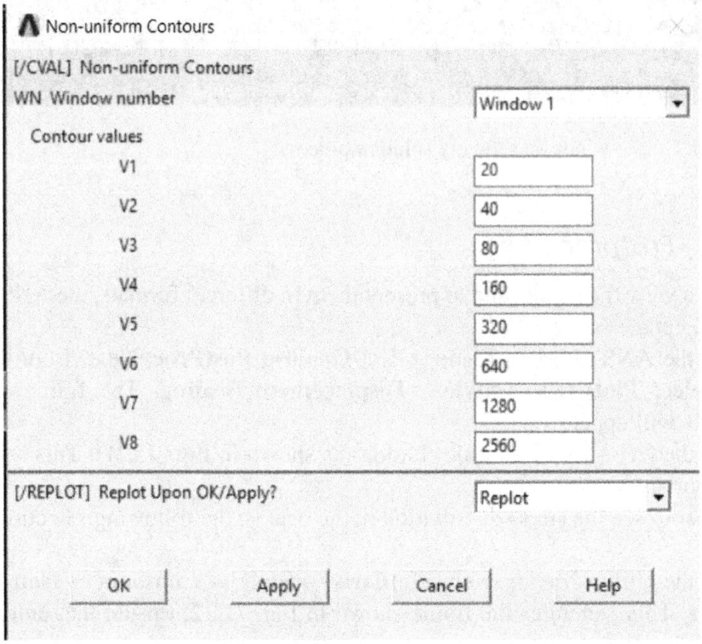

Fig. 7.222 Selection of nonuniform contours.

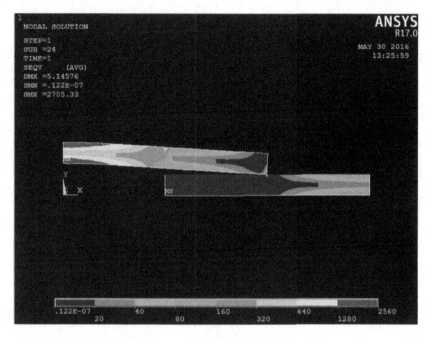

Fig. 7.223 View of the state of stresses in beams.

Next, select **General Postproc → Plot Results → Contour Plot → Nodal Solution → Stress → von Mises**. As a result, a frame showing stress distribution within both beams will be shown (see Fig. 7.223).

As can be seen in Fig. 7.223, the load on the upper beam caused it to deflect and come into contact with the lower beam, producing a state of stress in both of them.

Reference

[1] K.L. Johnson, Contact Mechanics, Cambridge University Press, Cambridge, 1985.

Appendix: Using ANSYS workbench for structural analysis

A.1 Introduction

The strength design of machines and structures has long been made based on the design by formula or rule approach, but it has been reported that the application of the design by analysis allows removing the unnecessary conservatism caused by applying the design by formula approach [1]. For structural components that have complex geometries and/or geometric discontinuities, i.e. stress raisers, the elementary theory of elasticity and/or the mechanics of materials is no longer effective for evaluating stress distributions in such components. Thus the design by formula approach tends to bring about excessive conservatism.

Owing to the astonishing development of computer technology, however, FEM has markedly increased its speed and performance to become a useful tool for strength design of machines and structures. In addition, commercial FEM software has become more and more user-friendly. Strength design engineers used to ask analysis specialists to make stress analyses for their strength designs, whereas they have now begun to use FEM as one of their design tools. They can use CAD data directly for FEM analyses. They can make combined analyses, e.g., fluid-structure coupling analyses, without much difficulty in their daily design activities.

ANSYS Workbench is powerful platform software that is the backbone for an integrated, comprehensive simulation system. This chapter describes how to operate ANSYS Workbench to carry out structural or stress analyses of a fundamental type of component parts for demonstrating elementary use of the software. The example problem to analyse in this chapter is not a multiphysics one, but it provides a good example for ANSYS Workbench Mechanical (AWM). The result obtained by the present analysis is compared with that by an empirical stress concentration factor diagram based on the photo-elasticity, showing a good agreement between the two.

A.2 Problem description: A stepped round bar subjected to tension

Perform an FEM analysis of a 3D stepped round bar subjected to an applied tension at the free end and clamped to the rigid wall at the other end, as described below.

Fig. A.1 Stepped round bar subjected to a uniform tensile stress σ_0.

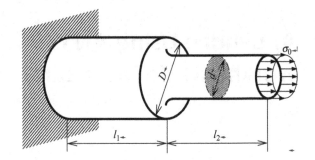

Calculate the stress concentration at the foot of the fillet of the bar and the tensile stress distribution on a cross section of the bar.

Geometry: lengths $l_1 = 50$ mm, $l_2 = 50$ mm, diameters $D = 40$ mm, $d = 20$ mm, fillet radius $\rho = 3$ mm

Material: mild steel having Young's modulus $E = 200$ GPa and Poisson's ratio $\nu = 0.3$.

The value of Young's modulus is the datum registered as that of a structural steel in the standard database of the ANSYS Workbench and is slightly lower than that used in Chapter 3.

Boundary conditions: A stepped bar is rigidly clamped to a wall at the left end and subjected to a uniform tensile stress of $\sigma_0 = 100$ MPa at the right free end, as shown in Fig. A.1.

A.3 Starting workbench

Double-click the **ANSYS Workbench** shortcut icon to start the workbench. The **Unsaved Project – Workbench** window as well as the **Getting Started** window open, as shown in Fig. A.2.

[COMMAND]   **Windows Start Menu** → **ANSYS 17.0** → **Workbench 17.0**

After reading 'Welcome to ANSYS Workbench!', click the close button in the upper right corner of the **Getting Started** window.

Fig. A.3 shows the components of the **Workbench** window. Click the **Static Structure** icon in the **Analysis Systems** toolbox to display the **Static Structural** window in the **Project Schematic** window, as shown in Fig. A.3. The symbols in the third row of the cells in the **Static Structural** window indicate the statuses of the cells as shown in Table A.1. The status symbol of the checkmark for the **Engineering Data** cell in Fig. A.4 indicates that the material of the stepped round bar to be analysed has been selected and is ready to be shared with other cells such as the **Model** and **Solution** cells.

Clicking the **Engineering Data** cell shown in Fig. A.4 launches the **A2: Engineering Data** screen, as shown in Fig. A.5. Click and highlight the A3 cell for the structural steel listed in the **Outline of Schematic A2: Engineering Data** window. The structural steel cell is selected as the material for the stepped round bar to be created and analysed.

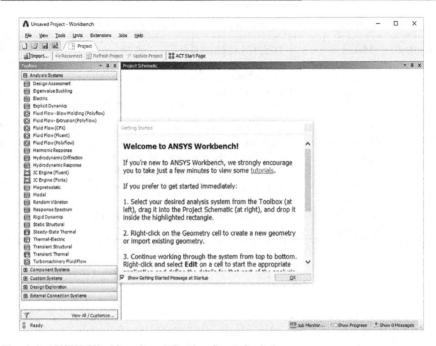

Fig. A.2 ANSYS Workbench and **Getting Started** windows.

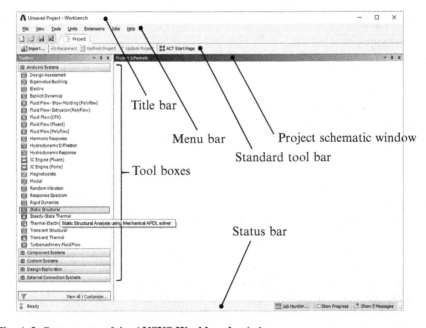

Fig. A.3 Components of the **ANSYS Workbench** window.

Table A.1 Symbols for cell statuses and their meanings

Symbol	Meaning	Detailed explanation
	Attention required	Immediate action is required for the cell, or a user cannot proceed further
	Unfulfilled	The previous cells do not have sufficient data
	Up to date	The cell is up to date and the data in the cell is ready to be shared with other cells
	Refresh required	The data in the previous cells have been changed since last update. The cell needs to be refreshed
	Update required	The input data of the cell has been changed and the output data needs to be updated
	Input changes pending	The cell is locally updated but may change

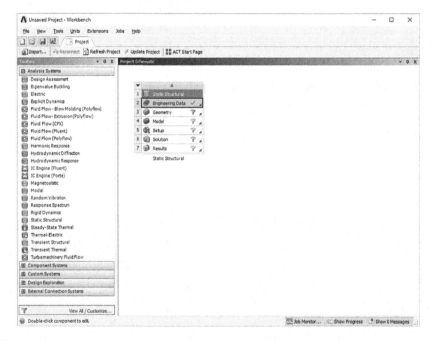

Fig. A.4 **Engineering Data** cells in the **Project Schematic** window.

The status symbol for the **Geometry** cell is the solid question mark as shown in Fig. A.6, meaning the geometry data of the model to carry out the analysis is required (see Table A.1). Geometry data can be created in two ways: (1) by creating the model in the **Design Modeler** (**DM**) window of ANSYS Workbench; or (2) by importing a CAD model created by commercial 3D CAD applications. Here we create the same stepped round bar model as that analysed in Chapter 3 in the **DM** window.

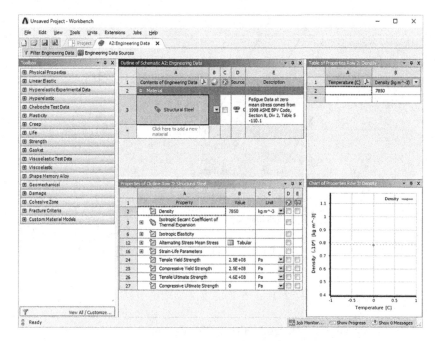

Fig. A.5 Tool boxes and property tables stored in the **A2: Engineering Data** screen.

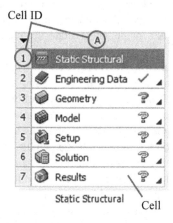

Fig. A.6 Project Schematic window displaying various standalone systems in it.

A.4 Creation of a 3D analysis model using the design modeller

Click the **Geometry** cell to display the **DM** window as shown in Fig. A.7; this depicts the components of the **DM** window. The **DM** window can be used in two basic modes

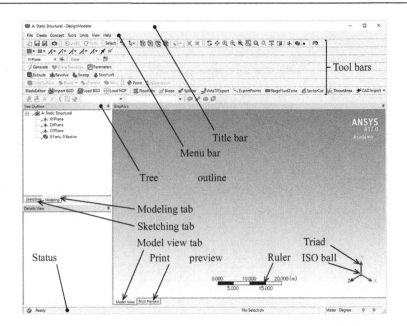

Fig. A.7 Components of the **Design Modeler (DM)** window.

of creating a model: (1) the sketching mode that is used to draw 2D sketches; and (2) the modelling mode that converts 2D sketches into 3D models. First, sketch a 2D stepped plate on, say, the XY plane, then rotate the plate about the X-axis to get the 3D stepped round bar.

Click the **XYPlanes** branch in the **Tree Outline** window to display the O-XYZ Cartesian coordinate system in the **Graphics** window, as shown in Fig. A.8.

Click the **Look At Face/Plane/Sketch** icon [A] in Fig. A.9, or the rightmost icon, in the top tool bar of the window to display the **XY plane**, as shown in Fig. A.9. Click the **Sketching** tab of the **Sketching Toolboxes** window to display menus, as shown in Fig. A.10. Click the **Grid** button in the **Settings** window and check the **Show in 2D** and the **Snap** boxes to display the grid in the **DM** window. Select **Millimeter** as the unit to use in the **Units** menu in the menu bar, as shown in Fig. A.11. The grid spacing is set to 10 mm. The spacing of the grid is automatically changed according to the scaling of the display. The display can be zoomed in and out by rolling the middle mouse wheel forward and backward, respectively. Click the **Pan** icon [A] in the tool bar, as shown in Fig. A.10, and the arrow cross appears in the **DM** window to execute the translation movement of the XY coordinates in all directions.

Click the **Draw** and the **Line** buttons. Click at the beginning of a horizontal line at a point of (0 mm, 20 mm) and click again at the end point of (50 mm, 20 mm), as shown in Fig. A.12. Continue drawing vertical and horizontal lines until the drawing of the outline of the stepped plate is completed, as shown in Fig. A.13.

[COMMAND] Sketching Toolboxes → Dimensions → Horizontal/Vertical

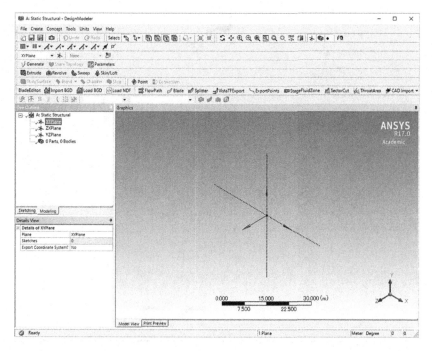

Fig. A.8 O-XYZ Cartesian coordinates in the **Graphics** window of the **DM**.

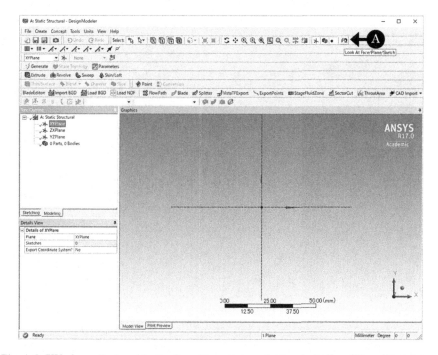

Fig. A.9 **XY plane** displayed in the **Graphics** window and **Look At Face/Plane/Sketch** icon [A] in the **Tool Bar** of the **Design Modeler** window.

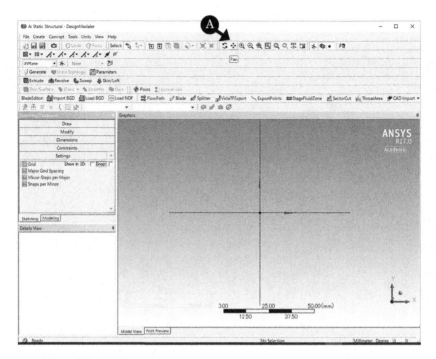

Fig. A.10 Menus in the **Settings** menu in the **Sketching Toolboxes** window.

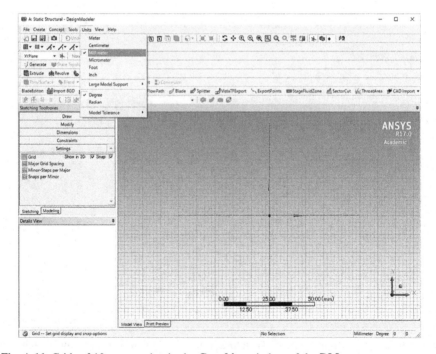

Fig. A.11 Grids of 10 mm spacing in the **Graphics** window of the **DM**.

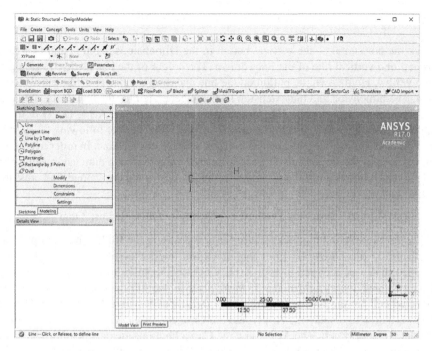

Fig. A.12 A horizontal line of 50 mm in length drawn on the **XY plane** in the **Graphics** window.

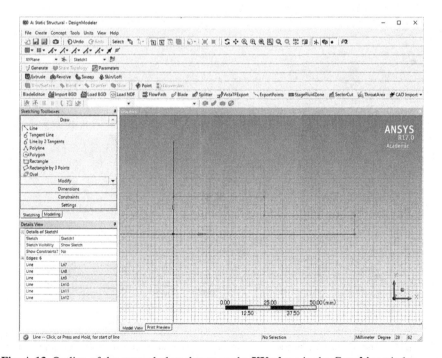

Fig. A.13 Outline of the stepped plate drawn on the **XY plane** in the **Graphics** window.

Click the horizontal or the vertical button in the **Dimensions** menu and click a horizontal or a vertical line of the lines drawn. The colour of the line clicked is changed into yellow. Hold down the button on the line and drag the line to a convenient location. The dimension and the extension lines are generated for the line designated and the dimension line is named **H1**, for example, as shown in Fig. A.14. The **H** or the **V** sign indicates that the line is horizontal or vertical. The number following the sign **H** or **V** indicates the order of drawing a horizontal or the vertical line. In order to edit the dimensions of the lines to give them the desired values, click a dimension value of a line in the **Details View** window as shown in Fig. A.14, enter the change, and press the return key.

[COMMAND]   Sketching Toolboxes → Modify → Fillet; Radius: 3 mm

Click the **Modify** and the **Fillet** buttons. Enter **3 mm** in the **Radius** box and pick the vertical and the horizontal lines that form the interior corner of the joint of the stepped plate, as depicted in Fig. A.15. Click the **Revolve** icon in one of the tool bars and pick the sketch of the stepped plate in the **DM** window. Click the **Not selected** cell for the **Geometry** cell of the **Details of Revolve 1** table in the **Details View** window to display the **Apply** and the **Cancel** buttons. Click the **Apply button** for the **Geometry** cell. The **Apply** and the **Cancel** buttons are changed into the name of the geometry, i.e., **Sketch 1** selected to revolve. Click the **Not selected** cell for

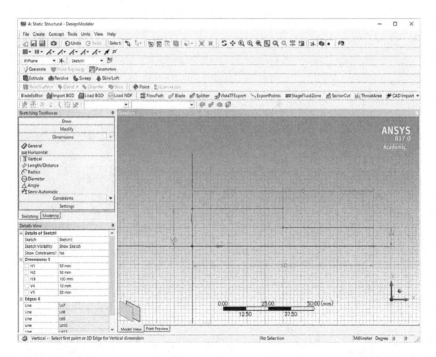

Fig. A.14 Dimension and extension lines for the stepped plate drawn on the **XY plane** in the **Graphics** window.

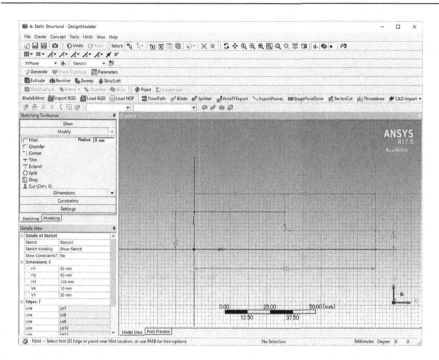

Fig. A.15 Circular fillet of 3 mm in radius set at the interior corner of the joint of the stepped plate.

the **Axis** cell to display the **Apply** and the **Cancel** buttons. Pick the x-axis about which the 2D stepped plate is to be revolved, as shown in Fig. A.16. Push the **Apply** button and the cell is changed into **2D Edge**. Click the **Generate** icon in one of the tool bars. The 2D stepped plate is transformed into the stepped circular cylindrical bar, as shown in Fig. A.17.

Click the **Workbench** icon in the task bar to return to the **Workbench** window to confirm that the solid question mark is changed for the **Up to date** mark, as shown in Fig. A.18. Double-click the **Model** cell marked by the **Refresh required** symbol, as shown in Fig. A.19. The **Static Structural – Mechanical** window appears, as shown in Fig. A.20. The stepped circular cylindrical bar is displayed in the window and the symbol by which the **Model** cell is marked turns into the **Update required** symbol, as shown in Fig. A.20. The software is switched from the **DM** to the **Mechanical** for the static structural analysis, or the static stress analysis of the stepped circular cylindrical bar. Fig. A.20 also shows the components of the **ANSYS Workbench Mechanical** window.

Note: If you already have a model to analyse with ANSYS Workbench, the modelling process described in Section A.4 can be omitted. Instead, the model to analyse is imported by right-clicking the **Geometry** cell, as shown in Fig. A.21. The image file formats that can be imported into the **ANSYS Workbench** are IGES, Parasolid, STEP, etc.

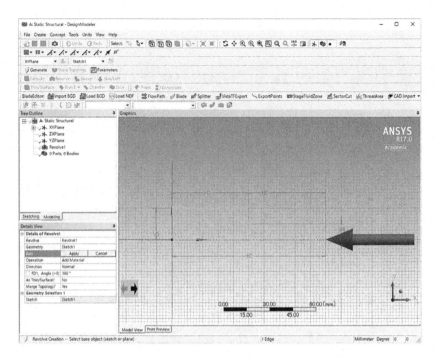

Fig. A.16 Giant 3D *arrow* indicating that the *x*-axis is the axis of revolution for creating the stepped round bar from the stepped plate.

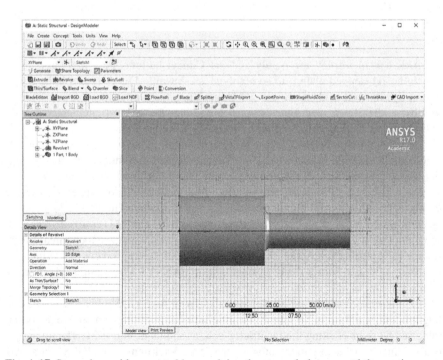

Fig. A.17 Stepped round bar created by revolving the stepped plate around the *x*-axis.

Fig. A.18 Selection of the **Geometry** cell in the **Project Schematic** window.

Fig. A.19 Selection of the **Model** cell in the **Project Schematic** window.

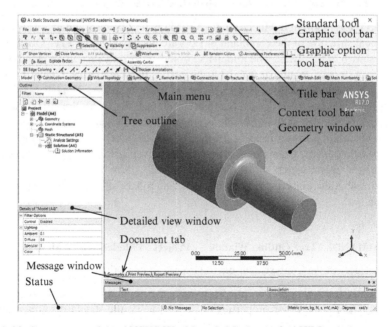

Fig. A.20 Components of the **ANSYS Workbench Mechanical (AWM)** window.

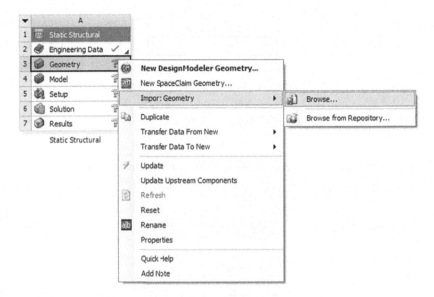

Fig. A.21 Hierarchical menu of the **Model** cell for importing a CAD file of the 3D stepped bar model.

A.5   Stress analysis of a stepped round bar subjected to uniform tension

A.5.1   Finite-element discretisation of the round bar volume

The **ANSYS Workbench Mechanical (AWM)** is an object-oriented software. The hierarchical structure of the command objects are displayed in the **Tree Outline** window, as shown in Fig. A.22. The folder-like object or the branch named '**Model (A4)**' represents a stepped round bar which has properties like **Geometry**, **Coordinate System**, **Mesh**, **Static Structural** responses, etc. as lower branches. The label '**A4**' in the parenthesis indicates the ID of the **Geometry** cell (see Fig. A.6). Specific contents of each property are stored in a respective branch. Thus, the analytical procedures are reduced to adding lower branches which store detailed contents of the above-mentioned properties. Changing a command involves just deleting the branch and adding a new object which stores desired contents.

Confirm that the units are mm, t, N, s, mV, and mA by clicking the **Units** menu in the menu bar.

(1) Meshing

    [COMMAND]   **Tree Outline Window → Project → Model (A4) → Mesh**

1. The category of the **Context Tool Bar** is changed from **Model** to **Mesh**.

Set the size of finite elements before discretising the stepped bar model as displayed in Fig. A.23. As for the other properties of the elements, the standard settings for the

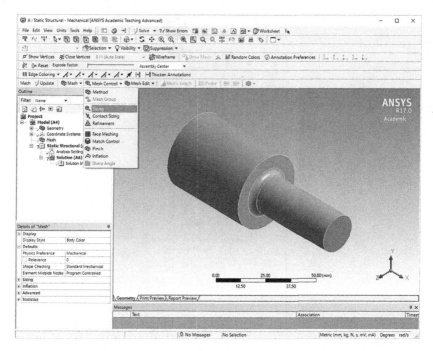

Fig. A.22 Hierarchical structure of the command objects displayed in the **Tree Outline** window.

elements used in the present analysis are adopted as described in the **Defaults** table in the **Details of 'Mesh'** window. Among the items listed in the table, for example, the description that the **Element Midside Nodes** are **Program Controlled** means that the higher order elements having midside nodes are used by default for linear elastic stress analyses. The type of elements need not be selected.

[COMMAND]   Context Tool Bar (Mesh) → Mesh Control → Sizing

1. The **Sizing** branch is created under the **Mesh** branch.
2. Enter **2** mm in the **Element Size** cell of the **Definition** table in the **Details of 'Sizing'** window, as shown in Fig. A.23.
3. Click the **Select Body** icon [A] in Fig. A.23 and place the mouse cursor on the stepped bar model and click the left button of the mouse. The whole body of the model is selected for element sizing, as shown in Fig. A.24.
4. Click the **No Selection** cell for the **Geometry** in the **Scope** table to display the **Apply** and the **Cancel** buttons.
5. Click the **Apply** button. The colour of the stepped bar model changes from green into purple to indicate that the whole body of the model is selected to be meshed, as shown in Fig. A.25.

[COMMAND]   Context Tool Bar (Mesh) → Mesh → Generate Mesh

1. The stepped round bar model is meshed, as shown in Fig. A.26.

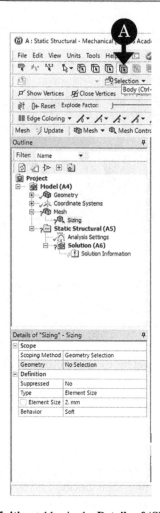

Fig. A.23 Scope and the **Definition** tables in the **Details of 'Sizing'** window.

A.5.2 Imposition of boundary conditions

(1) Imposing constraint conditions on the left-end face of the stepped round bar

    [COMMAND] Tree Outline Window → Project → Model (A4) → Static Structural (A5)

1. Select the **Static Structural (A5)** branch [A] in Fig. A.27, and the stepped bar model is changed into the volume of the bar model without mesh. The label 'A5' in the parenthesis indicates the ID of the **Setup** cell (see Fig. A.6).
2. The **Context Tool Bar** is changed from **Mesh** to **Environment**.
3. Click the **Rotate** icon [B] to rotate the model and display the left-end face to which the displacement constraint is to be applied, as shown in Fig. A.27.

    [COMMAND]   Context Tool Bar (Environment) → Supports → Fixed Support

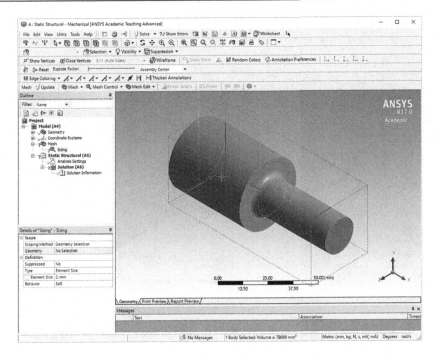

Fig. A.24 Whole body of the present stepped bar model selected for element sizing.

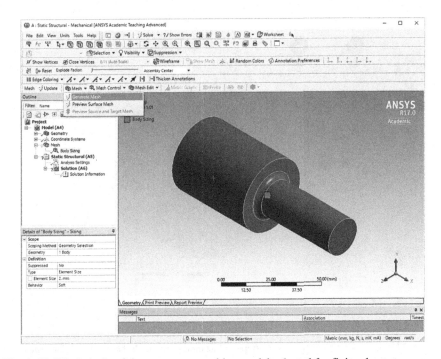

Fig. A.25 Whole body of the present stepped bar model selected for finite-element discretisation.

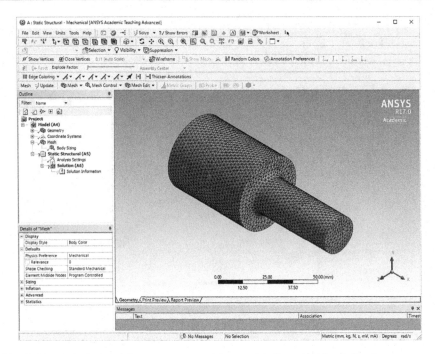

Fig. A.26 Whole body of the present stepped bar model discretised by finite elements.

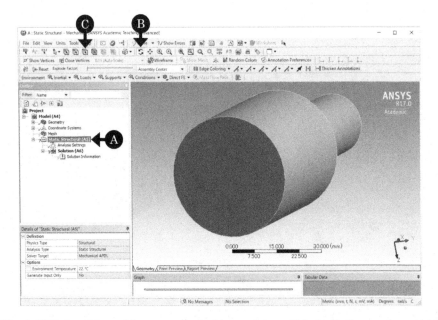

Fig. A.27 Left-end face of the present stepped bar model to which the displacement constraint is to be applied.

1. Click the **Select Face** icon [C], place the mouse cursor on the left-end face of the stepped bar model, and click the left button of the mouse. Select the left-end face of the model to be fixed, as shown in Fig. A.27.
2. The colour of the left-end face of the stepped bar model turns green.
3. Click the **1 Face** cell for the **Geometry**, and display the **Apply** and the **Cancel** buttons.
4. Click the **Apply** button in the **Geometry** cell of the **Scope** table in the **Details of 'Fixed Support'** window, as shown in Fig. A.28.
5. The colour of the left-end face of the stepped bar model changes from green into purple to indicate that the left-end face of the model is selected to be fixed.
6. The **Fixed Support** branch is created under the **Static Structural (A5)** branch.

(2) Applying uniform tensile stress to the right-end face of the stepped round bar model

[COMMAND]   Context Tool Bar (Environment) → Loads → Pressure

1. Click the **Rotate** icon to rotate the model and display the right-end face to apply a pressure of 100 MPa to the face, as shown in Fig. A.29.
2. Click the **Select Face** icon, place the mouse cursor on the right-end face of the stepped bar model, and click the left button of the mouse. Select the right-end face of the model to be subjected to pressure, as shown in Fig. A.29.
3. The colour of the right-end face of the stepped bar model turns green.
4. Enter **-100 MPa** in the **Magnitude** cell of the **Definition** table in the **Details of 'Pressure'** window, as shown in Fig. A.29.
5. Click the **1 Face** cell for the **Geometry**, and display the **Apply** and the **Cancel** buttons.

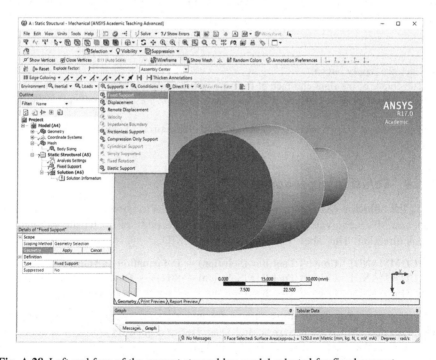

Fig. A.28 Left-end face of the present stepped bar model selected for fixed support.

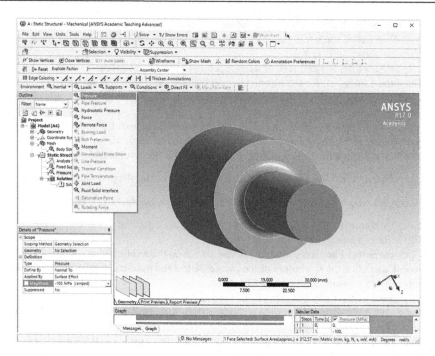

Fig. A.29 Right-end face of the present stepped bar model selected to be subjected to pressure, or uniform tensile stress.

6. Click the **Apply** button.
7. The colour of the right-end face of the stepped bar model changes from green into red to indicate that the right-end face of the model is subjected to pressure, or uniform tensile stress, as shown in Fig. A.30.
8. The **Pressure** branch is created under the **Static Structural (A5)** branch.

A.5.3 Solution procedures

Click the **Solution (A6)** branch to choose the solutions to obtain, e.g., total deformation, von Mises equivalent stress and a user defined result, or the tensile stress in the x-axis direction. The label 'A6' indicates the cell ID.

[COMMAND]   Context Tool Bar (Solution) → Deformation → Total
[COMMAND]   Context Tool Bar (Solution) → Stress → Equivalent (von Mises)
[COMMAND]   Context Tool Bar (Solution) → User Defined Result

After the three commands above are performed, the corresponding branches (**Total Deformation**, **Equivalent Stress**, and **User Defined Result**) are created under the **Solution (A6)** branch, as shown in Figs A.31–A.33, respectively.

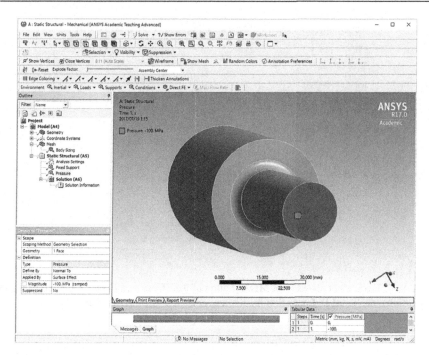

Fig. A.30 Right-end face of the present stepped bar model subjected to pressure, or uniform tensile stress.

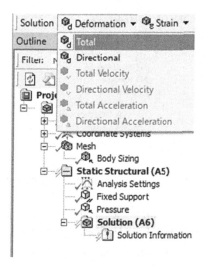

Fig. A.31 Total Deformation button of the **Deformation** menu in the **Context Tool Bar (Solution)**; the **Total Deformation** branch is not yet created under the **Solution (A6)** branch.

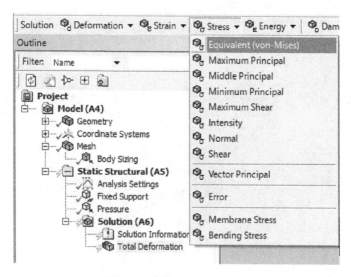

Fig. A.32 **Equivalent stress (von Mises)** button of the **Stress** menu in the **Context Tool Bar** (**Solution**); the **Total Deformation** branch was created under the **Solution (A6)** branch.

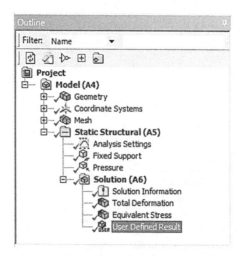

Fig. A.33 Designated quantity branches created under the **Solution (A6)** branch.

Select the **User Defined Result** branch, and enter **SX** in the **Expression** cell of the **Definition** table in the **Details of 'User Defined Result'** window, as shown in Fig. A.34.

[COMMAND]   Standard Tool Bar → Solve

The solution procedure starts. The **ANSYS Workbench Solution Status** window appears, displaying the status of the solution process, as shown in Fig. A.35. When the solution is finished, the solution status window disappears.

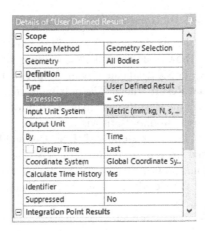

Fig. A.34 Expression **SX** entered in the **Expression** cell of the **Definition** table in the **Details of 'User Defined Result'** window.

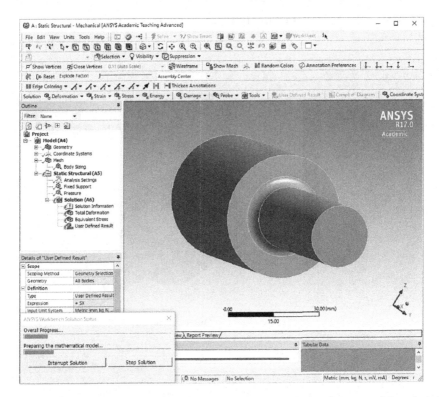

Fig. A.35 ANSYS Workbench Solution Status window displaying the status of the solution process.

A.5.4   Contour plots of the results

Contour plots of the results are displayed by clicking branches under the **Solution (A6)** branch. Fig. A.36 shows, for example, the contour plot of the normal stress in the x-axis direction σ_x, or SX designated by the **Expression** cell of the **Definition** table in the **User Defined Result** branch. The maximum and the minimum values of each result can be found in the **Results** table in the **Details** of the corresponding result and also in the table in the **Tabular Data** window.

The **Maximum** cell of the in the **Details of 'User Defined Result'** window shows that the maximum normal stress $\sigma_{x\max}$ obtained by AWM for the present stepped bar subjected to the applied uniform tensile stress σ_0 of 100 MPa is 160.08 MPa, as shown in the **Maximum** cell of the **Results** table in the **Details of 'User Defined Result'** window, as set out in Fig. A.36.

Note: The **User Defined Result** expressions available are listed in the table in the **Worksheet** window, as shown in Fig. A.37. The table can be accessed by clicking the **Solution** branch and then by selecting the **Worksheet** icon in the **Standard Tool Bar**. This icon does not become apparent until one or more solution quantities are designated.

Instead of using the **User Defined Result** command, six stress components can be obtained by the following command:

[COMMAND]   Context Tool Bar (Solution) → Stress → Normal/Shear

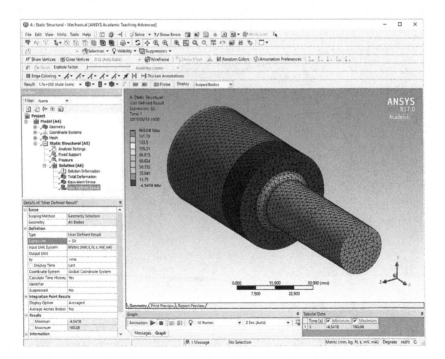

Fig. A.36 Contour plots of the normal stress in the x-axis direction σ_x, or SX showing that the maximum normal stress $\sigma_{x\max}$ obtained by the present analysis is 160.08 MPa.

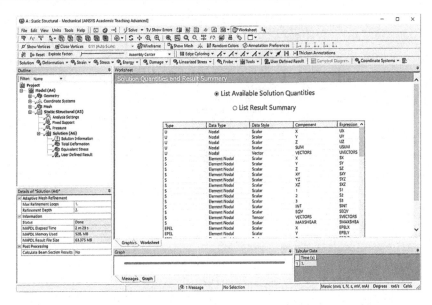

Fig. A.37 Available solution quantities and their expressions listed in the table in the **Worksheet** window.

After the above command is performed, the corresponding branch, i.e. the **Normal Stress** or the **Shear Stress** branch, is created under the **Solution (A6)** branch, as shown in Fig. A.38. The orientation of the normal or the shear stresses shall be designated as, for example, **X Axis** or **XY Plane** in the **Orientation** cell of the **Definition** table in the **Details of 'Normal Stress'** or **'Shear Stress'** window.

A.6 Discussion

As previously described, the maximum normal stress $\sigma_{x\max}$ obtained by ANSYS for the present stepped bar is 160.08 MPa. The tensile stress concentration factor is obtained as $\alpha = \sigma_{\max}/\sigma_0 \approx 1.78$ by the tensile stress concentration vs radius of curvature diagram [2] (also see Problem 3.19). The relative difference between the maximum stress obtained by AWM and that by the stress concentration factor is $(\sigma_{x\max} - \alpha\sigma_0)/(\alpha\sigma_0) \approx -10\%$. This difference is acceptable, but if even more accuracy is required, the operation of mesh refinement would be helpful. Mesh refinement can be achieved by following the procedures below.

(1) Mesh refinement for improving the solution

    [COMMAND]   **Tree Outline Window → Project → Model (A4) → Mesh**

1. The category of the **Context Tool Bar** is changed from **Model** to **Mesh**.

    [COMMAND]   **Context Tool Bar (Mesh) → Mesh Control → Refinement**

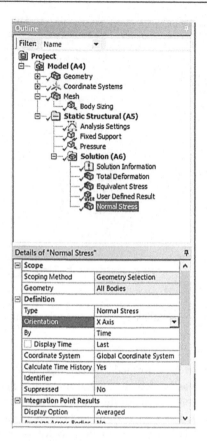

Fig. A.38 **User Defined Result** designated in the **Orientation** cell of the **Definition** table in the **Details of 'User Defined Result'** window.

1. The **Refinement** branch is created under the **Mesh** branch in the **Tree Outline** window, and the original finite element discretisation image of the stepped round bar in Fig. A.39 is changed into its blank image in Fig. A.40.
2. Click the **Select Face** icon and place the mouse cursor on the circumferential shoulder fillet region in the stepped bar model. Click the left button of the mouse, and the fillet region turns green, as shown in Fig. A.40.
3. Click the **Apply button** for the **Geometry** cell, and the colour of the fillet region is changed to blue.

[COMMAND]   Context Tool Bar (Mesh) → Mesh → Generate Mesh

1. The finite elements around the fillet in the present stepped round bar model are remeshed and refined, as shown in Fig. A.41.

[COMMAND]   Standard Tool Bar → Solve

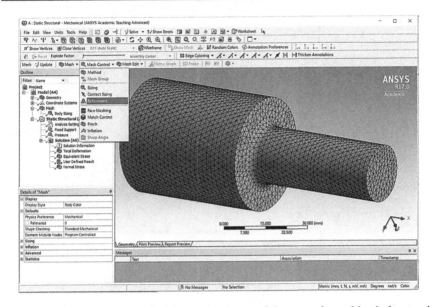

Fig. A.39 Original finite element discretisation image of the stepped round bar before mesh refinement.

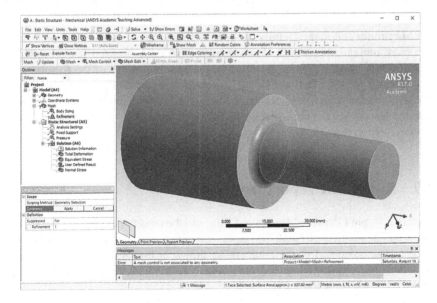

Fig. A.40 Fillet region selected to be remeshed for the solution improvement.

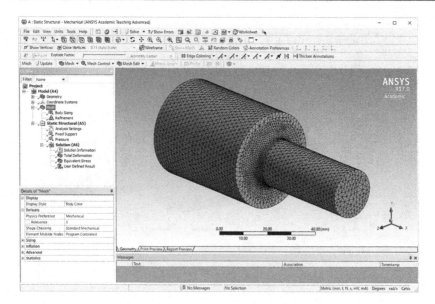

Fig. A.41 Remeshed and refined finite elements around the fillet region in the present stepped round bar model.

1. The solution procedure starts. The **ANSYS Workbench Solution Status** window appears, displaying the status of the solution process. The window disappears when the solution is finished.
2. Click, for example, the **User Defined Result** branch under the **Details of 'User Defined Result'** window and the contour plot of σ_x is displayed in the **Geometry** window.

The **Maximum** cell in the **Details of 'User Defined Result'** window shows that the maximum normal stress σ_{xmax} obtained by the remeshed stepped bar subjected to $\sigma_0 = 100$ MPa is 180.21 MPa, as shown in the **Maximum** cell of the **Results** table in the **Details of 'User Defined Result'** window, as set out in Fig. A.42. The relative difference between the maximum stress obtained by AWM after the mesh refinement and that by the stress concentration factor is about 1.24%, showing a significant improvement in the solution.

A.7   Automatic report generation

Click the **'Report Preview' Document Tab**, and the report of the present stress analysis is automatically generated. The report provides a summary of the stress analysis made and is composed of the contents, as shown in Fig. A.43. The report can be printed by the following command, published in the MHTML format, or exported into the Word/PowerPoint format.

   [COMMAND]   Context Tool Bar (Report Preview) → Print

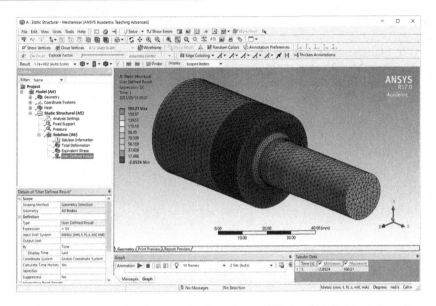

Fig. A.42 Contour plots of the improved value of σ_x, or SX showing that the value of σ_{xmax} obtained after the mesh refinement is 180.21 MPa.

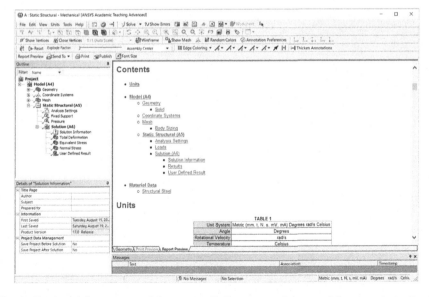

Fig. A.43 List of the contents of the automatically generated report of the present structural analysis displayed in the **Report Preview** window.

References

[1] U. T. Murtaza and M. J. Hyder, Proc. Inter. Multi-Conference of Engineers and Computer Scientists, IMECS, vol. II (2015).
[2] M. Nishida, Stress Concentration, Morikita Pub. Co., Ltd., Tokyo (1967) (in Japanese).

Index

Note: Page numbers followed by *f* indicate figures, and *t* indicate tables.

Printed in the United States
By Bookmasters